ICT Diffusion in Developing Countries

Ewa Lechman

ICT Diffusion in Developing Countries

Towards a New Concept of
Technological Takeoff

 Springer

Ewa Lechman
Gdańsk University of Technology
Faculty of Management and Economics
Gdańsk, Poland

ISBN 978-3-319-18253-7   ISBN 978-3-319-18254-4   (eBook)
DOI 10.1007/978-3-319-18254-4

Library of Congress Control Number: 2015940998

Springer Cham Heidelberg New York Dordrecht London

Springer International Publishing AG Switzerland is part of Springer Science+Business Media
(www.springer.com)

To my Daughter A.

Acknowledgments

While working on this book I was fortunate to obtain support and constructive criticism from many people. I am deeply indebted to all of them. My colleagues from the Faculty of Management and Economics at Gdansk University of Technology have been extremely generous in providing me sound and fresh ideas, arguing with me and discussing issues. This was a source of inspiration and encouraged me to further work. I am especially thankful to Professor Piotr Dominiak and Professor Jerzy Czesław Ossowski from the Faculty of Management and Economics at Gdansk University of Technology for being patient listeners, their intellectual support, and valuable suggestions. They both are responsible for a mountain of improvements throughout the manuscript; undoubtedly, without their support, this book would not exist. I also give enormous thanks for Professor Luciano Segreto from the University of Florence, who read every word of the preliminary concept of this book, showing me the right directions for its development. I am also very grateful to my parents for endless conversations concerning the book content, generous criticism, and stimulating questions, which guided me throughout my further research. Above all, many special thanks are owed to my beloved 10-year-old daughter for her enormous patience, giving me peace of mind and understanding while I was working on the manuscript. During these days, I could not spend with her as much time as I always wished to.

Contents

1 Introduction 1
  1.1 Background 1
  1.2 The Story 2
  1.3 Structure and Content 3
  References 6

**2 Technology, The Economy, and Society: Casting the
  Bridges—Introductory Notes** 7
  2.1 Introduction 7
  2.2 Technology: Ideas and Concepts 8
  2.3 Technology: A Timeless Value 11
  2.4 From Industrial to Information Revolution 17
  2.5 ICT: Opportunity Window for Developing Countries? ... 20
  References 23

3 Technology Diffusion 29
  3.1 Technology Diffusion: Theoretical Framework 29
  3.2 Technology Diffusion and Technological Substitution ... 33
      3.2.1 Technology Diffusion. Concepts and models 34
      3.2.2 Approximating Technology Diffusion Trajectories ... 41
      3.2.3 Technological Substitution 46
  3.3 The '*Critical Mass*': What Stands Behind? 49
      3.3.1 The '*Critical Mass*'. Explaining the Concept ... 49
      3.3.2 The '*Technological Take-Off*' and the '*Critical Mass*'.
             A Trial Conceptualisation 54
  3.4 Technology Convergence and Technology Convergence Clubs ... 61
      3.4.1 Convergence: Theoretical Specification 63
      3.4.2 Convergence Clubs Hypothesis 67
  References 71

**4 Information and Communication Technologies Diffusion Patterns in
  Developing Countries: Empirical Evidence** 83
  4.1 Introduction 84
  4.2 Data Explanation and Rationale 84

4.3   Information and Communication Technologies in Developing
       Countries: Preliminary Evidence   87
4.4   Shaping Country-Specific ICT Diffusion Trajectories   94
       4.4.1   Mobile Cellular Telephony Diffusion   95
       4.4.2   Internet Networks Diffusion and Internet Usage   107
4.5   Tracing the Technological Substitution   124
       4.5.1   Final Remarks   140
4.6   Technology Convergence, Divergence or Club Convergence?
       The Worldwide Evidence for the Period 2000–2012   141
       4.6.1   Technology Convergence   142
       4.6.2   Technology Convergence Clubs   153
       4.6.3   Brief Evidence on How Wireless Broadband Networks
                Expanded Worldwide Over the Period 2010–2012   159
4.7   Summary   162
References   163

5   What Matters for ICT Diffusion?   167
5.1   Introduction   168
5.2   Tracing the 'Technological Take-Off' and the 'Critical Mass'
       Effects   168
       5.2.1   The Data   169
       5.2.2   Ready for the 'Technological Take-Off'?   169
5.3   ICT Diffusion Determinants. A 'Traditional' Approach   200
       5.3.1   The Data   203
       5.3.2   Graphical Evidence   204
       5.3.3   Panel Regression Results   213
5.4   Summary   219
References   220

6   Conclusion, Recommendations and Implications   223
6.1   Introduction   223
6.2   Underlying Conclusions   224
       6.2.1   The First Perspective. Moving Ahead or Lagging
                Behind?   225
       6.2.2   The Second Perspective. Technological Substitution:
                Illusion or Fact?   226
       6.2.3   The Third Perspective. Digital Gaps Closing
                or Growing?   227
       6.2.4   The Fourth Perspective. Ready for the 'ICT
                Revolution'?   228
6.3   A Brief Look at ICT Policies in Developing Countries   231
       6.3.1   What Needs to Be Addressed? Some Recommendations . . .   231

6.3.2   A Few Words on ICT Policies and *e-Strategies*
        Implementation in Developing Countries 233
6.4   Toward the Great Escape. . .? 237
References 241

Appendix A. ICT Core Indicators: Definitions 247

**Appendix B. Core ICT Indicators. Low-Income and Lower-
Middle-Income Economies. Years 1975 and 2012** 249

**Appendix C. Countries Included in the Technology Convergence
Analysis. Core ICT Indicators. Period 2000–2012** 253

**Appendix D. ICTs Inequalities. 113 World Countries.
Period 2000–2012** 259

**Appendix E. ICTs Distributions. 113 World Countries.
Period 2000–2012** 261

**Appendix F. ICT Marginal Growths and Replication Coefficients.
17 Low-Income Economies. Period 1995–2012** 265

**Appendix G. ICT Marginal Growths and Replication Coefficients.
29 Lower-Middle-Income Economies. Period 1995–2012** 271

**Appendix H. Mobile Cellular Telephony and Internet Users Penetration
Rates: Determinants. Correlation Matrices. Low-Income and
Lower-Middle-Income Economies. Period 1997–2012** 279

**Appendix I. Mobile Cellular Telephony and Internet Users: Regression
Results. Low-Income and Lower-Middle-Income Economies. Period
1997–2012** 289

**Appendix J. 'Technological Take-Off' Conditions. Low-Income and
Lower-Middle-Income Economies** 301

Introduction

1

*(. . .) the information and communication revolution—
perhaps the most pervasive and global technological
revolution in recent human history*

Nagy H. Hanna (2010)

Abstract

This chapter presents the general purposes and aims of the book. It briefly discusses the conceptual and theoretical background, explains the major targets of the presented theoretical and empirical analysis. It also explains the structure of the book and the contents of its consecutive chapters.

Keywords

ICT • Technology diffusion • Critical mass • Technological takeoff

1.1 Background

For the last few decades, the world has witnessed unprecedented growth and diffusion of Information and Communication Technologies (ICT) in terms of speed and geographic coverage. It is difficult to determine the exact time when these tremendous changes began; however, many claim that the year 1971 was the turning point when the Technological (Information) Revolution emerged, giving rise to the new techno-economic paradigm (Dosi 1982; Freeman and Louca 2001; Perez 2009). Therefore, from 1971 onward, ICT has been gradually reshaping social and economic landscapes. As claimed by Hanna (2003), the ICT Revolution *'is so profound and pervasive that it challenges many traditional economic concepts that are rooted in incrementalist thinking'* (Hanna 2003).

Undeniably, technology and innovation have triggered in-depth transformation of societies throughout history, allowing for advances in overall well-being. Today, however, the ongoing Information Revolution is transforming socioeconomic systems more quickly and profoundly than any technological revolution has ever

© Springer International Publishing Switzerland 2015

E. Lechman, *ICT Diffusion in Developing Countries*,

DOI 10.1007/978-3-319-18254-4_1

done before, hence generating special attention and interest. The following example speaks for itself. In 1876, Graham Bell patented the analogue telephone, but in 2012 (after 136 years), a huge share of world society still lacked access to this form of communications. According to ITU (2013) statistics, the world average fixed telephony penetration rate was at approximately 21.2 per 100 inhabitants[1]). In contrast, mobile telephony over that 41-year period (between 1971 and 2012) diffused so rapidly that in 2012 it was accessible by nearly 100 % of the world's population[2]; this exhibits the unprecedented ability of ICT to spread at a high pace worldwide.

The Information Revolution introduced technological solutions, which are quickly distributable throughout societies, that overcome geographical, infrastructural and—to a point—financial constraints. Moreover, Information and Communication Technologies may be easily accessed and used, even by low-income and low-skilled people, regardless of their physical location, freeing them from mental, informational, technological and geographical isolation, and offering instead unlimited opportunities to benefit from global information and knowledge flows. Thus, technological peripheries are gradually disappearing from the world map.

To some extent, economically backward countries have been omitted from previous technological revolutions. This is not to say, however, that *no* type of technological progress has ever reached them; however, the spread and access to use of various technologies was extremely limited (see, e.g., low electrification rates or negligible access to railway networks). Today, economically backward countries, which 'traditionally' lag behind in terms of technology adoption, are rapidly heading towards broad deployment of ICT. This is, undoubtedly, the *revolutionary*, and one of the most striking facts in the development 'history' of economically backward countries.

1.2   The Story

This book tells a story about Information and Communication Technologies' diffusion in 46 economically backward countries[3] between 2000 and 2012, offering to the reader a fresh perspective on the issues discussed. It examines the spread of ICT from four broadly defined perspectives that highlight the major aims and scope of this work. These perspectives are:

- Explaining the ICTs diffusion patterns and the dynamics of the process itself;
- Detecting technological substitution;
- Examining technological convergence;

---

[1] Author's calculations.

[2] Ibid.

[3] To avoid varying terminologies, we use alternatively the term 'developing countries'.

- Identifying the 'critical conditions' that enhanced the emergence of the 'technological take-off.'

By convention, the central focus of this book is on developing countries, although this group of economies is extremely heterogeneous, and examining them is a challenging task. Per capita income varies significantly across the group; however, these countries also differ with respect to level of social development, economic performance, political regimes, dominant religion and, for example, population density. These differences matter, not only because they shape a country's individual features but also because they heavily predetermine a country's ability to develop in various ways and—as in our case—to assimilated ICT. Additionally, we argue that treating all 46 countries within the scope of this book as an aggregate may be misleading. Hence, we deliberately disaggregated the evidence and analysed each country individually. Such an approach allows the unveiling of significant differences among and unique characteristics of examined economies. Treating the countries as one homogeneous group would have resulted in a loss of information and an inability to present the above-mentioned differences and characteristics.

Why is it important to ask whether developing countries are gaining access to ICT? The key point is that ICT enables unbounded flows of information and knowledge that will undeniably have far-reaching consequences for reshaping social and economic systems. ICT as General Purpose Technologies pervasively affect societies (Bresnahan and Trajtenberg 1995; Helpman 1998; David and Wright 1999), accelerating economic growth and development, although the positive effects of ICT deployment may be visible in national accounts only on a long-term horizon, as technological change does not necessarily induce productivity shifts immediately following its arrival (David 1990). Arguably, ICT brings opportunities to accelerate economic growth and development, also inducing advances in human development, and opening the 'Opportunity Windows' for economically backward countries.

1.3   Structure and Content

This book comprises six logically structured chapters. The first chapter is introduction. Chapters 2 and 3 provide theoretical background and analytical framework for further analysis. Chapters 4 and 5 address the empirical objectives of the book and present major findings of the analysis. Finally, Chap. 6 contains conclusions and recommendations.

The following briefly explains the major issues and contents of each part of this book.

Chapter 1 constitutes the Introduction itself.

Chapter 2 addresses basic ideas and concepts related to technology, technological progress and technological revolutions. It is intended to explain why technological changes constitute prerequisites enabling advancements along the socioeconomic development pattern. Moreover, it introduces the terms Information

Revolution and information and communication technologies (ICT), placing them in a broad historical perspective. The chapter explains why information and communication technologies are labelled as general purpose technologies, demonstrating four major aspects underlying the advantage of ICT compared with other, 'old' technologies'. Along these lines, this chapter exhibits the special relevance of information and communication technologies when implemented in developing countries. Finally, it briefly discusses the potential channels through which information and communication technologies may contribute to socioeconomic development in economically backward countries.

Chapter 3 introduces the theoretical outline of technology diffusion, which is defined as a dynamic and time-attributed process involving the transfer of information, knowledge and innovations, and standing for a continuous and gradual spread of new ideas throughout large-scale and heterogeneous societies. First, it extensively discusses theoretical technology diffusion concepts and models, explaining the technology diffusion trajectories by the use of S-shaped curves. Second, it presents the fundamental ideas and models standing behind the idea of technological substitution. Third, there is demonstrated a novel methodological approach to identification of the 'technological take-off' interval and the 'critical mass' with respect to the dynamics of the technology diffusion process and its prerequisites. Finally, based on theoretical frameworks derived from economic growth theories, it shows conceptualizations of technology convergence and technology convergence clubs.

Chapter 4 portrays country-specific ICT diffusion patterns in 17 low-income and 29 lower-middle-income economies during the period 2000–2012. We propose using six ICT indicators extracted exclusively from the World Telecommunication/ICT Indicators database 2013 (17th Edition). These indicators include the following: Fixed telephone lines per 100 inhabitants, Mobile cellular telephone subscriptions per 100 inhabitants, Fixed Internet[4] subscriptions per 100 inhabitants, Fixed broadband Internet subscriptions per 100 inhabitants, Wireless-broadband subscriptions per 100 inhabitants, and number of Internet users. In this part, the concept of an S-shaped curve is adopted to examine the ICT diffusion trajectories. This enables learning about the dynamics of the process and distinguishing its characteristic phases. Additionally, the chapter refers to arguments raised by Landes (2003), who claims that '*each innovation seems to have a life span of its own, comprising periods of tentative youth, vigorous maturity, and declining old age. As its technological possibilities are realized, its marginal yield diminishes and it gives way to newer, more advantageous techniques*' (Landes 2003, p. 3). Along these lines, the chapter examines technology substitution effects regarding fixed-telephone lines *versus* mobile cellular telephony, and fixed-internet networks *versus* wireless-broadband networks. The final sections of this chapter report on technology convergence and trace technology club formation among 46 developing and 67 developed economies over the period 2000–2012. At this point, the focus

---

[4] Refers to narrowband network.

shifts to answer the prominent question of whether countries exhibit growing cohesion (decreasing digital gaps) in terms of their level of adoption and use of ICT. Put another way, we discover whether rapid diffusion of ICT is accompanied by the process of technology convergence worldwide or, instead, by a gradual technology divergence or even dual-divergence leading to emergence of specific technology convergence clubs.

The targets of Chap. 5 are twofold. First, adopting a newly developed approach, it traces the country-specific *technological take-off* intervals and the *critical mass* regarding ICT diffusion (Mobile Cellular Telephony and Internet) in 17 - low-income countries and 29 lower-middle-income countries over the period 2000–2012. To this end, it identifies *critical penetration rate of new technology* and country-specific conditions during the *technological take-off* intervals. This approach provides a broad perspective on seminal factors that influence ICT diffusion in economically backward countries. Second, the chapter provides additional evidence on ICT diffusion determinants in low-income and lower-middle-income countries during the analogous period. It empirically traces the potential effect of selected socioeconomic factors on ICT spread. The analysis covers 10 indicators, which are used to explain level of mobile cellular telephony penetration rates, and 9 indicators to explain the level of usage of Internet by individuals. Moreover, we have selected another 8 indicators to demonstrate general socioeconomic and infrastructural features of examined countries. All data used in the analysis were extracted from the World Telecommunication/ICT Indicators database 2013 (17th Edition) (International Telecommunication Union), World Development Indicators 2013 (World Bank), Human Development Reports 2005–2013 (United Nation Development Program) and Measuring the Information Society reports 2009–2013 (International Telecommunication Union). Additional data were derived from the CIA World Factbook 2014, Freedom House 2014, The Heritage Foundation 2014 and national telecommunication agencies.

Chapter 6 comprehensively describes major empirical findings that are mentioned throughout the book. It shows major ICT diffusion trends, demonstrates the main features of the technological substitution process, and shows the technological convergence dynamics. It also provides insight into seminal factors that accelerate—or, conversely, hinder—rapid ICT diffusion in the countries under discussion. Moreover, it briefly discusses ICT policies that aim to foster ICT deployment in economically backward countries. Finally, it sheds light on the potential role of ICT in boosting growth and development economically backward countries.

I am fully aware that the main findings of this broad study may differ slightly from what the reader might have initially expected. Above all, however, I have intended to separate facts from suppositions. The unconventional approach for identification *technological take-off* and *critical mass*, although conclusive and interpretive, may well not be the best way to analyse the problem. Finally, this trial approach may yield further modifications and adjustments. I am also convinced that the numerical results of the examined ICT diffusion process and ICT diffusion determinants are at best rough approximations, as extreme variability in the dynamics of technological progress accounts for a mountain of different factors that are not always easy to capture and isolate. All of these shall be borne in mind when drawing conclusions and formulating recommendations.

References

Bresnahan, T. F., & Trajtenberg, M. (1995). General purpose technologies 'Engines of growth'? *Journal of Econometrics, 65*(1), 83–108.

David, P. A. (1990). The dynamo and the computer: An historical perspective on the modern productivity paradox. *The American Economic Review, 80*(2), 355–361.

David, P. A., & Wright, G. (1999). *General purpose technologies and surges in productivity. Historical reflections on the future of the ICT revolution*. University of Oxford, Discussion Papers in Economic and Social History, No. 31.

Dosi, G. (1982). Technological paradigms and technological trajectories: A suggested interpretation of the determinants and directions of technical change. *Research Policy, 11*(3), 147–162.

Freeman, C., & Louca, F. (2001). *As time goes by: From the industrial revolution to the information revolution*. Oxford: Oxford University Press.

Hanna, N. K. (2003). *Why national strategies are needed for ICT-enabled development*. World Bank Staff Paper. Washington, DC: World Bank.

Helpman, E. (Ed.). (1998). *General purpose technologies and economic growth*. Cambridge: MIT Press.

ITU. (2013). *World telecommunication/ICT indicators database 2013* (17th ed.). ITU.

Landes, D. S. (2003). *The unbound prometheus: Technological change and industrial development in Western Europe from 1750 to the present*. New York: Cambridge University Press.

Perez, C. (2009). Technological revolutions and techno-economic paradigms. *Cambridge Journal of Economics, 33*, 185–202.

Technology, The Economy, and Society: Casting the Bridges—Introductory Notes

2

'ICT has been the fastest technological change in history'
Nagy K. Hanna (2003)

Abstract

This chapter is intended to provide basic ideas and concepts related to technology, technological progress and technological revolutions. It is designed to explain why technological changes constitute prerequisites enabling advancements along the socioeconomic development pattern. Moreover, it introduces the terms Information Revolution and information and communication technologies (ICT), placing them in a broad historical perspective. The chapter explains why information and communication technologies are labelled as general purpose technologies, demonstrating four major aspects underlying the advantage of ICT compared with other, 'old' technologies'. Along these lines, it exhibits the special relevance of information and communication technologies when implemented in developing countries. Finally, it briefly discusses the potential channels through which information and communication technologies may contribute to socioeconomic development in developing countries.

Keywords
Technology • Technological revolution • ICT • Developing countries

2.1 Introduction

Before the Industrial Revolution, economies were characterised by negligible rates of economic growth and development (Cipolla 1994) and were thus relatively stagnant. Similarly, Deane (1979) claims that although growth occurs in stagnant economies, that growth *'is either painfully slow or spasmodic, or is readily reversible'* (Deane 1979, p. 11). Moreover, Granato et al. (1996) argue that pre-industrial economies were the zero-sum systems *'characterized by little or no economic*

© Springer International Publishing Switzerland 2015
E. Lechman, *ICT Diffusion in Developing Countries*,
DOI 10.1007/978-3-319-18254-4_2

growth which implies that upward social mobility only comes at expense of someone else' (Granato et al. 1996, p. 609).

Still, prior to the 1750s, medieval European societies made *'path-breaking inventions'* (Mokyr 2005) and produced a multitude of goods and services. Those pre-industrial societies adopted a number of seminal inventions, such as paper and wind power; regardless, the impact of those inventions on long-term growth and development was barely detectable. This is not to say that those inventions were unimportant, but rather that knowledge of how and why those technologies worked was not widespread. Put in another way, one could argue that people in pre-1750s societies knew too little and were too poorly educated to ensure the intellectual foundations for the expansion of technology. Therefore, the dynamic spread of knowledge of how technologies work and how they can be used to generate benefits has emerged as a critical factor in fostering long-term technology-driven socio-economic development.

In his influential book *'A farewell to alms: a brief economic history of the world'*, Gregory Clark (2008) writes: *'(...) the average person in the world of 1800 was no better off that the average person of 100,000 BC'*. Fortunately, the 1750s brought the Industrial Revolution, which radically transformed social and economic life in Europe, shifting individuals from material subsistence as personal incomes began to grow (Landau and Rosenberg 1986). It is claimed that the Industrial Revolution enabled today's developed countries to escape from the Malthusian trap (Galor and Weil 2000), mainly due to enormous gains from increasing productivity fostered by the spread of technological progress.

The remainder of this chapter briefly outlines major themes associated with 'technology' and 'technological progress' broadly defined, intending to highlight their pervasive role in shifting and transforming socio-economic life. The latter term is at the core of many theoretical and empirical debates seeking to capture and understand the socio-economic interpretation of overwhelming technological change (Dosi 1997; Comin, Hobijn, et al. 2006). It is undeniable that technology and innovation have transformed the way we live and brought about changes in civilisation throughout history, but today, the information revolution is transforming socio-economic systems faster than any technological revolution has ever done before and thus commands special attention and interest.

2.2 Technology: Ideas and Concepts

The basic notion of technology has been systematically transforming over the last 200 years. It has always been difficult to rigidly define the term 'technology' due to its complexity, and its contemporaneous definitions largely depend on the adopted frame of reference. Singer and Williams (1954) provide a coherent definition of the term technology, defining it as *'how things are commonly made or done'* (Singer and Williams (1954), I:vii). Technology may also be defined as *'a manner of accomplishing a task especially using technical processes, methods and knowledge'* (Comin et al. 2006). Fagerberg et al. (2010) label technology as a unique subset of knowledge on how to produce and distribute goods and services. As proposed by Wilson and Heeks, technology is *'a purposeful, practical activity that involves the application of knowledge by organizations of human beings and their interaction with hardware'* (Wilson and Heeks in: p. 403). These perceptions of the technology encompass four different dimensions: purposeful activity, human-machine interaction, knowledge, and organisational issues. Stoneman (2002) understands technology as the means deployed to produce goods and services at the firm, industry or national level, while Gomulka (2006) argues that in a narrow sense, technology may be defined as a set of available techniques to produce goods. He further states that technology may be equivalent to the state of knowledge necessary to production processes. Following this approach, Gomulka (2006) claims that technological change may be perceived as the enlargement of existing technologies (available techniques). Similarly, Fagerberg et al. (2010) emphasize that technology constitutes a subset of knowledge on how to produce and distribute goods, a definition that opens a new conceptual window. Layton (1974) views technology in a traditional way, and treats it as 'systematic knowledge'. In the same vein, Mokyr in his seminal book 'The gifts of Athena: historical origins of knowledge economy' (2002) claims that *'technology is knowledge, even if not all knowledge is technological'* (Mokyr 2002, p. 2). He also states that knowledge is a non-rivalrous good that is instantaneously transmitted and shared among society members so that each individual can make effective use of it. Following the conceptual framework provided in the works of Law (1991) and Bijker and Law (1992), we may also say that technology encompasses various heterogeneous elements originating from human skills and knowledge. They also claim that these heterogeneous elements create networks among society members, leading to the construction of more complex socio-technological systems. The concept of technological progress defined as knowledge may also be traced in works of Solow (1956, 1957), and hereafter is cited by Fagerberg (1994).

Finally, the literature recognises a concept of technology that seems to combine these two approaches. Dosi (1982) underlines that the economic literature defines the term technology rather narrowly, as a set of factors, whose combinations contribute to overall productivity. At the same time, Dosi (1982) suggests that technology should be viewed more broadly and proposes defining it as a *'set of pieces of knowledge, know-how, methods, procedures, experience of successes and failures, and also, of course, physical devices and equipment'* (Dosi 1982, p. 151). Following Arrow (1962), Dosi (1988) claims that technology may be perceived as information that is applicable and perhaps easily reproduced by economic actors. Furthermore, Pavitt (1999) underlines that technology is *'specific, complex, partly tacit, and cumulative in its development'* (Pavitt 1999, p. 3).

Technology as such is intimately related to technological change. In 1943, Schumpeter claimed that technology and technological progress transform ways of doing things (Schumpeter 1934). Developing the Schumpeterian idea, Rosenberg (1976) argues that technological change covers a wide array of human activities and constitutes an important element of complex socio-economic systems whose effects usually appear over the long run. Moreover, technological change and the stock of scientific knowledge are inseparable (Rosenberg 1974); that is to say, technological change is a consequence of knowledge, and vice versa. Perez and Soete (1988) perceive technological change as a long-term disruptive process that alters social and economic structures. They also state that technological change is a *more or less continuous process*' (Perez and Soete 1988, p. 460) that emerges globally.

The conceptualisation proposed by Mokyr (2002) and Mokyr and Scherer (1990) yields the claim that, technology may be broadly defined as knowledge. This interpretation of term technology has far reaching implications. On the one hand, technology is a consequence, an outcome and a product of human thought and embodies human knowledge; on the other hand, technology serves as a tool to transmitting knowledge among individuals. Recently emerged Information and Communication Technologies (ICTs hereafter) are prominent examples of technologies that may be perceived as a product of human activity and knowledge that simultaneously constitute a channel of dissemination of all types of knowledge and information among society members.

Broadly defined, ICTs may be understood as an extension of Information Technologies (IT); however when referring to ICTs, the primary focus is on media enabling communication. According to the World Bank (2014)[1], ICTs encompass hardware, software, networks and media for the collection, storage, processing, transmission and presentation of information (e.g., voice or data) and related services. Put another way, ICTs stand for a unique set of activities that enable storage, processing, transmitting and displaying all types of information and knowledge by electronic means (Rodriguez and Wilson 2000). UNESCO (2002) provides a slightly different definition of ICTs, claiming that information and communication technologies are a combination of informatics technology with other related technologies, especially communication; while informatics techno-logy is defined as a society's technological applications (artifacts) (UNESCO 2002). Hargittai (1999) defines the ICTs mainly through the lens of the Internet, arguing that it is worldwide network of both computers and the people who use them. He also argues that ICTs enable people to acquire vast amounts of information. In the same vein, Kiiski and Pohjola (2002) emphasize that ICTs provide unbounded possibilities for delivering information and interacting with other people (network building), and also constitute a type of 'virtual' market place to buy and sell goods and services.

In a broad sense, ICTs refer to technologies that use electronic means to serve people by sharing, distributing and stocking all sorts of information and knowledge. Regardless of whether the term ICT refers to devices or applications, there is always significant emphasis on its role in supporting various spheres of socio-economic activities. The concept of ICTs encompasses all arrangements that foster flows of information and knowledge and facilitate different forms of communication.

---

[1] See the ICT Glossary Guide (100 ICT Concepts) at http://web.worldbank.org (accessed: September 2014).

Henceforth, ICTs are designed to serve people and are thus often perceived through the lens of their functionalities, applicability and usability.

As suggested by Bresnahan and Trajtenberg (1995) and Jovanovic and Rousseau (2005), ICTs may be classified as General Purpose Technologies. The term General Purpose Technologies (GPTs hereafter), initially proposed by Bresnahan and Trajtenberg (1995) signifies the technologies that deeply affects both societies and economies at the national and global level. It is widely agreed that GPTs have the potential to dramatically transform social structures, norms and attitudes, and their implementation exhibits far reaching consequences for economic growth and development. Jovanovic and Rousseau (2005) conclude that the major effects of the extensive use of general purpose technologies are exhibited through their impact on economic growth, the transformation of ways of doing business and shifting social structures. Bresnahan and Trajtenberg (1995; Bresnahan 2003) claim that general purpose (or generic) technologies have three major characteristics: *pervasiveness*, which means that they may be implemented in all sectors of the economy; *technological dynamism*, which shows their inherent potential to constant improvements and lowering use costs; and *innovation spawning*, which explains increases in sectorial productivity as GPTs contribute to ease of inventing new goods and products. In this line, Helpman (1998), David and Wright (1999) and Lipsey et al. (2005) write that GPTs have the potential to generate innovations across industries and that they induce discontinuities in the long-term development of technologies due to their pervasiveness. The importance of adopting GPTs has also been underlined by Helpman and Trajtenberg (1996), who state that *'both historical evidence and theoretical analysis have brought forth the notion that general purpose technologies may play a key role in economic growth'* (Helpman and Trajtenberg 1996, p. 1). Jovanovic and Rousseau (2005) go one step further and conclude that both electricity and information technology are the most important general purpose technologies ever invented.

2.3 Technology: A Timeless Value

(...) in a fundamental sense, the history of technological progress is inseparable from the history of civilization itself, dealing as it does with human efforts to raise productivity under an extremely divers range of environmental conditions
Nathan Rosenberg (1982)

(...) the Industrial Revolution marked a major turning point in man's history
David Landes (2003)

Technology has always been at the centre of human interest because technology makes human advancement possible. Technology, the economy and society are intimately interrelated and fundamentally inseparable (Rosenberg 1982). Kindleberger (1995) argued that socio-economic systems and technological progress are interdependent in a way that may be either positively or negatively influential. Moreover, the relationships between society, the economy and technology are linked by two-way causality. On the one hand, technological progress induces social and economic changes, but on the other, the speed and adoption of technologies is predetermined by socio-economic capabilities and performance. Technology and technological change are deeply rooted in broad social and economic contexts (Mokyr 1990; Fox 1996), and each society faces certain *technological facts* that have far reaching consequences for its dynamic performance (Bresnahan and Trajtenberg 1995). Mokyr argues that throughout the ages technology has '*revolutionized the structure of firms and households, it altered the way people look and feel, how long they live, how many children they have, and how they spend their time*' (Mokyr 2002, p. 2). Technological changes are often revolutionary in nature; that is to say, they are disruptive, continuous, and sometimes abrupt, causing deep and long-term changes in the social and economic status quo and becoming the primary engine of economic growth and development (Landes 2003). Similarly, Fagerberg and Verspagen (2002) restate that '*technology is a key factor shaping economic growth*' (Fagerberg and Verspagen 2002, p. 1294).

Thorstein Veblen (1915) was one of the first to examine the role of technological changes in economic development and the catch-up process. He argued that due to cross-country technology transfers, poor countries should inevitably enter a sustainable growth path and catch up with the developed economies. Early neoclassical models and concepts, e.g., those proposed by Solow (1956), treat technological advancement exogenously and highlight its seminal role in fostering long-run economic growth and development. A similar approach to the role of technology in shaping economic performance of countries may be traced in works of Schumpeter (1934, 1947) and Kaldor (1957), to name just two examples. Other significant theoretical and empirical contributions recognizing the links between technology and economic development were made by, *inter alia,* Uzawa (1965), Nelson and Phelps (1966) and Shell (1967). All the authors mentioned above stressed the importance of technological change and the permanent growth of technology as determinants of significant shifts in labour force skills and abilities that should influence positively national income growth rates. In addition to the previously cited authors, a remarkable literature has emerged that is concerned strictly with endogenous growth models. Examples include the works of Lucas (1988), Romer (1990), Grossman and Helpman (1991) and Aghion and Howitt (1992), in which role of technologies in fostering economic growth was highly emphasised. In line with the literature explaining technology as a factor of economic growth, another subset of work emerged in economic theory that combines the previous ideas with the hypothesis of catching-up in reference to developing countries. The idea of implementing technology in broad development theories in this sense was undertaken in works of Gerschenkron (1962), Findlay (1978) and Abramovitz (1986), to name a few. Gerschenkron claimed that developing countries mainly operate below the world technology (innovation) frontier, and by coping (imitating) the developed technologies gain the opportunity to converge (catch-up) with developed countries in terms of economic development. 'Technological congruence', meaning a lack of appropriate technology to enter the development path, has also been stressed in the works of Abramovitz (1994).

Gerschenkron (1962) writes that 'borrowed technology, so much and rightly stressed by Veblen (1915), was one of the primary factors assuring a high speed of development in a backward country'. Technology and innovation will foster the catch-up process of low-income countries mainly by enabling improvements in education, the diffusion of knowledge and shifts in labour productivity. The concepts cited above have also been extensively studied in empirical works by Castellacci (2006, 2008, 2011) and Ben-David (1998). Apart from the works cited above, there is a voluminous body of contemporary theoretical and empirical literature concentrating on identifying technology's role in economic growth and development. Evidence of this is traced in contributive and influential works by, *inter alia*, Romer (1990, 1993, 1994), Hewitt and Wield (1992), Mankiw et al. (1995), Savvides and Zachariadis (2005), Antonelli (2011), Nelson (2011) and Fukuda-Parr and Lopes (2013).

The positive outcomes of technological change may differ substantially across countries, as technological progress will only generate economic gains when it is accepted and assimilated by societies. Arguably, the process of adopting new technologies and their contribution to growth and development is far from automatic, and the pervasiveness and acceptance of new technology, as well as the speed at which technology diffuses, are attributed to complex social, institutional and economic forces. As underlined by Keller (1996) and Kostopoulos et al. (2011), the emergence of sustained benefits from extensive technology adoption is preconditioned by societies' absorptive capacities and their ability to deploy and use technology. Whether societies are able to rapidly adopt emerging technologies is essentially preconditioned by institutional environment, social attitudes, norms and values, and a wide array of economic or institutional incentives (Rosenberg 1982, 1994; Rosenberg et al. 2008). Certain societies may be endowed with poor education, low quality human capital, cultural constraints, an unfavourable institutional and legal environment, or simply geography, any of which may heavily impede the possibility of deploying new technologies. Similar arguments are raised by Soete and Verspagen (1993), who claim that societies assimilate new technologies by relying on their 'intellectual capital', namely, institutional and cultural prerequisites. On the contrary, better educated societies that exhibit little risk-aversion and a high propensity to adopt novelties, assimilate new technologies relatively easily and quickly. Put another way, the rate of diffusion and deployment of technologies depends on the absorptive capacities of the respective societies (see e.g., Baumol 1986; Perez and Soete 1988; Cohen and Levinthal 1990; Verspagen 1991; Criscuolo and Narula 2008). Some empirical evidence shows that the most prominent factors in a country's ability to adopt and effectively use new technologies are the education level and skills of its labour force (Baumol 1989; Benhabib and Spiegel 2005). Countries experiencing significant gaps in these areas may never be able to utilise the full potential of technological change. Various aspects of how societies progress technologically and are able to exploit the full potential of newly emerging technologies for economic benefit are discussed in works of, *inter alia*, Kim and Lee (2004) and Jensen et al. (2007), who write about the technological capabilities of societies. Others writing on this topic include

Lundvall (2010), Lall (1992) and Nelson (1993), who underline the role of innovations systems in technology adoption.

The key to a better understanding of the long-term impact of technological change on socio-economic development is to examine the issue from a wider historical perspective. A brief look at the last 200 years of economic development and technological progress sheds more light on their interdependency, and shows that technological revolutions are excellent examples of how technological change may shape societies and impact long-term economic growth and development. As stated by Dosi (1997) and Mowery and Rosenberg (1991), looking backward, long-term socio-economic development was essentially influenced by technological changes, which are especially well pronounced in disruptive technological revolutions. Landes (2003) perceives the technological revolution in terms of the First Industrial Revolution and claims that technological revolution may be regarded as technological innovations, which *'by substituting machines for human skills and inanimate power for human and animal force, brings about a shift from handicraft to manufacture and, so doing, gives birth to a modern economy'* (Landes 2003, p. 1). Perez (2009) offers a more general definition of technological revolution, defining it as a set of radical breakthroughs that give rise to set of new interrelated technologies. In the same work, Perez argues that technological revolutions have two main characteristics: newly emerged technologies are strongly interconnected and interdependent, and the technological revolution deeply transforms society and the economy. In the same vein, Hanna (2010) claims that technological revolutions may be defined as a cluster of newly emerged technologies that which dynamically diffuse across new industries and products and, when combined with infrastructural shifts, induce upswings in total productivity and hence economic development. Perceived broadly, the concept of technological revolution closely relates technological changes to socio-economic development. Arguably, due to the vast deployment of innovation and infrastructural and organisational improvements, technological revolution provides a solid background for increases in productivity, which, in turn, brings about a dramatic surge in economic performance and society's wealth.

From a historical perspective and especially in terms of the depth and pervasiveness of influence on society and the economy, the Industrial Revolution was the event that brought the most remarkable changes. In his influential book *'The Unbound Prometeus'*, Landes (2003) emphasises the great importance of the First Industrial Revolution, stating that 'the technological changes that we denote as the *'Industrial Revolution' implied a far more drastic break with the past that anything since the invention of the wheel'* (Landes 2003, p. 42). Moreover, he concludes that this revolution brought disruptive changes on the entrepreneurial side, enhanced shifts in investments, induced fundamental changes in the occupational role of the labour force, generated strong incentives for further re-organisations and restructuring in manufacturing and forced entrepreneurs to accept change by inducing them to increase their tolerance for risk in the hope of future gains. As an example of radical changes caused by the Industrial Revolution, Landes (2003) writes that in 1760, the British consumption of cotton was at approximately 2.5

million pounds, while most of labour associated with cotton production was done by hand mainly in workers' homes. Twenty-seven years later in 1787, cotton consumption had grown to 22 million pounds, and the cotton industry was second only to wool in Great Britain in terms of number of persons employed. After another 50-year period, the cotton industry was Britain's most important industry in terms of employment, invested capital and the value of the product, advances that were accompanied by a dramatic drop in the price of yarn. Similar claims regarding the Industrial Revolution may be traced in work of Ashton (1970). He argues that the intensity of the impact of technology and technological changes on society and the economy was helped by the dramatic changes that brought about the Industrial Revolution, which completely reshaped and profoundly transformed the socio-economic landscape (Ashton 1970). In the same line, Mokyr (1993) claims that *'in the past two centuries (...) output per capita have increased dramatically and in sustained manner, in a way they have never done before. It seems by now a consensus to term the start of this phenomenon 'the Industrial Revolution' although it is somewhat in dispute what precisely is meant by that term'*. The Industrial Revolution has undoubtedly had an essential impact in countries where it took place. Unquestionably, the Industrial Revolution has generated radical increases in per capita income across countries (Mokyr 2005); however, it is worth noting that *'(...) though the conventional date for the onset of the Industrial Revolution in Britain is given as the 1760s there is little sign of rapid growth of income per person until the decade of the 1860s'* (Clark 2008, p. 194). Mokyr (2002) makes similar claims, writing that *'It has become a consensus a view that economic growth as normally defined (...) was very slow during the Industrial Revolution, and that living standards barely nudged upward until the mid-1840s. (...). Yet it is also recognized that there are considerable time lags between the adoption of major technological breakthroughs (...) and their macroeconomic effects'* (Mokyr 2002, p. 30).

The Industrial Revolution and the systematic and continuous technological changes that it brought impacted not only economies but also societies (Deane 1979; Galor and Weil 2000). Increasing per capita output induced deep social changes (Deane 1979) that had far reaching consequences. The key point is that technological changes opened up a window of opportunities for entrepreneurs, who began trading successfully and investing in new companies, which dramatically shifted their incomes. The modernisation of production required improved skills from workers while newly emerging firms still demanded a workforce, which enhanced rural to urban migration. It is important to note that due to the growing demand for labour, women and children began actively participating in the labour market, which is, *inter alia*, interpreted as one of the causes of further declines in total fertility rates and the demographic revolution. An even more dramatic social and economic transformation occurred in the wake of the Second Industrial Revolution (approx. 1870s–1908[2]) (see below this Sect. 2.4) when mass production

---

[2] Or 1914—according to various sources.

became common, which tremendously increased the degree of urbanisation and population growth[3] (due to improving access to healthcare and the agricultural revolution). On the other hand, during the epoch of rapid industrialisation, visible social inequalities emerged in the form of the burgeoning middle class, whose members were relatively better off than the working class, who lived in dramatically worse conditions. Even so, the Industrial Revolutions brought about radical improvements in living standards across the board. People had access to better products at lower prices, began building increased education and skills, and experienced lowering death rates. Over the long-run, there are various rational reasons to believe that technological change undeniably forces steady increases in medical, social and economic outcomes.

Regarding the positive gains enhanced by technological change, it is important to note that the *real* impact of technology on either the economy or society is not always clearly demonstrated in the immediate aftermath of technological change. Roughly speaking, gains in quality of life brought by technological progress will not be fully apparent until a significant amount of time has passed. As noted by Mokyr (2005), the Industrial Revolution *itself* was not a period of rapid economic growth. Moreover, benefits from technological changes were not revealed immediately after the emergence of the Industrial Revolution, and early industrialised societies waited almost 100 years for economic growth to speed up. In support of Mokyr's supposition, Clark (2008) underlined that in Britain from the 1760s to the 1860s, the signs of rapid growth in per capita income were scarcely noted. Today, the phenomenon of substantial time lags in reaping the gains from technological progress throughout national accounts and in terms of per capita income growth is recognised as 'Solow's productivity paradox', an idea that was introduced to the economic literature by Solow himself in 1987. Through his famous claim that *'you can see the computer age everywhere but in the productivity statistics'*, Solow explains that rapid technological changes demonstrate slow gains in total productivity (David 1990). According to our intuition, this observed 'productivity paradox' may be at least partially explained by the dynamics to which the process of the diffusion, adoption and deployment of technological changes across a socio-economic system are attributed. Technological changes are being gradually installed and embodied within society and economy, moving through two characteristic phases (periods) (see Perez 1985, 2002; Cvetanović et al. 2012). The first period is recognised as the installation period, and the second as the deployment period. During the installation period, technological change spreads over society and the economy. Although diffusion is initially slow, once a critical mass of adopters emerges, diffusion accelerates and technological changes spread widely. The installation phase is critical, as it preconditions wide deployment and society's adaptation to technological change, which, in turn, induces structural shifts and economy-wide re-organisations. When leaving the installation phase and entering the deployment period, technological changes are adopted by a vast majority of

---

[3] According to Clark (2008), the English population tripled between 1770 and 1860.

society and are disruptive enough to induce shifts in productivity. By the end of the deployment period the positive gains from technological changes are gradually unveiled in growth statistics.[4]

2.4 From Industrial to Information Revolution

It is widely recognised that historically, technological revolutions typically occur every 40–60 years (Hanna 2010) and that each technological revolution is associated with radical changes in the techno-economic paradigm (regime). The concept of the techno-economic paradigm was originally proposed by Pérez (1986) and has since been adjusted and augmented in the works of Freeman and Perez (1986, 1988) and Perez (2002, 2003, 2009). Relying on conceptual foundations developed by Kuhn[5] (1962), Dosi (1982), Dosi[6] et al. (1988) and Freeman and Soete (1997), Pérez (1986) proposed the incorporating the term techno-economic paradigm into the broad analysis of technological revolutions. She also argued that each technological revolution induces its own techno-economic paradigm (regime) that constitutes a newly emerged technical and institutional best-practice frontier. The new institutional 'frontier' is perceived as a necessary organisational transformation that fosters shifts in sectorial productivity enabled by technological change. Freeman and Perez (1986, 1988) define the techno-economic paradigm as a set of technical and economic features of emerged technological solutions that are gradually incorporated into economic systems, eventually becoming integral. As redefined by Hanna (2010), *'a techno-economic paradigm articulates the technical and organisational model for taking the best advantage of the technological revolution and results in the rejuvenation of the whole productive structure'* (Hanna 2010, p. 31).

In this vein, we briefly discuss the major technological breakthroughs that have occurred over the last 200 years.[7] The First Industrial Revolution began around the 1770s in Britain and enabled the mechanisation of cotton industry, improvements in

---

[4]To a certain extent, this view coincides with the Kondratiev's concept of long-waves and its Schumpeterian interpretation regarding the role of technological progress in long-term growth. Both Kondratiev and Schumpeter attribute the emergence of long business cycles (approximately 50–60 years long) to the diffusion of technological progress. As successive technologies diffuse along logistic patterns, they gradually unveil their potential growth in productivity. The full potential of newly emerged technological changes, however, is exhibited once they are broadly adopted by society, which allows for the generation of gains such as growth in per capita income. A similar approach can be also found the work of Göransson and Söderberg (2005).

[5]Kuhn (1962) used the term *paradigm* to explain shifts in the theoretical perspectives in the development of sciences from historical perspective (Cvetanović et al. 2012).

[6]Dosi (1982) and Dosi et al. (1988) offered the definition of *technological paradigm* and state that it is a sphere of technology that hosts the search for innovations and is placed in a certain historical context. They also argue that it is a useful tool for analysing the role of technological change in the production of goods and services.

[7]We follow Freeman and Soete (1997), Perez (2002), Landes (2003) and Hanna (2010).

the water wheel (and thus more effective use of water energy) and the refinement of turnpike roads and canals. During the First Technological (also labelled Industrial Revolution), industry was organised mainly around small firms and individual entrepreneurs to which the rise of incomes and wealth was generally subjected. The period from 1829 onward is the time of Second Technological Revolution. This period is sometimes labelled as the 'Age of Stream and Railways', as further development was enabled by steam engines and steam-powered railways. The Second Technological Revolution is marked by the growing significance of railways, postal and telegraph services, ports and international sailing ships. Moreover, in those times, growing market competition and the emergence of large companies is observed. It is generally considered that the development of railways, postal and telegraph services gave rise to the increasing importance of networks and communication in economic development and social change. The period from 1875 to 1908 accounts for the Third Technological Revolution, which was the age of steel and electricity. The later perpetuated the further development of global railways and telegraph services, gave rise to analog telephony services and heavy and electrical engineering, and also saw the use of electricity for industrial purposes become common. The Third Technological Revolution was also marked by the emergence of giant companies, trusts and cartels, which, in turn, induced a growing number of legal anti-trust regulations. In 1908 the Fourth Technological Revolution began. This was the age of oil, mass production, and the dynamic development of roads, automobiles (Ford plants), ports and airports. Electricity was gradually deployed in a growing number of homes, which created electrical networks. The Fourth Technological Revolution was also a period of the global spread of analog modes of telecommunication (telephone, telegraph and cablegram). These changes mandated the further development of various types of networks, which began to constitute the prime engines of economic development while simultaneously deeply transforming social structures, norms and attitudes. Finally, beginning in the 1970s, the world has witnessed the Fifth Technological Revolution, broadly recognised as the age of information and telecommunications and encompassing microelectronics, software, computers and different forms of digital communications, including the Internet (Perez 2002; Freeman and Louca 2001). The Fifth Technological Revolution started with the introduction of the microprocessor to the public in 1971, the first personal computer in 1973 (by Intel) and mobile telephony by Motorola in the same year. Since then, both PCs and mobile telephony have diffused worldwide, profoundly reshaping societies and economic systems. According to Freeman and Louca (2001), Perez (2002) and Conceição and Heitor (2003), the development of information and communication technologies (ICTs) constitutes a newly emerged techno-economic paradigm, which has also been labelled the digital (ICT) paradigm. It is evident that in many ways, the Information Revolution is different from those in the past. The prime element and the most essential is that revolutionary changes are much faster and more pervasive compared to those generated by the previous revolutions, and, moreover, technological changes today are broadly embodied in various goods and services offered to the mass-market.

As has been argued in Sect. 2.3, information and communication technologies are recognised as contemporary general purpose technologies. ICTs are 'enabling technologies' that offer unbounded opportunities through their adoption and implementation on multiple social, institutional and economic grounds. The impact of ICTs on reshaping social and economic development is thought to be pervasive, and in the long-term perspective, likely to induce structural and organisational changes that lead to essential shifts in productivity. Hanna (2010) claims that the information and communication revolution is probably the most pervasive in recent human history, and has also argued that the timing of the Fifth Technological Revolution is mainly due to decentralisation and integration, network structures, adaptability, knowledge as capital and economies of scope (see Hanna 2010, p. 32).

There are at least four major aspects underlying the major advantages and importance of ICT compared to other 'old' technologies. First ICTs, like electricity or railways, create different types of networks (Shapiro and Varian 1999, 2013; Valente 1995, 1996; Castells et al. 2009; van den Berg et al. 2013). In late 1990s, Shapiro and Varian (1999) claimed that the major difference between the old and the new economies is crucial in that the old economies were predominantly driven by economies of scale, while the new economies are driven by economies of networks. Similarly, Servon (2008) argued that ICT has fundamentally reshaped societies and shifted many countries from the 'industrial age to a network age'. Following Katz and Shapiro (1985) and Economides (1996), network effects reveal the increasing utility derived from using a given good or service when accompanied by an increasing number of users of analogous goods or services. The revelation of the 'network effects' explains the value of potential connectivity, which tends to grow exponentially in heterogeneous societies. On the ground of the economy, it means that the growing number of links is potentially translated into real revenues (e.g., increasing GDP per capita). Second, the ease of creating different forms of networks through ICT adoption relies on the very nature of these technologies. Currently, ICT offers a wide array of services that are based on wireless solutions, which enables connectivity, data and voice transfer from any location. ICTs may be thus perceived as technologies that free people from geographic isolation and virtual marginalisation, shifting different activities to remote regions. Societies can use ICT regardless of their geographical location, so that the physical distance itself does not hinder the possibility of accessing ICT and thus accessing the ability to communicate and acquire information. Bearing the latter in mind, ICT may be defined as inclusive technologies that enable the 'death of distance' (Cairncross 2001). Similar arguments have been raised in works of Quah (2001), Venables (2001), Redding and Venables (2002) and Wresch and Fraser (2012), who emphasised the special role of information and communication technologies in various aspects of socio-economic development. Third, ICT enables and enhances massive flows of information across both individuals and entire societies. What is seminal here is that both information and knowledge sharing occurs rapidly and at a negligible cost, and thus becomes available even in low-income societies that have been traditionally left behind in terms of access to and use of various forms of technology. To a point, the widespread deployment of ICT allows for the gradual eradication of various forms of exclusion from access to knowledge and information. From a broader perspective, unbounded flows of knowledge and information have broad implications for socio-economic development and fostering growth. The key is that ICT provides a solid background for making knowledge work, thus helping knowledge to be transformed into long-term economic and social gains. Fourth, ICTs are recognised as general purpose technologies, and their key feature is therefore generality of purpose (Bresnahan and Trajtenberg 1995). For this reason, ICTs' influence on society and the economy is pervasive; they affects wide range of economic sectors, social structures and institutions, creating new frameworks within which all actors operate and interact. Many authors claim that at present, the deployment of ICT is sufficiently extensive that that no individual remains unaffected by the information revolution.

2.5   ICT: Opportunity Window for Developing Countries?

(. . .) the biggest beneficiary of the Industrial Revolution has so far been the unskilled
Gregory Clark (2008)

People living in economically backward countries have, to a certain extent, been omitted by previous technological revolutions. This is not to say that *no* type of technological progress has ever reached them, because certainly it has; however, the spread and access to use of that technological progress was extremely limited (e.g., low electrification rates or negligible access to railway networks). Many developing countries have never had any opportunity to adopt and effectively use the blessings of previous technological revolutions, and they 'traditionally' lag behind in terms of 'modern' technology adoption. This lag has obviously hindered their ability to develop rapidly or advance in overall well-being. Permanent inability to access and benefit from technological progress has resulted in major barriers to development having never been broken but instead persisting over time. In a way, societies in economically backward countries have never had an opportunity to 'consume' the technological changes that emerged over the last 200 years. Of particular importance, such societies have been unable to use technological progress as a driving force for socioeconomic development. Such an unfavourable situation was determined by, *first*, the potential to exploit past technological revolutions, which required essential financial resources and relatively well-developed hard infrastructure to be installed and used countrywide, and, *second*, that this potential required much more knowledge, skills and absorptive capacities to be deployed and then used effectively to induce scalable and long-term economic benefits. In short, economically backward countries have never been the real beneficiaries of past technological revolutions.

Luckily, in the early 1970s, the Information Revolution emerged, giving rise to new opportunities. Arguably, the contemporary Information Revolution is critical regarding the technological progress it induced because that progress can be accessed and adopted worldwide. This was not the case with previous revolutions.

Following Hanna (2010), it is correct to state that for low-income and slowly growing economies, the Information Revolution is more like *'a tsunami rather than a new technological wave'* (Hanna 2010, p. 32). This irresistibility is clearly the great advantage of the newly emerged Information Revolution.

Recently, much attention has been paid to recognition of the opportunities that ICT offers for developing economies, emphasizing their high relevance when deployed in low-income societies. With regard to that relevance, two seminal questions arise. The first question is, why would ICT exhibit high relevance when installed and adopted in economically backward countries? The second question is, what are the Opportunity Windows[8] through which ICT might affect economically backward countries? Identifying these Opportunity Windows may show channels through which ICT potentially affects societies and economies.

Question 1

Why would ICT exhibit high relevance when installed and adopted in economically backward countries?

There are several aspects to be discussed. *First*—ICT are installable relatively easily in permanently underserved, remote, rural and geographically isolated regions where the degree of development of backbone hard infrastructure is poor. This allows perceiving ICT as technological solutions that, to a point, go 'beyond geography', overcoming physical distances and infrastructure shortages. Regarding economically backward countries, this ICT feature is critical if they are to be deployed and adopted by low-income societies. Moreover, opportunities offered by the newest wireless networks yield special relevance in this respect, as they connect previously unconnected people with the outside world. *Second*—ICT may be bought at relatively low prices, essentially increasing their affordability even by low-income people. In economically backward countries, a vast majority of society usually permanently suffers from poverty and material deprivation; hence, a low price becomes a prerequisite for ICT to be afforded and adopted. The low price of ICT is especially relevant with regard to mobile cellular services, which are usually offered in pre-paid systems, allowing people with no regular income to use this type of communication. *Third*—ICT may be easily adopted and used even by low-educated, low-skilled or even illiterate people. This makes these technologies really *'for all'*. ICT do not require much knowledge from the final consumer to deliver benefits. *Fourth*—ICT are often adopted by traditional societies that, until now, have been left in social, cultural and economic isolation. It seems that ICT are acquired regardless of existing social norms and attitudes, and thus break communication barriers and enhance growing social cohesion and interactivity. *Fifth*—ICT are easily imitable and deliverable, which ensures historically unprecedented rapid diffusion to individuals and throughout societies. *Sixth*—the marginal cost of an additional user of ICT is negligible (even close to zero). Some ICT, especially the

---

[8] The term 'windows of opportunity' was introduced to the literature by Carlota Perez (see, e.g., Perez and Soete 1988).

Internet, are recognized as non-rivalrous goods; thus, their usage and consumption do not occur at the expense of any other individual.

All of the elements listed above are decisive to open wide Opportunity Windows and allow treating ICT as a highly favourable techno-economic paradigm compared with paradigms that emerged during previous technological revolutions, particularly with regard to the special characteristics of economically backward countries and the multiple constraints they face.

Question 2

What are the Opportunity Windows[9] through which ICT might affect economically backward countries?

The prime attribute of ICT is that they enable widespread communication and rapid and easy access to information and knowledge, which are critical prerequisites to providing solid foundations for long-term socioeconomic development. Indeed, all of the opportunities that ICT offer are closely related to the unbounded flows of information and knowledge that they foster (Hanna et al. 1995).

ICT may directly affect socioeconomic development through resource mobilizing and enforcing market activities. A crucial point is that ICT offer opportunities for greater involvement of resources in market activities. Through better access to financial markets (e.g., e-finance and mobile-finance solutions), ICT foster mobilization of savings and provide opportunities to convert them into investments, which has long-term positive consequences for market activity and economic growth. On the other hand, ICT enhance greater mobilization of the labour force, which has multidimensional consequences. First, growing participation in formal labour markets provides a solid fundament for obtaining regular income. This shifts people from subsistence, allows for gradual alleviation of poverty, reducing their vulnerability and risk exposure to external shocks. Undoubtedly, improving engagement in labour markets, both through growing employment or establishing new small firms, creates economic gains and allows for gradual eradication of various forms of socioeconomic deprivation. Most likely, increasing labour force engagement constitutes the first and the most important step through which the ICT potential may be exploited and exhibited in developing countries. Importantly, the application of ICT enables timely access to information, which facilitates removal of one of the fundamental barriers to the effective functioning of the market: information asymmetries. These two elements—shifting labour force participation and removing constraints to information access—if combined appropriately, force increases in number of transactions, enable participation in global trading markets, force drops in transaction costs and ensure worldwide visibility, which—in turn—from a long-term perspective, offer good prospects for economic growth and development.

---

[9] The term 'windows of opportunity' was introduced to the literature by Carlota Perez (see, e.g., Perez and Soete 1988).

Indirectly, ICT may affect socioeconomic development though improved access to education and knowledge, improved and more effective functioning of healthcare systems (mainly e-health and telemedicine applications) or the so-called e-government solution. All of these significantly foster increases in human capital and skills, contribute to social cohesion, enhance empowerment of all social groups (e.g., endogenous people), and ensure transparency and political inclusion. Obviously, the effect of ICT on, for example, educational or healthcare systems is qualitative in nature; the real gains are gradually demonstrated by advancements in social and economic aspects.

In summary, ICT allow for opening the Opportunity Windows by breaking barriers that have deprived societies of various social and economic activities, and offering instead all of these previously denied opportunities. *That is* the potential of ICT from the perspective of economically backward countries.

Obviously, ICT adoption and channels that affect socioeconomic systems are not limited to what was presented above, especially with respect to highly developed countries. Evidently, ICT applications, modes of usage and channels of effect are dramatically different in developing countries compared with those in highly developed economies. However, if economically backwards countries are considered, ICT are initially adapted to rather 'basic' activities that do not require knowledge and financial resources, before moving to more sophisticated applications and channels affecting socioeconomic systems.

These Opportunity Windows are not opened unconditionally, and the full exploitation of ICT potential is far from automatic. Favourable legal and institutional environments and a degree of telecommunication market competition all are obviously critical for adoption and usage of ICT. Many prerequisites emerge that help or hinder widespread implementation of ICT. Some claim that basic infrastructure must be assured, legal reforms are needed to allow for market competition in the telecommunication market and inflows of direct foreign investments are needed. On the other hand, effective ICT deployment requires continuous learning and growing social capabilities so that the ICT potential can be realized. ICT need to be promoted and supported to ease their economical implementation and adoption in society. If and *only* if fundamental preconditions enabling ICT spread and acquisition are ensured, ICT should diffuse and be gradually implemented in multiple fields, generate future social and economic gains, and convert societies 'information poor' into 'information rich'.

References

Abramovitz, M. (1986). Catching up, forging ahead, and falling behind. *Journal of Economic History, 46*(2), 385–406.

Abramowitz, M. (1994). Catch-up and convergence in the postwar growth boom and after. In W. Baumol, R. Nelson, & E. Wolff (Eds.), *Convergence and productivity: Cross-national studies and historical evidence* (pp. 86–125). Oxford: Oxford University Press.

Antonelli, C. (Ed.). (2011). *Handbook on the economic complexity of technological change.* Cheltenham: Edward Elgar Publishing.

Aghion, P., & Howitt, P. (1992). A model of growth through creative destruction. *Econometrica,* *60*(March), 323–351.

Arrow, K. (1962). Economic welfare and the allocation of resources for invention. In R. R. Nelson (Ed.), *The rate and direction of inventive activity: Economic and social factors* (pp. 609–626). Princeton, NJ: NBER.

Ashton, T. S. (1970). *The industrial revolution, 1760–1830* (Vol. 38). CUP Archive.

Baumol, W. (1986). Productivity growth, convergence, and welfare: What the long-run data show. *American Economic Review, 76*, 1072–1084.

Baumol, W. J. (1989). Reflections on modern economics: Review. *Cambridge Journal of Economics, 13*(2), 353–58.

Benhabib, J., & Spiegel, M. M. (2005). *Human capital and technology diffusion. Handbook of economic growth* (Vol. 1, pp. 935–966). Amsterdam: Elseiver.

Ben-David, D. (1998). Convergence clubs and subsistence economies. *Journal of Development Economics, 55*(1), 155–171.

Bijker, W. E., & Law, J. (1992). *Shaping technology/building society: Studies in sociotechnical change.* Cambridge: MIT Press.

Bresnahan, T. F. (2003). The mechanisms of information technology's contribution to economic growth. In: J.-P. Touffut (Ed.), *Institutions, innovation and growth: Selected economic papers. the centre Saint-Gobain for economic studies* (p. 116). Northampton, MA: Edward Elgar.

Bresnahan, T. F., & Trajtenberg, M. (1995). General purpose technologies 'Engines of growth'? *Journal of Econometrics, 65*(1), 83–108.

Cairncross, F. (2001). *The death of distance: How the communications revolution is changing our lives.* Boston: Harvard Business Press.

Castellacci F. (2006). *Convergence and divergence among technology clubs* (DRUID Working Paper No. 06–21).

Castellacci, F. (2008). Technology clubs, technology gaps and growth trajectories. *Structural Changes and Economic Dynamics, 19*(2008), 301–314.

Castellacci, F. (2011). Closing the technology gap? *Review of Development Economics, 15*(1), 189–197.

Castells, M., Fernandez-Ardevol, M., Qiu, J. L., & Sey, A. (2009). *Mobile communication and society: A global perspective.* Cambridge: MIT Press.

Cipolla, C. M. (1994). *Before the industrial revolution: European society and economy, 1000–1700.* New York: WW Norton & Company.

Clark, G. (2008). *A farewell to alms: A brief economic history of the world.* Princeton, NJ: Princeton University Press.

Cohen, W. M., & Levinthal, D. A. (1990). Absorptive capacity: A new perspective on learning and innovation. *Administrative Science Quarterly, 35*(1), 128–152.

Comin, D., Easterly, W., & Gong, E. (2006). *Was the wealth of nations determined in 1000 BC?* (No. w12657). National Bureau of Economic Research.

Comin, D., Hobijn, B., & Rovito, E. (2006). *Five facts you need to know about technology diffusion* (No. w11928). National Bureau of Economic Research.

Conceição, P., & Heitor, M. V. (2003). Techno-economic paradigms and latecomer industrialization. Mimeo.

Criscuolo, P., & Narula, R. (2008). A novel approach to national technological accumulation and absorptive capacity: Aggregating Cohen and Levinthal. *The European Journal of Development Research, 20*(1), 56–73.

Cvetanović, S., Despotović, D., & Mladenović, I. (2012). The concept of technological paradigm and the cyclical movements of the economy. *Facta universitatis-series: Economics and Organization, 9*(2), 149–159.

David, P. A. (1990). The dynamo and the computer: An historical perspective on the modern productivity paradox. *The American Economic Review, 80*(2), 355–361.

David, P. A., & Wright, G. (1999). *General purpose technologies and surges in productivity. Historical reflections on the future of the ICT revolution*. University of Oxford, Discussion Papers in Economic and Social History, No. 31.

Deane, P. M. (1979). *The first industrial revolution*. Cambridge: Cambridge University Press.

Dosi, G. (1982). Technological paradigms and technological trajectories: A suggested interpretation of the determinants and directions of technical change. *Research Policy, 11*(3), 147–162.

Dosi, G. (1988). Sources, procedures, and microeconomic effects of innovation. *Journal of Economic Literature, 26*, 1120–1171.

Dosi, G., et al. (Eds.). (1988). *Technical change and economic theory*. London: Columbia University Press.

Dosi, G. (1997). Opportunities, incentives and the collective patterns of technological change. *The Economic Journal, 107*(444), 1530–1547.

Economides, N. (1996). The economics of networks. *International Journal of Industrial Organization, 14*(6), 673–699.

Fagerberg, J. (1994). Technology and international differences in growth rates. *Journal of Economic Literature, 32*(3), 1147–1175.

Fagerberg, J., & Verspagen, B. (2002). Technology-gaps, innovation-diffusion and transformation: An evolutionary interpretation. *Research Policy, 31*(8), 1291–1304.

Fagerberg, J., Srholec, M., & Verspagen, B. (2010). Innovation and economic development. *Handbook of the Economics of Innovation, 2*, 833–872.

Findlay, R. (1978). Relative backwardness, direct foreign investment, and the transfer of technology: A simple dynamic model. *The Quarterly Journal of Economics, 92*(1), 1–16.

Fox, R. (Ed.). (1996). *Technological change: Methods and themes in the history of technology* (Vol. 1). New York: Psychology Press.

Freeman, C., & Perez, C. (1986). *The diffusion of technical innovations and changes of techno-economic paradigm*. Science Policy Research Unit, University of Sussex.

Freeman, C., & Perez, C. (1988). *Structural crises of adjustment, business cycles and investment behaviour*. London: Pinter.

Freeman, C., & Soete, L. (1997). *The economics of industrial innovation*. London: Pinter.

Freeman, C., & Louca, F. (2001). *As time goes by: From the industrial revolution to the information revolution*. Oxford: Oxford University Press.

Fukuda-Parr, S., & Lopes, C. (Eds.). (2013). *Capacity for development: New solutions to old problems*. London: Routledge.

Galor, O., & Weil, D. N. (2000). Population, technology, and growth: From Malthusian stagnation to the demographic transition and beyond. *American Economic Review, 90*(4), 806–828

Gerschenkron, A. (1962). *Economic backwardness in economic perspective*. Cambridge: Belknap.

Gomulka, S. (2006). *The theory of technological change and economic growth*. London: Routledge.

Göransson, B., & Söderberg, J. (2005). Long waves and information technologies—On the transition towards the information society. *Technovation, 25*(3), 203–211.

Granato, J., Inglehart, R., & Leblang, D. (1996). The effect of cultural values on economic development: Theory, hypotheses, and some empirical tests. *American Journal of Political Science, 40*, 607–631.

Grossman, G. M., & Helpman, E. (1991). *Innovation and growth in the global economy*. Cambridge, MA: Cambridge MIT Press.

Hanna, N. K., Guy, K., & Arnold, E. (1995). *Information technology diffusion: Experience of industrial countries and lessons for developing countries* (World Bank Staff Working Paper). Washington DC: The World Bank.

Hanna, N. K. (2003). *Why national strategies are needed for ICT-enabled development* (World Bank Staff Paper). Washington, DC: World Bank.

Hanna, N. K. (2010). *Transforming government and building the information society: Challenges and opportunities for the developing world*. New York: Springer.

Hargittai, E. (1999). Weaving the Western Web: Explaining differences in Internet connectivity among OECD countries. *Telecommunications Policy, 23*(10), 701–718.

Helpman, E., & Trajtenberg, M. (1996). *Diffusion of general purpose technologies* (No. w5773). National Bureau of Economic Research.

Helpman, E. (Ed.). (1998). *General purpose technologies and economic growth*. Cambridge: MIT Press.

Hewitt, T., & Wield, D. (1992). *Industrialization and development*. Oxford: Oxford University Press.

Jensen, M. B., Johnson, B., Lorenz, E., & Lundvall, B. Å. (2007). Forms of knowledge and modes of innovation. *Research Policy, 36*(5), 680–693.

Jovanovic, B., & Rousseau, P. L. (2005). General purpose technologies. *Handbook of Economic Growth, 1*, 1181–1224.

Kaldor, N. (1957). A model of economic growth. *The Economic Journal, 67*, 591–624.

Katz, M. L., & Shapiro, C. (1985). Network externalities, competition, and compatibility. *The American Economic Review, 75*(3), 424–440.

Keller, W. (1996). Absorptive capacity: On the creation and acquisition of technology in development. *Journal of Development Economics, 49*(1), 199–227.

Kiiski, S., & Pohjola, M. (2002). Cross-country diffusion of the Internet. *Information Economics and Policy, 14*(2), 297–310.

Kim, S., & Lee, H. (2004). Organizational factors affecting knowledge sharing capabilities in e-government: An empirical study. In *Knowledge Management in Electronic Government* (pp. 281–293). Berlin: Springer.

Kindleberger, C. P. (1995). Technological diffusion: European experience to 1850. *Journal of Evolutionary Economics, 5*(3), 229–242.

Kostopoulos, K., Papalexandris, A., Papachroni, M., & Ioannou, G. (2011). Absorptive capacity, innovation, and financial performance. *Journal of Business Research, 64*(12), 1335–1343.

Kuhn, T. (1962). *The structure of scientific revolutions*. Chicago: University of Chicago Press.

Lall, S. (1992). Technological capabilities and industrialization. *World Development, 20*(2), 165–186.

Landau, R., & Rosenberg, N. (Eds.). (1986). *The positive sum strategy: Harnessing technology for economic growth*. Washington, DC: National Academies Press.

Landes, D. S. (2003). *The unbound Prometheus: Technological change and industrial development in Western Europe from 1750 to the present*. New York: Cambridge University Press.

Layton, E. T., Jr. (1974). Technology as knowledge. *Technology and Culture, 15*(1), 31–41.

Law, J. (Ed.). (1991). *A sociology of monsters: Essays on power, technology and domination* (Vol. 171). London: Routledge.

Lipsey, R. G., Carlaw, K. I., & Bekar, C. T. (2005). *Economic transformations: General purpose technologies and long-term economic growth: General purpose technologies and long-term economic growth*. New York: Oxford University Press.

Lucas, R. (1988). On the mechanics of economic development. *Journal of Monetary Economics, 22*, 3–42.

Lundvall, B. Å. (Ed.). (2010). *National systems of innovation: Toward a theory of innovation and interactive learning* (Vol. 2). London: Anthem Press.

Mankiw, N. G., Phelps, E. S., & Romer, P. M. (1995). The growth of nations. *Brookings Papers on Economic Activity, 1*, 275–326.

Mokyr, J. (1990). *The lever of riches: Technological creativity and economic progress*. New York: Oxford University Press.

Mokyr, J., & Scherer, F. M. (1990). *Twenty five centuries of technological change: An historical survey* (Vol. 35). New York: Taylor & Francis.

Mokyr, J. (Ed.). (1993). *The British industrial revolution: An economic perspective*. Boulder: Westview Press.

Mokyr, J. (2002). *The gifts of Athena: Historical origins of the knowledge economy*. Princeton, NJ: Princeton University Press.

Mokyr, J. (2005). Long-term economic growth and the history of technology. *Handbook of Economic Growth, 1*, 1113–1180.

Mowery, D. C., & Rosenberg, N. (1991). *Technology and the pursuit of economic growth.* Cambridge: Cambridge University Press.

Nelson, R. R. (1993). *National innovation systems: A comparative analysis.* University of Illinois at Urbana-Champaign's Academy for Entrepreneurial Leadership Historical Research Reference in Entrepreneurship.

Nelson, R. R. (2011). Technology, institutions and economic development. *Techno-Economic Paradigms: Essays in Honour of Carlota Perez*, 269.

Nelson, R. R., & Phelps, E. S. (1966). Investment in humans, technological diffusion, and economic growth. *The American Economic Review, 56*, 69–75.

Pavitt, K. (1999). *Technology, management and systems of innovation.* Cheltenham: Edward Elgar Publishing.

Perez, C. (1985). Microelectronics, long waves and world structural change: New perspectives for developing countries. *World Development, 13*(3), 441–463.

Pérez, C. (1986). Las nuevas tecnologías: una visión de conjunto. *Estudios Internacionales, 76*, 420–459.

Perez, C. (2002). *Technological revolutions and financial capital: The dynamics of bubbles and golden ages.* Northampton, MA: Edward Elgar Publishing.

Pérez, C. (2003). Technological change and opportunities for development as a moving target. *Trade and development: Directions for the 21st century*, 100.

Perez, C. (2009). Technological revolutions and techno-economic paradigms. *Cambridge Journal of Economics, 34*(1), 185–202.

Perez, C., & Soete, L. (1988). Catching up in technology: Entry barriers and windows of opportunity. In G. Dosi et al. (Eds.), *Technical change and economic theory* (pp. 458–479). Francis Pinter: London.

Quah, D. (2001). ICT clusters in development: Theory and evidence. *EIB Papers, 6*(1), 85–100.

Redding, S., & Venables, A. J. (2002). The economics of isolation and distance. *Nordic Journal of Political Economy, 28*, 93–108.

Rodriguez F., & Wilson, E. (2000). Are poor countries losing the information revolution? (infoDev Working Paper). Washington, DC: World Bank.

Romer, P. (1990). Endogenous technological change. *Journal of Political Economy, 98*(Oct), S71–S102.

Romer, P. (1993). Idea gaps and object gaps in economic development. *Journal of Monetary Economics, 32*(3), 543–573.

Romer, P. M. (1994). The origins of endogenous growth. *The Journal of Economic Perspectives, 8*(1), 3–22.

Rosenberg, N. (1974). Science, invention and economic growth. *The Economic Journal, 84*, 90–108.

Rosenberg, N. (1976). *Perspectives on technology.* CUP Archive.

Rosenberg, N. (1982). *Inside the black box: Technology and economics.* Cambridge, MA: Cambridge University Press.

Rosenberg, N. (1994). *Exploring the black box: Technology, economics, and history.* Cambridge: Cambridge University Press.

Rosenberg, N., & Birdzell, L. E., Jr. (2008). *How the West grew rich: The economic transformation of the industrial world.* New York: Basic Books.

Savvides, A., & Zachariadis, M. (2005). International technology diffusion and the growth of TFP in the manufacturing sector of developing economies. *Review of Development Economics, 9*(4), 482–501.

Servon, L. J. (2008). *Bridging the digital divide: Technology, community and public policy.* New York: Wiley.

Schumpeter, J. A. (1934). *Theory of economic development.* Brunswick, NJ: Transaction Publishers.

Schumpeter, J. A. (1947). The creative responses in economic history. *Journal of Economic History, 7*, 149–159.

Shapiro, C., & Varian, H. (1999). *Information rules: A strategic guide to the network economy.* Boston: Harvard Business School Press. [Kurzweil, R. (1999) *The age of spiritual machines: when computers exceed human intelligence.* New York: Viking].

Shapiro, C., & Varian, H. R. (2013). *Information rules: A strategic guide to the network economy.* Boston: Harvard Business Press.

Shell, K. (1967). A model of innovative activity and capital accumulation. In K. Shell (Ed.), *Essays on the theory of optimal economic growth* (pp. 67–85). Cambridge: MIT Press.

Singer, C. J., & Williams, T. I. (1954). *A history of technology.* Oxford: Clarendon.

Soete, L., & Verspagen, B. (1993). Technology and growth: The complex dynamics of catching-up, falling behind and taking over. In A. Szirmai (Ed.), *Explaining economic growth.* Amsterdam: Elsevier.

Solow, R. M. (1956). A contribution to the theory of economic growth. *The Quarterly Journal of Economics, 70*(1), 65–94.

Solow, R. M. (1957). Technical change and the aggregate production function. *The Review of Economics and Statistics, 39*(3), 312–320.

Stoneman, P. (Ed.). (2002). *The economics of technological diffusion.* Oxford: Blackwell Publishing.

Thorstein, V. (1915). *Imperial Germany and the industrial revolution.* New York: Macmillan.

UNESCO. (2002). In J. Anderson (Ed.). *Information and communication technology in education: A curriculum for schools and programme of teacher development.* Paris: UNESCO.

Uzawa, H. (1965). Optimum technical change in an aggregate model of economic growth. *International Economic Review, 6*, 18–31.

Wresch, W., & Fraser, S. (2012). ICT–enabled market freedoms and their impacts in developing countries: Opportunities, frustrations, and surprises. *Information Technology for Development, 18*(1), 76–86.

Valente, T. W. (1995). *Network models of the diffusion of innovations* (Vol. 2, No. 2). Cresskill, NJ: Hampton Press.

Valente, T. W. (1996). Social network thresholds in the diffusion of innovations. *Social Networks, 18*(1), 69–89.

van den Berg, P., Arentze, T., & Timmermans, H. (2013). A path analysis of social networks, telecommunication and social activity–travel patterns. *Transportation Research Part C: Emerging Technologies, 26*, 256–268.

Venables, A. J. (2001). Geography and international inequalities: The impact of new technologies. *Journal of Industry, Competition and Trade, 1*(2), 135–159.

Verspagen, B. (1991). A new empirical approach to catching up or falling behind. *Structural Change and Economic Dynamics, 2*(2), 359–380.

World Bank. (2014). *The little data book on information and communication technology 2014.* Washington, DC: Author.

Technology Diffusion

3

*(. . .) diffusion concerns issues that are among the more
difficult to analyze adequately. Time is involved. Uncertainty
is inherent. Change is the major topic. Imperfect markets
abound*

Paul Stoneman (2002)

Abstract

This chapter provides a theoretical framework of technology diffusion, which is
defined as a dynamic and time-attributed process involving the transfer of information, knowledge and innovations, and standing for a continuous and gradual
spread of new ideas throughout large-scale and heterogeneous societies. First, it
extensively discusses theoretical technology diffusion concepts and models,
explaining the technology diffusion trajectories by the use of S-shaped curves.
Second, it presents the fundamental ideas and models standing behind the idea of
technological substitution. Third, there is demonstrated a novel approach to
identification of the 'technological take-off' and 'critical mass' effects with respect
to the dynamics of the technology diffusion process and its prerequisites. Finally,
based on theoretical frameworks derived from economic growth theories, it shows
conceptualizations of technology convergence and technology convergence clubs.

Keywords

Technology diffusion • S-curve • Technological substitution • Technology
convergence • Technology convergence clubs • Critical mass • Technological
takeoff

3.1 Technology Diffusion: Theoretical Framework

The term '*diffusion*' originates from the Latin nouns '*diffusio*' and '*diffusionis*', and
the verb '*diffundere*'. By definition, it refers to the process of spread, expansion,
dissemination, propagation or generalization.

© Springer International Publishing Switzerland 2015   29
E. Lechman, *ICT Diffusion in Developing Countries*,
DOI 10.1007/978-3-319-18254-4_3

Diffusion is a dynamic and time-related process, involving the transfer of information, knowledge and innovations. It stands for a continuous and gradual spread of new ideas and concepts, over large-scale and heterogeneous societies (Gray 1973). Therefore, from the socio-economic perspective, the diffusion of innovations, new technologies and new ideas is of seminal importance, as it provokes profound changes in society and economy, impacting shifts in productivity and education, and transforming markets and organizations, among other things.

The concept of diffusion of innovation developed by Everett Rogers (2010), and extensively described in his touchstone book 'Diffusion of innovation'[1], constitutes a starting point for a great variety of discussions on technology diffusion. Rogers (2010) defines technology diffusion as *'the process by which an innovation is communicated through certain channels over time among members of a social system'*. Mansfield (1961, 1968, 1971), following Rogers, emphasizes the unique role of 'two-step' communication in diffusion processes, which enables the exchange of knowledge between 'users' and 'non-users' about the advantages of new technologies.

Gray (1973) calls the process of diffusion the spread of innovations, which depends on the effectiveness of communication channels and social attitudes. Davies (1979) defines technology diffusion in a *strict* economic sense, claiming that the process can be seen as passing from an equilibrium state, determined by the use of 'old' technology, to another equilibrium where the whole society adopts the 'new' technology. This approach suggests that shifting from one technology to another, over a diffusion time path, implies that the process is marked by constantly emerging disequilibria. Nathan Rosenberg (1982) in his seminal book *'Inside the black box: technology and economics'*, underlines that diffusion introduces inventions into economy and society, and thus is perceived as being of seminal importance for further development. On the same lines, Mansfield (1986) recognizes diffusion as a process of transfer of innovation which hugely affects national economies. Following Rogers's concept, Mahajan and Peterson (1985) claim that technology diffusion stands for the spread of ideas over time among society members. Paul David (1986) argues that through diffusion channels new technologies randomly reach new users; however, considering the socio-economic environment, the process is less hazardous as agents are driven by the anticipated profitability of new technologies. A broader perspective on the perception of diffusion was proposed by John S. Metcalfe (1997), who considers diffusion as flows of a multitude of technological improvements which—despite the fact that they spread instantaneously—bring crucial changes to technological, social and economic progress. Stoneman (1995) argues that the process of diffusion involves increases in the number of adopters of new technologies, which results in a growing number of users, while Sarkar (1998) states that *'technological diffusion can be*

---

[1] In his work 'Diffusion of Innovation', E. Rogers presents 508 different case studies explaining the diffusion of different innovations adopted by both companies and individuals in rural areas (see Rogers and Havens 1962).

defined as a mechanism that spreads 'successful' varieties of products and processes through an economic structure and displaces wholly or partly the existing 'inferior' varieties'. Stoneman (2002) also suggests that the process of diffusion of innovation explains the constant expansion of newly emerging technologies which are being gradually adopted and used by individuals and/or companies[2]. Following the logic of Metcalfe, Saviotti (2002) argues that technology diffusion brings a wide array of new products to markets, and thus is perceived by societies as highly desirable. Apart from the contributions mentioned above, there exists a substantial body of literature discussing conceptual issues associated with technology diffusion. Various aspects of technology diffusion are studied in the works of Kindleberger (1995), Bell and Pavitt (1995, 1997), Geroski (1990, 2000), Reinganum (1981a, 1989), Castellacci (2006b, 2007), Helpman (1998), Findlay (1978a, b), Battisti (2008), Stoneman and Battisti (2010), Ireland and Stoneman (1986), Karshenas and Stoneman (1993, 1995), Fagerberg and Verspagen (2002), Kapur (1995, 2001), Gomulka (2006), Kubielas (2009), Antonelli (1986, 1991), Dosi and Nelson (1994), Dosi (1991), Soete and Turner (1984), Comin and Hobijn (2006).

The contemporary qualitative and quantitative conceptualization of technology diffusion is deeply rooted in the evolutionary paradigm of Charles Darwin (1968) and his pioneering work on natural growth and the spatial diffusion of species; but it also refers to the theories of natural selection developed by Fisher (1930). Darwin (1968) predicted the unique ability of species to multiply at exponential growth rates, and to compete for survival in the environment in which they live. This concept was then gradually adjusted for multipurpose use in the economic sciences, rigidly assuming that 'species' are various variables (e.g. national income, technology or products) which tend to grow over time. Today, technology diffusion theories are designed to explain the spread of new ideas, innovations and technologies within societies. Thus the process itself is strongly related to time and its speed depends on the unique characteristics of people (Rogers and Shoemaker 1971; Metcalfe 1997). Moreover, technology diffusion theories allow for detecting patterns in the spread of new ideas, discovering regularities that the process depends upon, and identifying factors stimulating or impeding it. Difficulties associated with the elaboration of diffusion trajectories of newly emerging technologies reflect the heterogeneity of the social and economic environment (Rosenberg 1972). People rarely make their decisions interdependently (Geroski 2000), their cognitive capacities are limited, and various reference points are referred to before accepting or rejecting new technology (Dosi 1991). People's behaviour is driven by customs, culture, traditions and moral attitudes (Simon 1972; Silverberg 1994). Moreover, an individual's decision on the adoption of a new technology is made under uncertainty (Keller 2004; Ward and Pede 2013) and through cost-benefit analysis. Risk-averse people will adopt innovations once they notice that a 'new' one brings relatively greater

---

[2] Agents.

advantages compared to the 'old' one, and consequently the 'old' is replaced by the 'new' and better technology (Hall and Khan 2003). However, it is important to note that diverse personal characteristics determine the diffusion time path, illuminating the strength of the *domino effect* which perpetuates the spread of new ideas. Rogers (2010) claims that the diffusion process encompasses four major elements: (1) innovation; (2) communication channels; (3) time; (4) a social system. He defines an innovation as a new idea (i.e. product) which is desirably adopted by market agents, and the process happens over time. As diffusion is time-related, the rate of diffusion[3]—explaining the speed at which individuals in heterogeneous societies adopt new ideas—is recognized as its most prominent feature. The speed of diffusion is, however, heavily conditioned by social system absorptive and learning capabilities, as well as by the propensity and ability to adopt novelties (Cohen and Levinthal 1990; Keller 1996; Castellacci and Natera 2013; Lall 1992). This implies that existing communication channels (means and forms of communication and information dissemination) and social systems (defined as sets of social norms, formal and informal institutions) precondition both diffusion itself and its speed (Rogers 1976).

Despite potential disruptions, discontinuities and permanent uncertainty (Ehrnberg 1995), the phenomenon of rises and falls of new technologies is well described by simple logistic growth models and S-shaped curves that are generated by plotting the technology's behaviour over time, as this unique shape allows for a straightforward explanation of the characteristic phases of the diffusion process. Simply plotting the total number of adopters of new technology *versus* time generates the sigmoid curve, and the special shape of this sigmoid pattern[4] explains the characteristic phases of the diffusion process. It is slow initially, then it accelerates (the *domino effect* is revealed), and finally slows down, heading for the stabilization phase as the population approaches full saturation regarding the new technology (Jaber 2011). Rogers (2010) uses a derivative of the sigmoid curve—the bell-shaped curve (Nakicenovic 1991; Van den Bulte and Stremersch 2004)—to show five types of adopters: (1) innovators, (2) early adopters, (3) the early majority, (4) the late majority and (5) laggards. The group of 'innovators' introduces new technologies to societies, while the 'early adopters' are those who acquire novelties quickly and demonstrate little risk-aversion. The 'early majority' group follows the 'early adopters' and, *prior* to decisions made by the 'early majority' decide to adopt the new technologies expecting benefits. The last two groups—the 'late majority' and the 'laggards'—are those who are generally uncomfortable with new technologies and lag behind in their broad adoption[5].

---

[3] The rate of diffusion is additionally associated with the concept of *critical mass* and it reveals 'network effects'—explained in Sect. 3.3.

[4] The unique characteristics and basic mathematics related to sigmoid curves are explained in Sect. 3.2.

[5] Goeffrey Moore, in his book 'Crossing the Chasm' (1991), proposes a modified version of Roger's bell-curve. He emphasizes the role of 'disruptive innovations' that generate the chasm

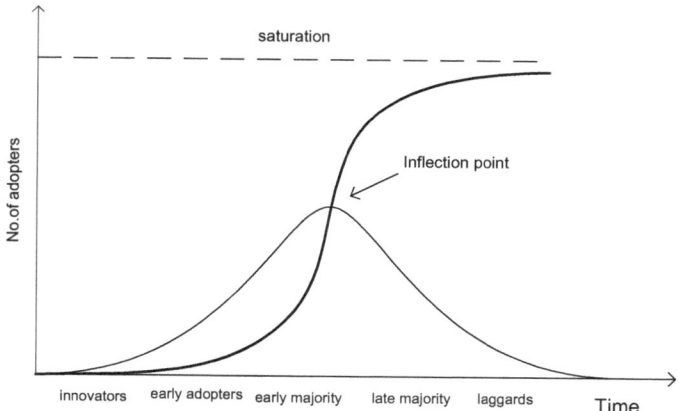

Fig. 3.1 Diffusion and innovation expansion curves. Theoretical specification

Figure 3.1 presents the cumulative sigmoid curve, approximating the new technology diffusion time path, and its derivative—the bell-curve.

The bell curve explains knowledge accumulation (or expansion of innovation)
that is generated by the gradual diffusion of new technology through society. The
slope of the bell curve decreases systematically as the cumulative number of
adopters grows, and its maximum coincides with the inflection point of the
S-shaped pattern.

The logic and basic mathematics used to formalize the phenomenon of the
diffusion of technologies and the process of shifting from 'old' to 'new' ones is
explained by technology diffusion and technological substitution models, discussed
in the following Sect. 3.2.

3.2 Technology Diffusion and Technological Substitution

Models are abstractions and simplifications of reality. Useful models capture the essence of
reality in a way that enhances the understanding of phenomena
Frank M. Bass (2004)

The technology diffusion process is formalized in a wide array of 'technology
diffusion models' describing how novel emerging technologies tend to spread
through societies. Most of these models are well grounded in mathematics, which
allows *ex-post* diffusion trajectories to be approximated; and, relying on rigid
assumptions, future development scenarios and forecasts to be draw up.

(gap, discontinuities) between the group of innovators and the early adopters and the group of the
early majority, the late majority and the laggards.

3.2.1 Technology Diffusion. Concepts and models

> For people who attempt to forecast the future, there is a continuing need for simple models
> that describe the course of unfolding events. Each such model should be based upon easily
> understood assumptions that are not susceptible to unconscious or invisible tampering by
> the forecaster in his efforts to make the future what he wants it to be. The model should be
> easy to apply to a wide variety of circumstances, and should be easy to interpret
> Fisher and Pry (1972)

As clarified in Sect. 3.1, the term 'diffusion' has multiple meanings. However, despite the diversity, it refers to the process of the physical spread of ideas, products and many other things in the human environment. Time plays a central role in most empirical studies that concern technology diffusion, as regardless of the source of an innovation, its type and the cost of acquiring it, it is always a time-consuming process for innovations to spread through societies and to be fully adopted and used. A great part of the theoretical and empirical literature on technology diffusion is mainly concerned with first the identification of factors that determine (enhance or hinder) the diffusion process, and second tracing causal links between the technology diffusion dynamics and its determinants. Put another way, diffusion models allow projections of how fast the technology will expand, and when (or *if*) the total population will be saturated with the new technology.

As discussed in Karshenas and Stoneman (1993), theories of technology diffusion can be classified into four general categories: epidemic models, rank (probit) models, order models, and stock models. The theoretical specifications falling within each of the four categories exclusively analyze technology diffusion from the demand-side perspective, and refer to stand-alone technologies, assuming that uncertainty does not emerge. In this book, to meet the general goals of our empirical analysis, we concentrate on technology diffusion models originating from 'epidemic models', as they well suit the major aims of our research, although the other theories and models are briefly discussed in this section.

Theoretically, technology diffusion process is analogous to the spread of information over society. Thus, the growing *'mass'* of those who get the information depends on intensity and the number of contacts that facilitate further information spread and acquisition. In a broad sense, this assumption yields the adoption of 'epidemic models' to explain technology diffusion dynamics and trajectories, while the adjustment of 'epidemic models' to the needs of the formal analysis of technology diffusion leads to incorporating the concept of the logistic growth curve,[6] which allows for approximating diffusion trajectories.

Originally, the concept of 'epidemic models' was derived from an analogy between the spread of contagious diseases and that of technological innovation (Sarkar 1998; Kumar and Krishnan 2002). The general logic behind the epidemic model is following. Suppose we have an area where a population of hypothetical agents (adopters, users) lives that tends to acquire new technologies as they emerge.

---

[6] The concepts and mathematics underlying logistic growth are explained in Sect. 3.2.2.

Moreover, the number of these potential adopters is constant over time. Initially, the groups of 'users' and 'non-users' coexist, but the 'non-users' imitate those who already use new technologies and are gradually '*contaminated*'. Hence, the '*contamination effect*' arises (Gray 1973) as agents are involved in personal contacts, which perpetuate the process of further diffusion. It is assumed that the probability of '*contamination*' is time invariant and 'non-users' convert into 'users' once the two get in touch; thus, the 'adopters' ('users') influence social systems in such a way that the total number of 'adopters' increases. In systems where innovation spreads, information and interpersonal contacts are perceived as significant driving forces of diffusion processes, which inevitably leads to a growing number of 'users' (Stoneman 2002). The concept of epidemics, adjusted to the needs of technology diffusion analysis, can be formalized as follows. Suppose that N denotes the total number of potential users of a new technology, and $n(t)$ stands for the actual number of those who have already adopted the new technology at time t. We assume that new adopters arrive as they get information on newly emerging technologies, and the process of transmitting information is not disrupted by any external factor. φ represents the probability of getting '*contaminated*' and acquiring new technology, so that the total number of users at a certain point in time t is expressed as (Stoneman 2002):

$$\frac{dn(t)}{dt} = \frac{\tau \cdot n(t)}{N(N - n(t))}, \tag{3.1}$$

where $\tau = \varphi \cdot \vartheta$, and ϑ stands for the probability that the contact between a 'user' and 'non-user' will be effective and lead to the adoption of the new technology. If Eq. (3.1) is a class of first-order differential equations, its solution can be formally written as:

$$n(t) = \frac{N}{(1 + \exp\{-\beta - \alpha t\})}. \tag{3.2}$$

In Eq. (3.2), τ from Eq. (3.1), is replaced by α. Equation (3.2) is the classical formula for a logistic curve with imposed growth limits (N), where β denotes the initial year of diffusion, and α is the rate (speed) of diffusion.

Starting from the late 1950s, many contributions in the field of technology diffusion studies were made. Extensive empirical analyses of technological diffusion both within and between countries were conducted (see, e.g., the works of Griliches 1957; Mansfield 1961, 1968), which resulted in the elaboration of diffusion models that provided theoretical frameworks for more sophisticated formal analysis of technology diffusion. The oldest and probably the most influential model of technology diffusion, strictly basing itself on the concept of 'epidemics', was proposed by Edwin Mansfield (1961). His pioneering works gave a solid background for future studies of technology diffusion and its economic consequences (Metcalfe 2004). In his works, Mansfield, strongly incorporates evolutionary ideas (Darwin 1968; Fisher 1930) into technology diffusion theories, which *inter alia*, induced a broad adoption of logistic curves into the analysis of the dynamics of innovation spread. The idea of incorporating logistic laws of evolution (Darwin 1968; Fisher 1930) into formalized concepts of technology diffusion was provoked by the analogies observed between the evolutionary paths of natural and social systems. The dynamics of evolving populations is significantly driven by a competitive selection process (Dosi and Nelson 1994; Silverberg and Verspagen 1995; Metcalfe 2004) that is often reported in economic processes. Social systems or market structures tend to evolve along time paths, and logistic laws can successfully approximate the dynamics of the evolutionary process. In the literature, Mansfield's prime technology diffusion model is classified as an evolutionary disequilibrium model (Srivastava and Rao 1990). It relies on four fundamental assumptions (Mahajan and Peterson 1985; Sarkar 1998): (1) adopters are rational; (2) adopters do not necessarily head for profit maximization that would be potentially obtained from new technology acquisition; (3) technology diffusion is self-perpetuating, and thus endogenous; and (4) the technology diffusion process might not be continuous and is disequilibrating in its nature. Even if it is assumed that the equilibrium is represented by N (in Eq. (3.2)), the level of use of the technology at time $t \, (\rightarrow n(t))$ is always below N. Diffusion trajectories can, however, be explained as processes of constant adjustment of the level of $n(t)$, which is approaching N. To capture the process of new technology spread, Mansfield suggests adopting a logistic growth equation to explain the phenomenon. Additionally, he introduces the '*word of mouth*' effect (Geroski 2000; Lee et al. 2010) to the formal model. This emerges once potential adopters of the new technology tend to communicate among themselves, which transmits knowledge of the advantages of new technologies[7]. Put another way, Mansfield's model assumes that the technology diffusion process is pre-determined by previous users, as they are the main source of information about new technologies.

Equation (3.3), below, summarizes Mansfield's technology diffusion concept. Assume that each 'user' of a new technology freely contacts a 'non-user', which leads to the adoption of the new technology by the latter, and the probability of an 'effective' contact is denoted as ϑ. If the total number of 'users' increases by Δt, and $\Delta t \rightarrow 0$, the time path for technology diffusion yields:

$$n(t) = N/(1 + \vartheta \, \exp[-\mu t])^{-1}, \tag{3.3}$$

or alternatively:

$$n(t) = \frac{N}{(1 + \vartheta \, \exp[-\mu t])}, \tag{3.4}$$

where $n(t)$ is the number of 'users' at time t, and N the potential number of total 'users'. Following Geroski (2000), for Eqs. (3.3 and 3.4) we assume that $\mu \equiv \vartheta N$ and

---

[7] '*Word of mouth*' models are also labelled '*contact*' or '*disease*' models.

$\vartheta \equiv (N - n(0))/(n(0))$, where $n(0)$ stands for the number of 'users' in the initial year of technology diffusion. Mansfield's model of technology diffusion explains the process as long as it is purely imitative. Thus, it explains the diffusion exclusively by the internal influence (Turk and Trkman 2012) of earlier adopters who, due to the 'word of mouth' effect, transmit information to later adopters. However, if we relax the assumption of strictly endogenous determinants of technology diffusion among 'non-users', and incorporate exogenous (external) factors which influence the diffusion process (Lee et al. 2010), Eq. (3.3) can be expressed in an adjusted form. Frank Bass (1969, 1974, 1980, 2004; Bass and Parsons 1969) in the late 1960s developed an extended version of the Mansfield model by incorporating a new 'innovator perspective'. The Bass model relies on the assumption that technology diffusion is determined not only by 'imitators' but also by 'innovators' (those who intend to try new technologies), who *massively* influence the decisions made by their peers (Satoh 2001). The Bass specification is also recognized as a 'mixed-information-source' model, as it assumes that 'users' of new technology differentiate their decision 'to adopt or not' according to information obtained from various sources. In the Bass diffusion model, it is assumed that the speed (rate) of diffusion is shaped by imitation and innovation determinants. If this is true, then, following the logic of the Bass model, we can propose that the final outcome of new technology diffusion can be easily decomposed into an 'innovation effect' and an 'imitation effect'. The basic linear specification of the Bass formula (1969) is as follows:

$$S(t) = p + \frac{q}{\kappa}(N(t)), \tag{3.5}$$

where $S(t)$ specifies the likelihood of adoption of the new technology by a 'non-user' at time t, p is the imitation coefficient, q is the innovation coefficient, and $N(t)$ is the cumulative adoption of the new technology (product) at time t. By differentiating Eq. (3.5), we obtain (Satoh 2001):

$$\frac{dN(t)}{dt} = \left(p + \frac{q}{\kappa}N(t) \right) \times (\kappa - N(t)). \tag{3.6}$$

In Eq. (3.6), p and q are parameters, $N(t)$ explains the same as in Eq. (3.5), while κ is the total potential number of users of the new technology (product). Imposing that $F(t)$ is the fraction of potential 'users' who have adopted the new technology at time t, so that $F(t) = N(t)/\kappa$, we rewrite Eq. (3.6) as:

$$\frac{dF(t)}{dt} = (p + qF(t)) \times (1 - F(t)). \tag{3.7}$$

The time path for new technology diffusion following the Bass specification is:

$$N(t) = \kappa \left(\frac{1 - e^{-(p+q)t}}{1 + \frac{q}{p}e^{-(p+q)t}} \right), \tag{3.8}$$

with notation analogous to that in Eqs. (3.5–3.7). Estimation of Eq. (3.8) returns predictions on the growth in the number of users of the new technology (product). The inflection point in the diffusion time path is at:

$$N(t^*) = \kappa \left(\frac{1}{2} - \frac{p}{2q} \right), \tag{3.9}$$

if $t^* = -\frac{1}{p+q}ln\frac{p}{q}$, and under the condition that $N(t = t_0 = 0) = 0$.

Today, the Bass model is broadly applied in marketing, mainly in predictions of the dynamics of purchases of new products by consumers, or in forecasting potential scenarios for future market exploitation.

Undoubtedly, the theoretical approaches to technology diffusion analysis have certain shortcomings and limitations. 'Epidemic models' have been widely criticized for their oversimplifying assumptions and weak theoretical background. The approach is 'blind' to societal, demographic, cultural, educational and institutional prerequisites which condition the rate of adoption of a product and its effective use. Additionally, in systems in which the spread of technologies is supposed to be highly homogenous, agents acquire perfect information on new technologies through interpersonal contacts, and the process of diffusion stops only in the case that all the members of society use the new technology. Moreover, as is stressed by Karshenas and Stoneman (1993), the 'epidemic model' assumes that agents' decisions on acquiring—or not—new technology are free of risk. However, omitting risk can be misleading, especially when predicting the development of future technologies, and risk should definitely not be ignored in the long-term perspective. Applying an *explicit* or *implicit* 'epidemic' analogy to the theoretical concepts explaining the technology diffusion process, to a point, was criticized by the next two prominent authors, Paul David (1969) and Stephen Davies (1979), who made significant contributions to the theory of technology diffusion. Davies (1979) points out that 'blind' acceptance of the assumption that the diffusion process is well approximated by logistic growth equations leads to another unrealistic assumption—of a constant diffusion rate. If we relax the assumption of a time-invariant diffusion rate, the logistic pattern is not generated. In addition, many authors claim (see Griliches 1957; Mansfield 1968; Romeo 1977; Davies 1979; Metcalfe 1987; Karshenas and Stoneman 1995; Stoneman 2001; Stoneman and Battisti 2005) that these models fully and correctly explain the process of the systematic adoption of new technologies by societies.

However, despite obvious limitations of the theoretical approaches to technology diffusion, their contribution to diffusion analysis is pervasive and unquestionable. As claimed and proved in multiple empirical studies, this approach, despite its drawbacks, approximates time diffusion paths and the dynamics of the process relatively well. Systems, despite being attributed to various features, tend to develop in a similar way. S-shaped curves (logistics growth patterns), which are what are 'generated' from the Mansfield and Bass models, allow for broad intuitive interpretations, describing and forecasting the growth of various technologies (products) (Bass 2004). As growth trajectories have generally similar features, classical S-time path analysis creates the possibility of 'guessing by analogy' (Bass 2004) with the growth histories of past technologies. 'Epidemic models' are simple, clearly describe and explain the diffusion trajectories of new technologies, and allow the prediction with little uncertainty of future development paths.

The next paragraphs briefly discuses alternative approaches to the conceptualization of technology diffusion: probit (rank), stock and order models. The probit (or rank) approach, mostly developed and explained by Paul A. David (1969) and Stephen Davies (1979), is based on two major assumptions: the behaviour of agents (individuals or firms) is rational; and they head toward utility maximization. This specification contains elements of rational choice theory (e.g. Rawls 1999; Foley 2009). In the probit approach, it is assumed that technology diffusion is attributed to unique features of agents (in the case of companies, these can be the size, geographical location and production profile of firms), risk aversion to new technologies, or just the opposite—risk acceptance, the relative prices of alternative technologies to be potentially acquired, and the variety of substitutes for the technology (product) in question. In other words, the rank approach relies on a supposition that technologies spread in heterogeneous societies, and potential users of new technology condition their decisions on cost/benefit analysis (Davies 1979; Stoneman 2002). If the cost of technology acquisition at time t is defined as $C(t)$, while the benefits[8] generated from effective use of it are $B(t)$, then an individual decides to buy the technology only if $B(t) > C(t)$ is satisfied. The model, however, although more sophisticated than simple 'epidemic' models, includes multiple latent factors (e.g. consumer expectations) that determine agents' final decisions on new technology acquisition, and which heavily disrupt quantitative specification. The rank models of technology diffusion fall into the equilibrium model category. The equilibrium, referring to the actual numbers of users of a particular technology, can be established along the diffusion path for each period of time. Once, due to some external (exogenous) factors, the number of users changes, so the equilibrium is disrupted and the system heads toward another equilibrium state.

The stock[9] models include three main approaches to technology diffusion: those of Reinganum (1981b) and Schumpeter (1984) and a last stream which is based on an evolutionary approach. Both Reinganum (1981b) and Schumpeter's (1984) specifications may be classified as equilibrium class models. Both concepts are

---

[8] The benefits from the adoption of new technology are mainly associated with introducing 'process innovation' that underlies company performance. This can be conditioned, *inter alia*, by prospective profitability, expected risk, organizational structure and other factors which may impact outcomes for a company.

[9] The models, are labelled 'stock', as diffusion in time $(t + 1)$ depends on the stock (number) of given technology users in period 't'.

deeply rooted in neoclassical theories. Thus, the technology diffusion path is characterized by a sequence of equilibria in each time period. The consecutive equilibria are generated as agents, driven by infinite rationality and having access to full information, make decisions on new technology acquisition. The Reinganum approach assumes that firms tend to buy new technology when they expect a reduction in cost, so that the cost generated by the 'old technology' $C_{old}(t)$, is greater than $C_{new}(t)$ in a given time period. If $C_{old}(t) > C_{new}(t)$, then positive externalities, accounted as increases in profits, are expected. The Schumpeterian approach is similar in its logic to Reinganum's. However, the Schumpeterian concept (Soete and Turner 1984; Aghion et al. 2013) of technology diffusion is conceptually placed in a broader macroeconomic perspective, and it accounts for spillovers as new technologies expand and are gradually acquired by new users. Finally, the evolutionary approach offers a similar explanation of the technology diffusion process to the two just discussed. However, the main difference between the Reinganum and Schumpeterian explanations of technology diffusion and the approach argued by the evolutionary school lies in the basic assumptions that the models rely on. Evolutionary concepts reject assumptions on perfect information and perfect market competition, as is the case in the Reinganum and Schumpeterian models, and they relax the assumption on profit maximization and the infinite rationality of agents. To a point, evolutionary models are similar to those based on 'epidemic' concepts, as they claim that technology diffusion is *not* a self-equilibrating process. In evolutionary models, the process of technology diffusion is also defined as self-perpetuating, and the individual features of agents are assumed to be endogenous. If companies (individuals) get profits from newly acquired technologies, i.e. if $B(t) > C(t)$, then new users arrive and the diffusion proceeds. Generally, models developed under evolutionary economics are recognized as being more open-ended and more real-world-oriented, providing a more suitable insight into the nature and dynamics of the process. Following Allen (1988a, b), Sarkar (1998) argues that '(...) *diffusion* (...) *of innovations and technological changes has been considered in neoclassical economics, abstracted from history, culture, social structure* (...). (...) *such abstraction may have rendered equilibrium models simpler,* (...) *but having very low economic plausibility of [their] assumptions, thereby making it difficult to test these models rigorously for falsification'*.

In the late 1980s, 'order' models were developed (see, e.g., Fudenberg and Tirole 1985). Also classified as equilibrium models, these rely on the same assumptions under which the stock models operate. However, the 'order' approaches emphasize that the order of adoption of new technologies matters for the diffusion process. Order models relax the assumption that each agent (user) gets equal profit from new technology acquisition, and assume that a user that adopts a new technology first (first in order) enjoys higher profits compared to those who acquire new technologies later on. Hence, along the diffusion path $B(t_i) > B(t_{(i+1)})$, where $B(t)$ explains the profits gained by a user of a new technology.

3.2.2 Approximating Technology Diffusion Trajectories

A deep insight into the dynamics of technology diffusion was provided in the influential works of, *inter alia*, Mansfield (1968), Griliches (1957), and Nelson (1982), who analyzed the phenomenon adopting the evolutionary dynamics concept. This resulted in the introduction to economic studies of the logistics law, which is broadly applied in natural science to describe the path dependence of biological growth (Verhulst 1838; Pearl and Reed 1922). According to the logistic law of growth, systems tend to grow exponentially. In 1838, inspired by the Malthusian growth model, the Belgian mathematician Pierre-Francois Verhulst (1838)[10] formalized logistic growth and introduced the logistic function. In a generic sense, the function that Verhulst proposed is a logistic equation. also known as a simple sigmoid asymptotic function, and it produces an S-shaped curve once empirical data on diffusion (growth) is plotted over time. The growth curve can be divided into two specific parts by the inflection point: first (before the inflection point), it is a downward powers function; second (after the inflection point), it is a logarithmic function. The ubiquitous family of S-shaped curves (also recognized as: S-curves, logistic curves, S-shaped patterns, S-shaped paths, S-shaped trajectories, S-shaped time paths, Gompertz[11] curve, Foster's curve, sigmoid curves) allow for the visualization of the logistic growth process and its intuitive interpretation (Modis 2007). Mathematically, the logistic growth function originates from the exponential growth model, and if written as an ordinary differential equation is as follows (Meyer et al. 1999):

$$\frac{dY_x(t)}{dt} = \alpha Y_x(t). \tag{3.10}$$

If $Y(t)$ denotes the level of variable x, (t) is time, and α is a constant growth rate, then Eq. (3.10) explains the time path of $Y(t)$. If we introduce e[12] to Eq. (3.10), it can be reformulated as:

$$Y_x(t) = \beta e^{\alpha t}, \tag{3.11}$$

or alternatively:

$$Y_x(t) = \alpha \exp \beta t, \tag{3.12}$$

with notation analogous to Eq. (3.10) and β representing the initial value of x at $t = 0$.

---

[10] The logistic equation is also recognized as the Verhulst-Pearl equation, as Pearl and Reed (1922), in the early 1920s already adopted similar formulas in the biological sciences.

[11] Referring to Benjamin Gompertz (1825) and his 'law of mortality', which is a mathematical specification to model time-series (Gompertz model, Gompertz growth).

[12] Base of naatural logarithms.

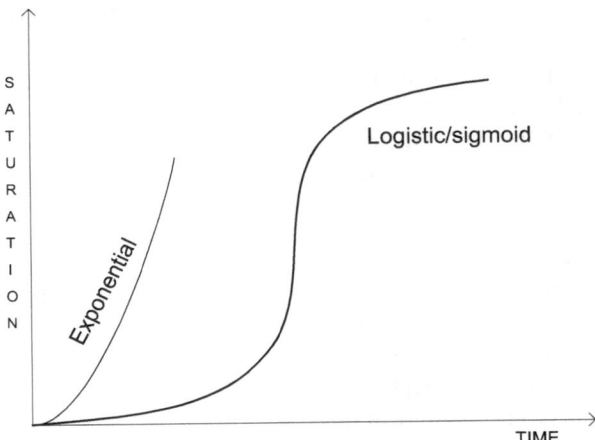

Fig. 3.2 Exponential versus logistic (sigmoid) curve specification. Theoretical specification

By convention, the simple growth model is pre-defined as exponential. Thus, if left to itself x will grow infinitely in geometric progression. But, indiscriminate extrapolation of $Y_x(t)$ generated by an exponential growth model would lead to unrealistic predictions, as due to various constraints, systems do not grow infinitely (Stone 1980; Kingsland 1982; Meyer 1994; Coontz 2013). Therefore, it is reasonable to impose growth boundaries to the original model. To solve the problem of 'infinite growth', the 'resistance' parameter (Meyer et al. 1999; Banks 1994; Cramer 2003; Kwasnicki 2013) was added to Eq. (3.10). This modification introduces an upper 'limit' to the exponential growth model, which instead gives the original exponential growth curve a sigmoid shape (Fig. 3.2).

Formally, the modified version of Eq. (3.10) is the logistic differential function, defined as:

$$\frac{dY(t)}{dt} = \alpha Y(t)\left(1 - \frac{Y(t)}{\kappa}\right), \tag{3.13}$$

where the parameter κ denotes the imposed upper asymptote that arbitrarily limits the growth of Y. As already mentioned, adding the slowing-down parameter to exponential growth generates an S-shaped trajectory[13] (see Fig. 3.3).

The three-parameter[14] logistic differential equation, Eq. (3.13), can be re-written as a logistic growth function, taking non-negative values throughout its path:

---

[13] Following Meyer et al. (1999), we define $\left(1 - \frac{Y(t)}{\kappa}\right)$ as a 'slowing term' ('negative feedback'), which is close to 1 as $Y(t) \ll \kappa$, but if $Y(t) \to \kappa$ then $\left(1 - \frac{Y(t)}{\kappa}\right) \to 0$.

[14] For estimates of the asymmetric responses 5-parameter logistic functions (5PL) are applied. A standard 5PL is as follows (Gottschalk and Dunn 2005): $y = f(x; p) = d + \frac{(a-d)}{\left[1 + \left[\frac{x}{c}\right]^b\right]^g}$, where

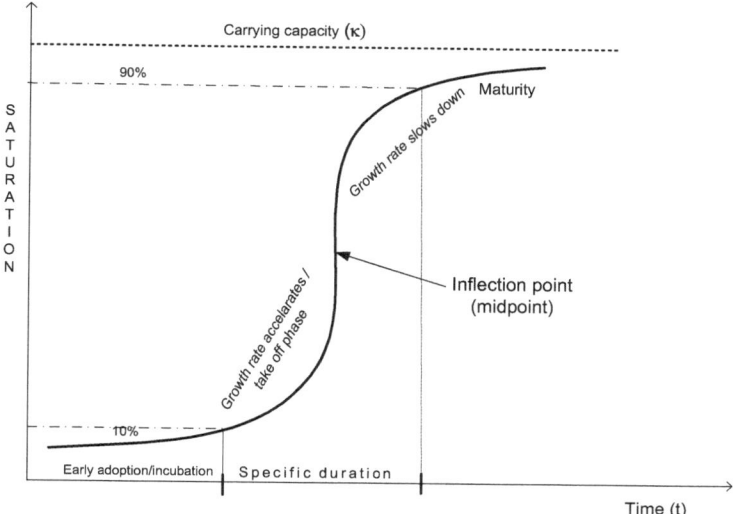

Fig. 3.3 S-shaped time path. Theoretical specification. Note: the logistic function follows the S-path if plotted on an absolute and linear scale. Once the Fisher-Pry (The Fisher-Pry (Fisher and Pry 1972) transform yields: $\equiv {^{N(t)}}/_{\kappa}$. Thus $^{F}/_{(1-F)} = e^{at+\beta}$.) transform is applied, the logistic curve can be plotted linearly.

$$N_x(t) = \frac{\kappa}{1 + e^{-at-\beta}}, \tag{3.14}$$

or, alternatively:

$$N_x(t) = \frac{\kappa}{1 + \exp(-\alpha(t - \beta))}, \tag{3.15}$$

where $N_x(t)$ stands for the value of variable x in time period t. The parameters in Eqs. (3.14 and 3.15)[15] explain the following:

- κ—upper asymptote, which determines the limit of growth ($N(t) \rightarrow \kappa$), also labelled 'carrying capacity' or 'saturation';

$p = (a, b, c, d, g)$, $c > 0$ and $g > 0$. If we restrict $g = 1$, a 4-parameter logistic function is generated.

[15] The parameters in Eqs. (3.14 and 3.15) can be estimated by applying ordinary least squares (OLS), maximum likelihood (MLE), algebraic estimation (AE), or nonlinear least squares (NLS). As Satoh and Yamada (2002) suggests, NLS returns the relatively best predictions, as the estimates of standard errors (of κ, β, α) are more valid than those returned from estimation using other methods. Adoption of NLS allows avoiding time-interval biases, which are revealed in the case of OLS estimates (Srinivasan and Mason 1986). However, the main disadvantage of the NLS procedure is that estimates of the parameters may be sensitive to the initial values in the time-series adopted.

- α—growth rate, which determines the speed of diffusion;
- β—midpoint, which determines the exact time (T_m) when the logistic pattern reaches 0.5κ.

The growth rate α additionally determines the 'steepness'[16] of the S-shaped curve. However to facilitate interpretation[17], it is useful to replace α with a 'specific duration'[18] parameter, defined as $\Delta t = \frac{\ln(81)}{\alpha}$. Having Δt, it is easy to approximate the time needed for x to grow from 10 to 90 % κ. The midpoint (β) describes the point in time at which the logistic growth starts to level off. Mathematically, the midpoint stands for the inflection point of the logistic curve. Incorporating Δt and (T_m) into Eq. (3.15), entails:

$$N_x(t) = \frac{\kappa}{1 + exp\left[-\frac{\ln(81)}{\Delta t}(t - T_m)\right]}. \tag{3.16}$$

A generalized version of the logistic function (Kudryashov 2013) including more than one explanatory variable of $N_x(t)$, is as follows:

$$N(Z) = \frac{expZ}{1 + expZ} = \frac{1}{1 + exp^{-Z}}, \tag{3.17}$$

with $Z = x^T \gamma$, where x stands for all covariates and γ is the coefficient of x.

Given that different growth processes are decomposable into *sub*-process, the model in Eq. (3.15) can easily be transformed into a multiple growth 'pulses' model. Assuming we are dealing with just two recognizable 'pulses' (*sub*-processes of growth), this gives rise to the expression:

$$N_x(t) = N_1(t) + N_2(t). \tag{3.18}$$

Hence, $N_1(t)$ and $N_2(t)$ yield: $\left[\frac{\kappa_1}{1+exp\left(\frac{\ln(81)}{\Delta t_1}\left(t-T_{m_1}\right)\right)}\right]$ and $\left[\frac{\kappa_2}{1+exp\left(\frac{\ln(81)}{\Delta t_2}\left(t-T_{m_2}\right)\right)}\right]$

respectively. The model defined in Eq. (3.18), is commonly known as a *bi*-logistic growth equation. The generalized version of Eq. (3.18) for multiple ($\rightarrow$'z') logistic growth *sub*-processes follows the *z-component* logistic growth model:

---

[16] Also labelled 'width'.

[17] The parameter α as such, is not economically interpretable, thus it is exclusively estimated to calculate the 'specific duration'.

[18] Also labelled 'characteristic duration' or 'specific time'.

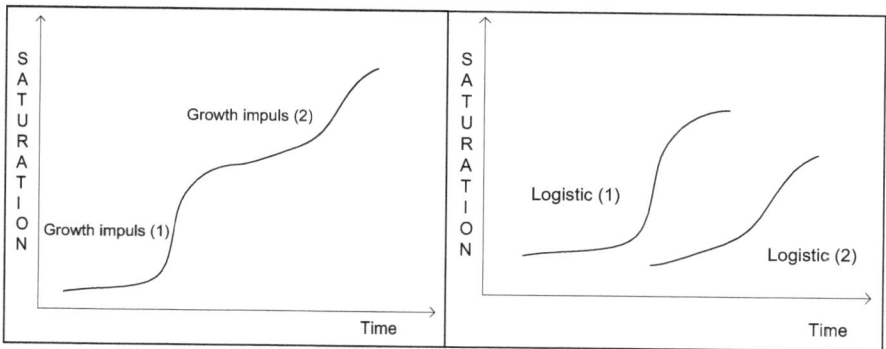

Fig. 3.4 Component logistic model decomposition into *bi*-logistic growth. Theoretical specification

$$N(t) = \left[\frac{\kappa_1}{1 + exp\left(\frac{\ln(81)}{\Delta t_1}\left(t - T_{m_1}\right)\right)}\right] + \ldots + \left[\frac{\kappa_i}{1 + exp\left(\frac{\ln(81)}{\Delta t_i}\left(t - T_{m_i}\right)\right)}\right]$$

$$= \sum_{i=1}^{z} N_i(t), \tag{3.19}$$

if:

$$N_i(t) = \frac{\kappa_i}{1 + \exp(-\alpha_i(t - \beta_i))}. \tag{3.20}$$

The concept formalized in Eq. (3.18) is graphically displayed in Fig. 3.4 (see below).

The left-hand side of Fig. 3.4, shows a component logistic curve with two clearly distinguishable growth phases (growth impulse (1) and growth impulse (2)). The left-hand curve is the approximated sum of two discrete 'wavelets' (Meyer et al. 1999), and can be decomposed into two separate three-parameter logistic functions. The curves on the right, instead, present the two distinct growth *sub*-impulses. Such decomposition allows for detailed analysis of the behaviour of the relevant technology in each phase of growth.[19]

Most technology diffusion models deal with strictly one technology (innovation) and describe its in-time behaviour. However, if another technology arrives there emerges a competition between the 'old' and 'new' technologies. Hence, the technological substitution process is revealed, which explains the life cycle of certain technologies, distinguishing certain phases of growth and decline. Here-after, Sect. 3.2.3 briefly describes technological substitution theories and models.

---

[19] If a Fisher-Pry transform is applied for normalization, then the logistic curves become linear, which additionally facilitates further analysis of growth *sub*-phases.

3.2.3 Technological Substitution

Technologies rise, saturate and finally decline when new and better ones emerge. The process of continuous replacement of 'old' technologies by 'new' technologies is labelled technological substitution, and can easily be encountered in various systems and under different circumstances (Fisher and Pry 1972). Technological substitution is evolutionary or revolutionary in its nature. It brings significant changes to societies (Kucharavy and De Guio 2011), and it may be perceived as a consequence of technology development marked by a stream of 'discontinuities' (Miranda and Lima 2013), and leading to replacements of 'old' technologies by 'new' ones. Generically, the process of technological replacement resembles competition between the 'old' and 'new' technology, in which the 'old' technology is initially a dominant competitor in the market and the 'new' 'invading' one fights for a growing market share (Morris and Pratt 2003).

By definition, the technological substitution model (also labelled logistic substitution model) explains the competitors' changing market shares (fractions) along the competition process, which is attributed to time. Technological replacement is gradual (Wang and Lan 2007), and, as broadly observed, the time behaviour of competing technologies follows a logistics trajectory. In a competitive system, each technology passes through three characteristic phases: a logistic growth phase (P1)—the prime phase of growth, when initially growth rates are slow, but they then enter an exponential growth phase (this results in fast diffusion of the technology); a saturation phase (P2)—the technology reaches the maximum of its market share and thus follows a non-logistic pattern; and a logistic decline phase (P3)—the technology is fading away from the market, its market share is gradually declining as it is substituted by new technology, which is in the logistic growth phase.

Most contemporary empirical works considering the process of gradual substitution between two competing technologies[20] can be traced back to the influential models proposed by Fisher-Pry (1972), Marchetti and Nakicenovic (1980), and Nakicenovic (1987). The Fisher-Pry model of technological substitution is based on three general assumptions (Fisher and Pry 1972; Bhargava 1995; Kumar and Kumar 1992): (1) many technological advances can be considered competitive substitutions of one method of satisfying a need with another; (2) if a substitution has progressed as far as a few percent, it will proceed to completion; (3) the rate of fractional substitution of new for old is proportional to the remaining amount of the old left to be substituted.

Technically, the technological substitution model explains changing shares of the market that competitors take over, and it relies on the assumption that the total

---

[20] Conceptually, technological substitution models refer to the seminal works of Alfred Lotka (1920) and Vito Volterra (1926), who were the first to introduce a generalized version of the logistic growth equation. They developed a model of competition among different species in biological systems (Voltera) and chemical chain reactions (Lotka). Today, the Volterra-Lotka competition equation is widely adopted for qualitative analysis of technological substitution if at least two competing technologies are involved.

sum of users of the two competing technologies is fixed.[21] Blackman (1971) and
Marchetti and Nakicenovic (1980) formalize the original technological substitution
model developed by Fisher and Pry, and develop a three-parameter logistic substitution model describing the behaviour of two competitors along the time path. The
technological substitution model is based on the following assumptions:

- There are n competing technologies;
- Once a 'new' technology has invaded the market, it grows at logistic rates;
- The 'old' technology fades away also at logistic rates, but the speed of decline is
  predominantly affected by the speed of diffusion of the 'new' technology[22];
- It is possible for only one technology (out of two or more competitors) to be in
  the saturation phase at a given point of time;
- A technology in the saturation phase follows a non-logistic pattern.

Let us assume a competitive system and consider the technology substitution
model with two technologies replacing each other. Assume that N is the total
population, where N_i represent the users of the two technologies, so that the share of
the population using i-technology at time t is:

$$f_i(t) = \frac{N_i(t)}{N}. \tag{3.21}$$

To avoid unrealistic estimates, it is presumed that the number of users is fixed and
each deploys one out of the two available technologies (Morris and Pratt 2003),
which implies an obvious constraint like:

$$f_i(t) + f_j(t) = 1, \tag{3.22}$$

where $'i'$ and $'j'$ are competing technologies. By convention, the technologies
follow a logistic growth trajectory (Kwasnicki 1999) defined as:

---

[21] Relaxing the assumption of a fixed total number of users would allow the system to grow
infinitely, which is not the case in real-data based empirical studies.

[22] Theodore Modis (2003) distinguishes six ways that two competitors can affect the growth rate in
a competitive system. These are: (1) pure competition (competitors need to fight to survive in the
same environment, as they use the same resources, which are limited); (2) predator-prey competition (one competitor is labelled prey and the second the predator—the 'predator' population grows
as there are abundant 'preys'; this kind of competition generates cyclical growths and declines in
populations of 'predators' and 'preys'. Lotka-Volterra equations are applied to describe this kind
of competition; (3) symbiosis (competitors are interrelated as the existence of the first is totally
dependent on the existence of the second); (4) parasitic (the first competitor benefits from the
second, but is does not affect the latter's existence, also labelled 'win-impervious' competition);
(5) symbiotic (the first competitor benefits from the second, but the latter is negatively affected by
the competition but remains indifferent to the loses, also labelled 'loss-indifferent'); (6) no
competition (the two competitors are not overlapping each other as they use different resources
to survive.

$$f_i(t) = \frac{1}{1 + \exp(-a - bt)},$$ (3.23)

where value a is defined for the initial year $(t = 0)$. To indicate the market share $(y_i(t))$ possessed by technology $'i'$ (either a declining or growing technology), we adopt a Fisher-Pry transform (1972) so that Eq. (3.23) yields $y_i(t) = ln\left[\frac{f_i(t)}{1 - f_i(t)}\right]$. Respecting the assumption defined in Eq. (3.22), we find that:

$$y_i(t) + y_j(t) = 1.$$ (3.24)

If Eq. (3.24) is satisfied, the market share of technology $'j'$ in the non-logistic saturation phase (P2) is given by:

$$f_j(t) = 1 - \sum_{j \neq i} f_i(t).$$ (3.25)

Thus, the share of the market possessed by technology $'i'$ is strictly subject to the share of the market possessed by technology $'j'$.

For an economic interpretation of the process of technological substitution, it is essential to determine the point in time when certain phases of substitution begin or end. Following Meyer et al. (1999), the estimate of the point in time when the saturation phase stops is given by:

$$\frac{y_i''(t)}{y_i'(t)} \to min.$$ (3.26)

Having y_i and thus y_i', it is possible to estimate the two parameters of the logistic curve for technology $'i'$, which can be mathematically expressed as:

$$\Delta t_i = \frac{\ln(81)}{y_i'(t)},$$ (3.27)

and:

$$T_{m_i} = ln\left[\frac{\left(y_i(t) - \frac{\ln(81)}{\Delta t}\right)}{\frac{\ln(81)}{\Delta t}}\right].$$ (3.28)

Δt_i is labelled 'takeover' (Fisher and Pry 1972) and it indicates the time needed for technology $'i'$ to increase its market share from $y_i(t) = 0.1$ to $y_i(t) = 0.9$. $T_{m_{is}}$ explains the specific point in time (e.g. year) when the substitution process between the competing technologies is half-complete; thus $y_i(t) = y_j(t) = 0.5$.

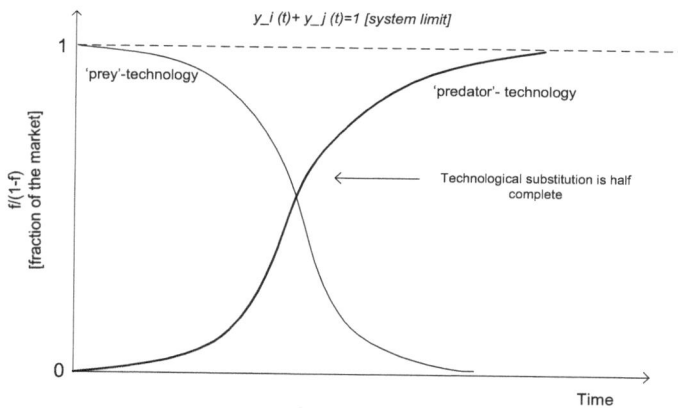

Fig. 3.5 Technological substitution process. Theoretical specification

Figure 3.5 graphically presents the mechanism of technological substitution, which combines two substitution curves with deterministic asymptotic behaviour.

Figure 3.5 shows the life cycles of both the 'predator' and 'prey' technologies, and three distinct phases are detectable: logistic growth, saturation and logistic decline. It is easy to note that once the 'predator' technology is in its logistic growth phase, the 'prey' technology follows a logistic decline. The intersection point depicts the specific time (i.e. the year) when the technological substitution process is half complete. Thus both the 'predator' and 'prey' control 50 % of the total market ($\rightarrow y_i(t) = y_j(t) = 0.5$).

3.3 The *Critical Mass*: What Stands Behind?

3.3.1 The *Critical Mass*. Explaining the Concept

Technology diffusion is strictly attributed to network externalities (network effects), which emerge as positive feedback from random contacts between society members, giving rise to exponential growth of the network itself. Carrington et al. (2005) and Villasis (2008) argue that 'network' stands for an interconnected chain or group, while the 'social network' '*is a social structure made of nodes tied by one or more types of relations (...)*'. If social networks give positive feedback, then network effects (externalities) may emerge, showing the value of potential connectivity exponentially increasing with the number of users of a new technology (Economides and Himmelberg 1995a, b; Villasis 2008). Katz and Shapiro (1985) and Shapiro and Varian (1998) define network effects as an increasing utility of using the product when the absolute number of users of this product increases. However, the positive effects of networks may arise only if the social system achieves a certain '*critical mass*', ensuring a further sustainable multiplication of users (Katz and Shapiro 1985, 1986, 1992; Markus 1987; Oliver et al. 1985).

In other words, a positive re-alimentation schema of revealing network effects is conditioned on the society reaching a certain '*critical mass*'. The notion of '*critical mass*' might be confusing, since it has multiple meanings. It originates from physics, and in its generic sense denotes the amount of radioactive material necessary for nuclear fission to take place (Oliver et al. 1985). Mancur Olson (1965) was the first to introduce the concept of '*critical mass*' to the social sciences, and he defines '*critical mass*' as the *critical* number of early adopters which is necessary to lead the rest of the population in collective actions.[23] Rephrasing this, '*critical mass*' theory leads to the *critical* (threshold) conditions for collective actions to emerge, and then continue as self-perpetuating[24] and profitable[25] (Marwell and Oliver 1993; Molina et al. 2001; Puumalainen et al. 2011).

To a certain extent, the '*critical mass*' concept has also been discussed in the literature on technology diffusion. The process of technology diffusion follows the third-order (S-shaped) time path, and so the main emphasis in analyzing the diffusion process is put on estimating the inflection point of the curve. By definition, the inflection point on an S-shape trajectory denotes the specific time period when saturation reaches 50 % of the population and the rate of diffusion starts to slow down. However, when considering the '*critical mass*' concept in reference to the diffusion process, it might be relevant to identify the *critical* (threshold) level of saturation of a given technology, at which the further process of diffusion becomes self-perpetuating. Rogers (2010) argues that at the '*critical mass*' 'diffusion becomes self-sustaining'. However, the concept of '*critical mass*' that Rogers (2010) uses is based on the assumption that the diffusion process will continue endogenously at exponential rates, finally reaching the stabilization phase once the '*critical mass*' of users is achieved (see Fig. 3.6) This therefore relaxes the assumption that the diffusion process of, e.g., a new product is determined by changes in relative prices or shifts in quality.

Similar explanations of significance of the '*critical mass*'[26] in the continuous diffusion process characterized by multiple equilibria states are given by Cabral (1990, 2006), Economides and Himmelberg (1995a, b), and Evans and Schmalensee (2010). For instance, Economides and Himmelberg (1995a, b) propose that the '*critical mass*' constitutes the smallest possible (minimal non-zero) equilibrium assuring the stability of a further diffusion process[27] at an exponential rate, while

---

[23] Oliver et al. (1985) recall that the *critical mass* effect is also known as the 'snob and bandwagon effect', the 'free rider problem' or the 'tragedy of commons'.

[24] Self-sustaining.

[25] Many claim (see, e.g. Bonacich et al. 1976; Frohlich et al. 1971; or Hardin 1982) that Olson's concept of *critical mass* was too general and unconditional and so it did not allow for any mathematical formalization. Additionally, their experiments have proved that Olson's concepts was not correct, as in many cases people's real behaviour does not confirm Olson's assumptions.

[26] The notion of *critical mass* is also known as 'installed base' (Grajek and Kretschmer 2012).

[27] In fact, they precondition the value of *critical mass* on prices, arguing that lower prices require lower *critical mass*, to assure sustainability of the diffusion process.

Fig. 3.6 *'Critical mass'* on the S-shaped diffusion trajectory

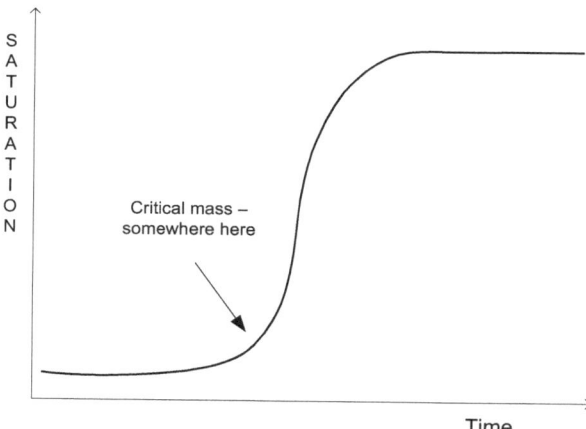

Evans and Schmalensee (2010) show that the level and diffusion of the *'critical mass'* are heavily determined by the nature of networks and individual consumer preferences.

In analyzing the phenomenon of *'critical mass'* as proposed by Rogers (2010), the theory of the diffusion of innovation becomes an obvious conceptual background. In diffusion theory, the very first adopters (innovators) of a new product do so because they benefit from the new product. Whether the rest of the society members will follow them or not usually depends on a *threshold,* defined as the number of people who have already adopted the new product. The central presumption of diffusion theory is that the process of diffusion follows a sigmoid pattern. Hence, identification of the *'critical mass'* might be strictly related to examining when and at what saturation level diffusion accelerates and the 'take-off' emerges. It is thus possible to state that diffusion accelerates once the *'critical mass'* is reached (Allen 1988a; Rogers 2010; Schoder 2000). Cabral (1990, 2006) claims that *'critical mass'* occurs if network effects are sufficiently strong and diffusion is endogenously driven. He also states that the *'critical mass'* point depicts the 'catastrophe point' on the diffusion time path, which corresponds to low-level equilibrium. Loch and Huberman (1999), in their work *'A Punctuated-Equilibrium Model of Technology Diffusion'*, propose an evolutionary model where two competing technologies (old and new) are available. Assuming that both technologies demonstrate network externalities and generate benefits from their use, consumers will switch to the new technology only if the technology diffuses at high speed. They also presume that other factors like, e.g., uncertainty, cultural 'openness' or personal preferences play a crucial role in the diffusion process, being strong incentives or barriers for new technologies to reach a *'critical mass'* and spread throughout society.

The works of Lim et al. (2003) and Kim and Kim (2007) attempt to identify the *'critical mass'* from the S-shaped diffusion pattern. Implementing the Bass diffusion model,[28] they develop the concepts of 'early take off' (Kim and Kim 2007) and

---

[28] For the formal specification, see Sect. 3.2.

'late take off' (Lim et al. 2003) with respect to diffusion studies. Adopting the formal specification of non-cumulative and cumulative adoption curves, they calculate the specific periods of time indicating the beginnings of the 'early take off' and 'late take off' phases. Assuming that t is time, that $(C(t))$ describes the cumulative curve and $(nonC(t))$ is the non-cumulative one, then:

$$C(t) = \kappa \frac{1 - e^{-(p+q)t}}{1 + \frac{q}{p}e^{-(p+q)t}}, \qquad (3.29)$$

and:

$$nonC(t) = \kappa \frac{p(p+q)^2 e^{-(p+q)t}}{\left(p + qe^{-(p+q)t}\right)^2}, \qquad (3.30)$$

where κ is the saturation level, and p and q explain the external and internal influence respectively. Mathematically (Kim and Kim 2007), the inflection points of the curves specified in Eqs. (3.29–3.39) correspond to:

$$t_{infl(C(t))} = -\frac{1}{p+q}\ln\frac{p}{q}, \qquad (3.31)$$

and:

$$t_{infl(nonC(t))} = -\frac{1}{p+q}\ln[(2 + \sqrt{3} frac\, pq)]. \qquad (3.32)$$

The inflection point defined as in Eq. (3.31) denotes entry into the exponential growth phase on the S-time path. By convention, by using the value of the inflection point we can determine the number of adoptions, which refers to $t_{infl(C(t))}$. Therefore, the number of adoptions at $t_{infl(C(t))}$ would presumably determine the level of the 'critical mass'. However, as Valente (1996, 2005), Mahler and Rogers (1999), and Lim et al. (2003) argue, it may be highly controversial whether the point $t_{infl(C(t))}$ unquestionably denotes the 'critical mass'. The question is whether, after passing the $t_{infl(C(t))}$ point, the diffusion turns out to be a self-sustaining process or not. If not, there is no justification for treating $t_{infl(C(t))}$ as the 'critical mass' point. Thus, the conviction that the 'critical mass' is easily detectable might be misleading and confusing.

Different approaches to the identification of 'critical mass' are offered by Grajek (2003, 2010), Grajek and Kretschmer (2011, 2012), Baraldi (2004, 2012), Arroyo-Barrigüete et al. (2010) and Villasis (2008). To quantify 'critical mass', Grajek and Kretschmer (2012) define it as a function of the installed base and price. Following, e.g., Cabral (1990, 2006), they presume that due to the installed base effect the diffusion of products should continue even if prices remain unchanged. Consequently, Grajek and Kretschmer (2012) develop a structural model of demand with installed base effects. To estimate the threshold level of the 'critical mass', they suggest that diffusion is highly endogenous and the process as such can be identified as multi-equilibrating. Their seminal findings, examining the case of the global cellular telephony market over the period 1998–2007,[29] suggest that the 'critical mass' can be predominantly attributed to the size of the installed base, prices and the market size. Strong network effects allow for a lower installed base and higher prices to assure the sustainability of further diffusion, and the opposite is true in the case of weak network effects. Additionally, Grajek and Kretschmer (2012) report that the 'critical mass' phenomenon is only revealed in the case of emerging (pioneering markets). The model they propose for the identification of the 'critical mass' combines an installed base effect, the current installed base and prices. If the 'critical mass' occurs under certain threshold conditions, then the diffusion becomes self-sustaining. In this spirit, they define the 'critical mass' point as a combination of the three factors previously listed. Considering the assumptions in the Grajek and Kretschmer (2012) model, in certain societies (countries) the diffusion of innovation will never occur unless the 'critical mass' is reached. This would imply that some societies might be stuck in a 'low-equilibrium trap' and unable to 'take-off'. Baraldi (2012) provides new insights into the estimation of the *size* of the 'critical mass' rather than concentrating exclusively on its determinants (recall the works of, e.g., Grajek and Kretschmer 2012). Baraldi (2012) argues that the size of the 'critical mass' is determined by the strength of network effects. To detect the strength of the network effects, she adopts a concave demand curve. Hence, the occurrence of the 'critical mass' (regardless of the price of the new product) takes place the sooner the stronger the network effect is and the opposite otherwise.[30]

Similar to Baraldi (2012), Arroyo-Barrigüete et al. (2010) offer a conceptualization of the 'critical mass'. They use a convex demand curve to depict the 'critical mass' point. Arroyo-Barrigüete et al. (2010) follow Oren et al.'s (1982) concept of 'critical mass', arguing that it explains the minimum size of the network that encourages new users to join the network and adopt the new product. Once the 'critical mass' of users is achieved, the process of diffusion is self-perpetuating. Following Katz and Shapiro (1985), who define the network effects as an increasing

---

[29] Similar evidence on the role of the installed base is offered by Gruber and Verboven (2001), Koski and Kretschmer (2005), and Grajek (2010).

[30] Baraldi (2012) specifies the network effects as: $X_{i,t} = f\left[\left(\frac{GDP}{population}\right)_{i,t}, p_{i,t}, g(X_{i,t-1})\right]$, where

i denotes country, and t the time period. $X_{i,t}$ is thus the installed base, $p_{i,t}$ is price, $\left(\frac{GDP}{population}\right)_{i,t}$ is GDP *per capita*, and $g(X_{i,t-1})$ reveals network externalities in country i at time t. To control for concavity, $g(X_{i,t-1})$ includes a squared term for the lagged installed base. To estimate the size of the *critical mass*, Baraldi (2012) follows Rohlfs (1974), Katz and Shapiro (1985) and Economides and Himmelberg (1995a) and formalizes the inverse demand function as $p_{i,t} = \alpha + \beta_1 base_{i,t} + \beta_2 \ln(base_{i,t-1}) + \beta_3 X_{i,t} + \varepsilon_{i,t}$, where $X_{i,t}$ captures control variables. To assure concavity, $\beta_2 > 0$, and $\beta_1 < 0$ must be satisfied. If $\beta_2 > 0$ and $\beta_2 > \beta_1$, the network externalities are revealed and the upward slope of the demand curve emerges. The higher β_2, the sooner the *critical mass* point is reached.

utility of using a product as the total number of users grows, Arroyo-Barrigüete et al. (2010) suggest that new users will arrive once the utility obtained from the product is higher than its price.[31] However, Arroyo-Barrigüete et al. (2010) claim that direct estimation of the 'critical mass' point is hardly possible, as the process of diffusion of new products is preconditioned by individual choices (not always rational) and preferences, market structure, legal conditions and other unquantifiable effects.

It is worth underlining that despite a relatively well-developed theoretical framework and conceptual background aiming to explain the 'critical mass'-like phenomenon, the number of empirical works seeking a quantitative assessment of it is very limited. This may be a consequence of the great heterogeneity of the proposed theoretical specifications without any clear and well-established definition of 'critical mass'.

Few empirical works provide quantitative identification of the critical mass in its generic sense. Some examples are the works of Mahler and Rogers (1999), who study telecommunication services in 392 German banks, and Cool et al. (1997), who analyze the diffusion of innovation in an intra-organizational context, providing evidence on the threshold share of the population that has already adopted the new product which can ensure the further process of diffusion is self-sustaining. Mahler and Rogers (1999) suggest that keeping diffusion at very low levels makes it impossible to reach the 'critical mass', which hinders the broad spread of innovations. Cool et al. (1997) find that the 'critical mass' can be reached in different organizational regimes. They also underline that before reaching the 'critical mass' point, diffusion is predominantly driven by supply factors, while after passing the 'critical mass' point further diffusion is mainly pushed by growing demand.

Most presented concepts of the 'critical mass' consider the phenomenon in a microeconomic rather than a macroeconomic perspective. This is a serious limitation, as reaching a 'critical mass' might be strongly affected by social, economic, institutional, cultural or legal prerequisites.

The following Sect. 3.3.2 is intended to explain a novel conceptualization of the 'critical mass' regarding technology diffusion process.

3.3.2 The *'Technological Take-Off'* and the *'Critical Mass'*. A Trial Conceptualisation

As was previously discussed, the 'critical mass', may be defined as the minimal necessary number of user of new technology, which ensure the emergence of the 'take-off' period along the diffusion trajectory, at which the further process of diffusion becomes self-perpetuating (see Fig. 3.3).

---

[31] The condition follows: $U = a + b(n^e) > P$, where U is the utility function and P is the product price.

The term 'take-off' itself, however has been originally introduced to the economic literature by Walt Rostow, who, in his founding paper '*The take-off into self-sustaining growth*' (1956), claimed that the process of economic growth is characterised by discontinuity '*centering on a relatively brief time interval of two or three decades when the economy and the society of which it is a part transforms themselves in such ways that economic growth is, subsequently, more or less automatic*' (Rostow 1956, p. 1). He labelled this transformation the '*take-off*'. Rostow (1956, 1963, 1990) also wrote that identifying the '*take-off*' entails seeking to isolate the specific period (interval) in which '*the scale of productive activity reaches a critical level, (…) which leads to a massive and progressive structural transformation in economic, better viewed as change in kind than a merely in degree*' (Rostow 1956, p. 16). The concept of the '*take-off*' was then developed and implemented in the works of, e.g., Hoselitz (1957), Ranis and Fei (1961), Bertram (1963), Azariadis and Drazen (1990), Becker et al. (1994), Evans (1995), Baldwin et al. (2001), and Easterly (2006). In most of the cited works, the notion of the 'take-off' was, however, combined with Rosenstein-Rodan's (1943) 'Big Push' doctrine, which was predominantly applied to describing and explaining the stages, patterns and determinants of economic development and growth.

Similar to economic growth, the process of technology diffusion may well be approximated by easily distinguishable phases (stages) (see Fig. 3.2). During the initial phase, the process of diffusion slows, whereas subsequently, under favourable circumstances, it accelerates and proceeds at an exponential growth rate, ultimately approaching relative stabilisation (maturity) when the growth rates gradually diminish.

In Sect. 3.3.2, we propose a novel trial conceptualisation of how to identify the '*take-off*' period and the '*critical mass*' regarding technology diffusion process. The presented throughout the Sect. 3.3.2 theoretical framework has been developed based on the previously run empirical analysis which outcomes are extensively discussed in Chap. 5 (see also Appendices F and G for detailed calculations).

To meet the objective of this work, we adjust the conceptual background provided by Rostow (1956, 1990) and develop the term '*technological take-off*' and define it the time interval when the nature of the diffusion process is radically transformed due to shifting the rate of diffusion and forcing the transition from condition of stagnation into dynamic and self-sustaining growth (diffusion) of new technology. In this sense, the emergence of the '*technological take-off*' is essential for ensuring the sustainability of technology diffusion and enabling the widespread adoption of new technology throughout society. Generally, before the '*technological take-off*', diffusion proceeds slowly, but once the '*technological take-off*' is achieved, diffusion proceeds more rapidly and the number of new technology adopters begins to expand fast, typically at an exponential rate. Finally, in the maturity phase, the number of new technology users reaches system carrying capacity (saturation) and stabilises. To remain in line with the previous, the long-term process of technology diffusion may be arbitrarily divided into four separate phases (stages). Firstly, the initial (early) phase is when the technology diffusion is initiated, but the annual growth and penetration rates are typically negligible. In the early stage of diffusion, the preconditions for the '*technological take-off*' are also established. The second phase constitutes the '*technological take-off*' itself; then, in the third phase—'*post technological take-off*'—the increase in users of the new technology is self-perpetuating and becomes a normal condition in a given economy. Finally, the fourth phase occurs when diffusion significantly slows down, approaching saturation (maturity).

However, the emergence of the '*technological take-off*' is intimately related to and preconditioned by achieving the '*critical mass*', which has yet to be defined. With this aim, we develop the following terms: the technology replication coefficient ($\Phi_{i,y}$) (hereafter, the replication coefficient), marginal growth in technology adoption ($\Omega_{i,y}$) (hereafter, marginal growth), critical year ($Y_{crit,i,y}$), and critical penetration rate ($critICT_{i,y}$), where i denotes country and y year.

Assume that for a given country (i) and a given technology (ICT), the term $N_{i,y}$ stands for the level of technology (ICT) adoption in y year. By definition, $N_{i,y} > 0$, because negative adoption is not possible, and if $N_y = 0$, the diffusion process is not reported. Along this line, the technology replication coefficient ($\Phi_{i,y}$) follows:

$$\Phi_{i,y} = \frac{N_{i,y}}{N_{(i,y-1)}}, \tag{3.33}$$

then:

$$N_{i,y} = \Phi_{i,y} \left[N_{(i,y-1)} \right], \tag{3.34}$$

if $N_{i,y} > 0$ and $N_{(i,y-1)} > 0$, and $\Phi_{i,y} \in (0;\infty)$. The replication coefficient for respective technology (ICT) explains the multiplication of technology users that occurs because of the emerging '*word of mouth*' effect (Geroski 2000; Lee et al. 2010). Suppose that for y year, the $\Phi_{i,y} = 3$. This shows that in $(y-1)$ year, each user of the given technology has 'generated' two **additional** new users of the new technology. In this sense, the replication is the cornerstone of the diffusion process itself. Figure 3.7 illustrates how respective values of $\Phi_{i,y}$ determine $N_{i,y}$ over time.

If $\Phi_{i,y} > 1$, it implies that in each consecutive year, the number of users of new technology increases, so that $N_{i,y} > N_{i,y-1}$. This indicates that the values of $\Phi_{i,y}$ must be higher than 1 to ensure diffusion. If $\Phi_{i,y} = 1$, the number of new technology users is constant over time, and thus $N_t = N_{(t+1)} = \ldots = N_{(t+n)}$ and no diffusion is reported. Finally, $\Phi_{i,y} < 1$ would imply that the number of users of new technology is decreasing over time, so that $N_{i,y-1} > N_{i,y}$. It may be argued that the replication coefficient ($\Phi_{i,y}$) exhibits the dynamics of the diffusion process and—to some degree—demonstrates the strength of the network effects that enhance the spread of new technology over society.

As was already claimed, if $\Phi_{i,y} > 1$, the number of new technology users is constantly increasing, so that $N_{i,y} > N_{i,y-1}$. Based on the latter, we propose the

Fig. 3.7 Changes is number of technology users *versus* replication coefficients. Source: Author's elaboration

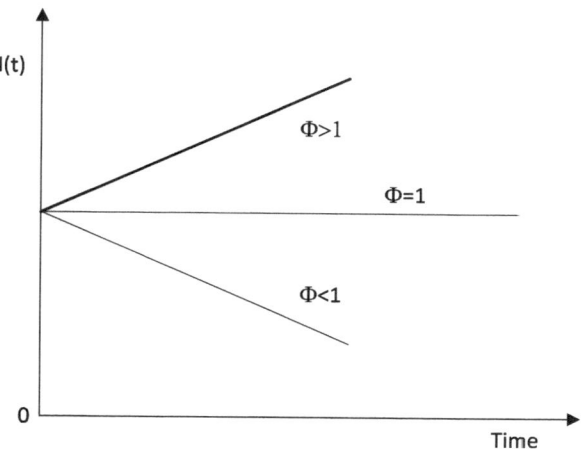

term 'marginal' growth in technology adoption ($\Omega_{i,y}$), which formally may be expressed as:

$$\Omega_{i,y} = N_{i,y} - N_{i,y-1}, \tag{3.35}$$

under the conditions that $N_{i,y} > 0$ and $N_{i,y-1} > 0$. The value of $\Omega_{i,y}$ expresses the change in the total number of users[32] of new technology over two consecutive years.

It is easily observed that these two coefficients—$\Phi_{i,y}$ and $\Omega_{i,y}$, are closely interrelated. Assuming that $\Phi_y > 1$, the level of marginal growth in i country and in y year is:

$$\Omega_{i,y} = N_{(i,y-1)}\left[\Phi_{i,y} - 1\right], \tag{3.36}$$

or:

$$\Omega_{i,y} = -N_{(i,y-1)}\left[1 - \Phi_{i,y}\right]. \tag{3.37}$$

Simply transforming Eq. (3.35) yields:

$$\frac{\Omega_{i,y}}{N_{i,y-1}} = \left[\Phi_{i,y} - 1\right]. \tag{3.38}$$

Generally, the $\Omega_{i,y}$ depends directly on the strength of the replication process that is expressed through the $\Phi_{i,y}$.

Examining the $\Phi_{i,y}$ and $\Omega_{i,y}$ simultaneously, it is easy to conclude that:

---

[32] In our case, expressed as number of users per 100 inhabitants.

1. If ($\Phi_{i,y} > 1$ then $\Omega_{i,y} > 0$), the replication process is sufficiently strong and the diffusion proceeds, which is demonstrated in the increasing number of new technology users $\left(N_{i,y} < N_{(i,y+1)}\right)$;
2. If ($\Phi_{i,y} = 1$ then $\Omega_{i,y} = 0$), the diffusion does not proceed, which results in a constant number of users of new technology $\left(N_{i,y} = N_{(i,y+1)} = \ldots = N_{(i,y+n)}\right)$;
3. If ($\Phi_{i,y} < 1$ then $\Omega_{i,y} < 0$), the replication process is so weak that the diffusion is limited, and there will be a decreasing number of users of new technology $\left(N_{i,y} > N_{(i,y+1)}\right)$.

If the replication coefficient is constant over time ($\Phi_{i,y} = \Phi_{i,y+1} \ldots = \Phi_{i,y+n}$), then in each consecutive period, the marginal growths in technology adoption are equal $\left(\Omega_{i,y} = \Omega_{i,y+1} \ldots = \Omega_{i,y+n}\right)$; and the diffusion proceeds linearly. However, as was already discussed in Sect. 3.2, the technology diffusion process is far from linear but rather follows an S-shaped trajectory instead.

In this vein, we intend to examine the behaviour of respective coefficients—$\Phi_{i,y}$ and $\Omega_{i,y}$—along the sigmoid technology diffusion pattern (for visualisation, see Fig. 3.8), which allows for determining the critical year ($Y_{crit,i,y}$) and critical penetration rate ($critICT_{i,y}$), and finally for identifying the 'technological take-off' interval.

In the early (initial) diffusion phase, the replication coefficient tends to be higher than marginal growth $\left(\Phi_{i,y} > \Omega_{i,y}\right)$, and thus, a gap emerges between $\Phi_{i,y}$ and $\Omega_{i,y}$. However, as the diffusion proceeds and the replication process is gains strength (so that $\Phi_{i,y} > 1$ and $\Omega_{i,y} > 0$), the $\Omega_{i,y}$ ultimately increases gradually while the $\Phi_{i,y}$ decreases in consecutive years, which will inevitably lead to closing the gap between $\Phi_{i,y}$ and $\Omega_{i,y}$ (the paths that show the changes in $\Phi_{i,y}$ and $\Omega_{i,y}$ are converging; see Fig. 3.8). If the latter is satisfied, the paths that show changes in $\Phi_{i,y}$ and $\Omega_{i,y}$ finally intersect (the gap between $\Phi_{i,y}$ and $\Omega_{i,y}$ is closed), so that in the next years, the replication coefficients are *lower* than marginal growth $\left(\Phi_{i,y} < \Omega_{i,y}\right)$, and the paths that show changes in $\Phi_{i,y}$ and $\Omega_{i,y}$ diverge. The specific time when the gap between $\Phi_{i,y}$ and $\Omega_{i,y}$ is closed (theoretically, $\Phi_{i,y} = \Omega_{i,y}$) we label the critical year ($Y_{crit,i,y}$); meanwhile, the penetration rate of new technology in $Y_{crit,i,y}$ we name the critical penetration rate ($critICT_{i,y}$). Technically, the critical year denotes the specific time period when the dynamic of the diffusion process is transformed, as the early diffusion phase is left behind and the new technology begins to diffuse exponentially; the 'critical penetration rate' we define as the *threshold* that, once passed, provokes the diffusion to become self-perpetuating, which implies overcoming the 'resistance to steady growth' (Rostow 1990). The 'critical penetration rate' traces the number of individuals—'innovators'—who demonstrate little risk aversion and high propensity to acquire novelties and who thus are the first new technology adopters and the ones who propagate its further diffusion throughout society. Finally, we argue that the 'critical penetration rate' approximates the 'critical mass' of new technology adopters, which preconditions

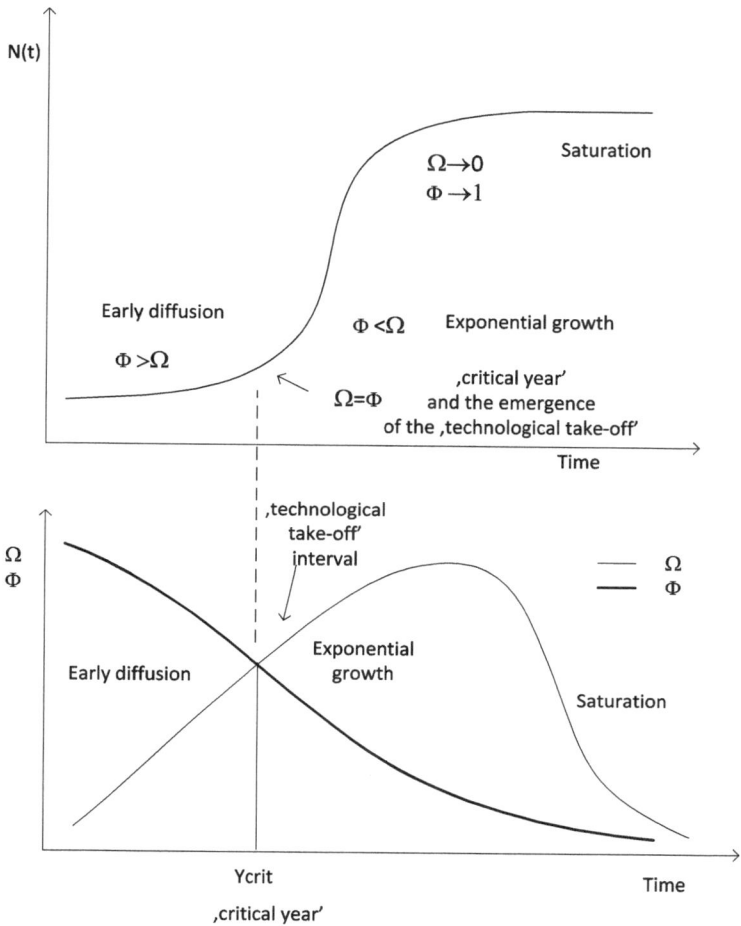

Fig. 3.8 Relationships between technology replication coefficient $(_{i,y})$, 'marginal' growth in technology adoption $(_{i,y})$, critical year $(Y_{crit,i,y})$ along the S-shaped technology diffusion trajectory. Source: Author's elaboration

the further spread of technology and forces the emergence of the '*technological take-off*'.

It is important to note that following this procedure would yield rigid identification of the exact date when $\Phi_{i,y} = \Omega_{i,y}$. However, to satisfy the latter, daily data on new technology penetration rates would be required, which for obvious reasons is scarcely possible. To challenge this obstacle, we choose to treat as the critical year $(Y_{crit,i,y})$ the first year when $\Phi_{i,y} < \Omega_{i,y}$, **if** in the previous year, the $\Phi_{i,y-1} > \Omega_{i,y-1}$ was reported. As was already mentioned, once it passes the $Y_{crit,i,y}$, the new technology begins to diffuse at an exponential rate, which is exhibited in the increasing values of $\Omega_{i,y}$. Finally, the process of diffusion slows and inevitably approaches the maturity phase when the desired saturation ($N_{y,i}$) is achieved. The slow-down and maturity phase $\Phi_y \to 1$ and $\Omega_y \to 0$ determines the termination of the diffusion process.

Finally, we propose labelling the 2-year interval right after the $Y_{crit,i,y}$ as the '*technological take-off*', which, as was previously defined, denotes the time period when the nature of the diffusion process is transformed because the diffusion rate shifts and forces the transition from stagnation to the dynamic and self-sustaining growth (diffusion) of the new technology.

Presuming that y stands for $Y_{crit,i,y}$ and to address the assumption that the '*technological take-off*' is the period during which the rate of diffusion is radically shifted, we suggest the following formalization of the conditions under which the '*technological take-off*' emerges:

$$\begin{cases} \Omega_{i,(y+1)} > 0 \\ \Omega_{i,(y+2)} > 0 \\ \Omega_{i,(y+1)} > \Omega_{i,(y)} \\ \Omega_{i,(y+2)} > \Omega_{i,(y)} \end{cases}. \tag{3.39}$$

Following Eq. (3.38) we argue that if y stands for $Y_{crit,i,y}$, the '*technological take-off*' interval occurs during the period $< y + 1; y + 2 >$.

If the critical year ($Y_{crit,i,y}$) is not identified, the conditions specified in Eq. (3.39) are also not satisfied, and this implies that the emergence of the '*technological take-off*' has been restricted. Technically, the previous indicates that during the initial diffusion phase, the replication lacked the strength to ensure gradual increases in $\Omega_{i,y}$, which would allow for closing the gap between $\Phi_{i,y}$ and $\Omega_{i,y}$ (see Fig. 3.9). As result, the paths that show the changes in $\Phi_{i,y}$ and $\Omega_{i,y}$ diverge rather than converge, and the critical year does not emerge. If $\Phi_{i,y} = 1$ or $\Phi_{i,y} < 1$, the situation is similar, and the technology diffusion is impeded. The countries where the $Y_{Crit,i,y}$ has not been identified are those where the process of entering exponential growth has been restrained and they remained virtually locked in the 'low-level-technology' trap, becoming latecomers in this respect.

Finally, we strongly argue that the '*critical year*', the '*critical penetration rate*' and the '*technological take-off*' do not emerge unconditionally or in isolation but are heavily predetermined by multiple social, economic and instructional prerequisites. The '*technological take-off*' is preconditioned and induced by strong stimuli that are typically well-established in the early diffusion phase. In this vein, we claim that the analysis of the '*critical mass*' should be considered in a broad context that allows for capturing a broad array of factors that could potentially foster or impede the '*technological take-off*'. We suggest that identifying both the critical penetration rate and the '*technological take-off*' interval should be complemented by broad analysis of the socio-economic and institutional conditions under which the '*technological take-off*' emerged. This approach places the purely numerical analysis in the broad macroeconomic perspective and is essential for capturing those factors that potentially foster or hinder the emergence of the

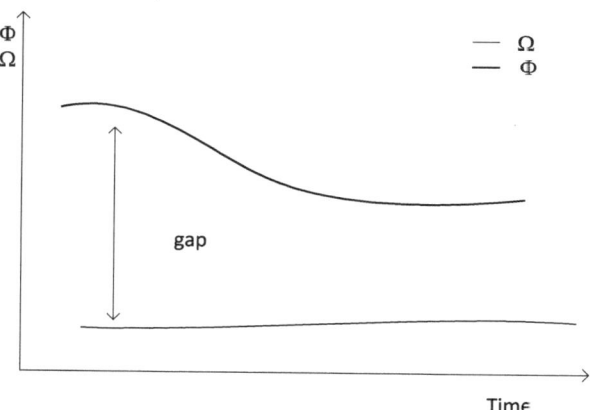

Fig. 3.9 The 'low-level-technology' trap. Source: Author's elaboration

'technological take-off'; and proposed broadening of the *'critical mass'* analysis sheds light on countries' socio-economic and institutional characteristics and situates the analysis in a broad macroeconomic perspective. These preconditions generally combine institutional change, economic performance, political regimes, social norms and attitudes, and the state of development of any backbone infrastructure. In a broad sense, the *'technological take-off'* requires that a society and an economy be prepared to actively respond to newly emerging possibilities (Rostow 1956). If these requirements are not sufficiently fulfilled, the *'technological take-off'* will not occur. Our concept of *'critical mass'* is, to a point, related to what was stressed in the works of Baumol (1986), Perez and Soete (1988), and Verspagen (1991), that a country's ability to adopt new technologies is preconditioned by a wide array of factors. Societies assess and assimilate technological novelties by relying upon 'intellectual' capital (Soete and Verspagen 1994) and institutional, governmental and cultural conditions. Some empirical evidence shows that the most prominent factors in a country's ability to adopt and effectively use new technologies are education and the skills of the labour force (Baumol 1986). Countries that experience significant lacks in these factors will likely never be able to ensure the widespread use of new technologies and use the full potential of technological change. As a result, they will never catch up with richer countries and will continue to lag behind as technologically disadvantaged regions.

3.4 Technology Convergence and Technology Convergence Clubs

Dynamic technology diffusion, accompanied by fundamental shifts in technology adoption and use, should inevitably lead to a significant reduction in cross-country technology gaps and growing cohesion. In other words, if countries experience growing levels of technology adoption, cross-country convergence should be exhibited. In this vein, we define 'technology convergence'[33] as a process leading to the 'technology gap' narrowing, and eradicating different forms of exclusion from access to and use of basic ICTs (Lechman 2012a, b). In this sense, technology convergence should fundamentally decrease cross-country inequalities in access to and use of ICTs,[34] as countries which are initially technologically-poorer shall exhibit relatively higher average annual growth rates of ICTs adoption, compared to countries which are initially better off with this respect. We intentionally encourage technology convergence unconditionally, leaving aside all factors which hypothetically might enhance or hinder the process. Still, our main attention shifts to providing an analytical framework to answer the prominent question of whether countries exhibit growing cohesion (decreasing technology gaps) in terms of their level of adoption and use of ICTs. So far, the approach to technology convergence analysis that we suggest is not commonly recognized, and the empirical evidence in the field remains relatively poor. Some evidence can be traced in the works of Comin and Hobijn (2004, 2011), Comin et al. (2006), Castellacci (2006a, 2008), Castellacci and Archibugi (2008), Castellacci (2011) and Lechman (2012a, b). Comin and Hobijn (2004) provide extensive analysis of technology convergence over the period 1788–2001. Their study covers 20 technologies in 23 different countries and tests the convergence hypothesis applying beta- and sigma-convergence procedures. Comin et al. (2006) perform similar exercises to Comin and Hobijn (2004). They test beta- and sigma-convergence using the CHAT (Cross-Country Historical Adoption of Technology) dataset, additionally separating within-technology and across-technologies effects. Castellacci (2006a, 2008) and

---

[33] In the literature discussing 'technological catching-up', the term is often confused with 'technology convergence'. In effect, it is misleading to use these two terms alternatively. Technological catching-up is the process through which countries benefit from the stock of knowledge available in the rest of the developed world, and goes far beyond simple technology convergence (Rogers 2010). The technological catching-up theories instead seek to answer how technologically backward countries may benefit from their underdevelopment and by diminishing the relative gap in Total Factor Productivity (TFP) experience economic growth (Soete and Turner 1984). The idea of incorporating different aspects of 'technology' into growth models traces back to pioneering works by Veblen (1915), Nurkse (1955), Gerschenkron (1962), Rostow (1971), Schumpeter (1984). Nelson and Phelps (1966) were the first to formalize the Veblen-Gerschenkron 'relative backwardness' idea and they introduced the idea of the function of technological catching-up depending on human capital and its absorptive capabilities (also argued by Abramovitz (1986)): $\frac{dA}{dt} / A = \varnothing(.)\left(\frac{T-A}{A}\right)$, where T stands for the level of the best practice technology, A is the level of technology in a backward country, and $\varnothing(.)$ is the function of absorptive capacities. Recently the literature treating international technological catching-up, and technology diffusion and transfer as factors contributing to rapid economic growth is pervasive. The most prominent evidence can be found in works by, *inter alia*, Fagerberg (1987, 1994), Perez and Soete (1988), Verspagen (1994), Dowrick (1992), Ben-David (1993), Coe and Helpman (1995), Barro and Sala-i-Martin (1990), Keller (1996), Bassanini et al. (2000), Dowrick and Rogers (2002), Castellacci (2002, 2006a, b, 2008, 2011), Liebig (2012), Stokey (2012), Shin (2013) and Serranito (2013).

[34] Apart from some empirical evidence on 'technology convergence' with respect to ICTs, there exist numerous studies where an analogous problem is tackled, but is labelled 'closing the digital divide' (see e.g. Servon 2008; James 2003, 2011; Vicente and López 2011).

Castellacci and Archibugi (2008) detect technology convergence clubs along with technology convergence testing. Castellacci (2008) reports on technology convergence and technology convergence clubs for 149 countries over the period 1990–2000. He additionally tests for 'technological capabilities' which may enhance or hinder the process of closing cross-country technology gaps. Additional evidence on the process of closing technology gaps is also reported by Castellacci (2011). Castellacci and Archibugi (2008), using data from the ArCo database (Archibugi and Coco 2004a, b, 2005) provide similar evidence over an analogous time period but they include 131 countries in their analysis. The empirical analysis found in works by Lechman (2012a, 2012b) reports on technology convergence exclusively for Information and Communication Technologies, for 145 countries over the period 2000–2010, and the technology convergence is tested adopting beta-, sigma-, and quantile-convergence approaches.

Originally, the concept of 'convergence' referred to growing cross-country cohesion in terms of economic development, approximated by *per capita* income level. Thus, the conceptual background for technology convergence analysis is derived from endogenous growth theories. These are explained in Sects. 3.4.1 and 3.4.2.

3.4.1 Convergence: Theoretical Specification

Following neoclassical growth theory (Solow 1956), countries follow a convergence pattern heading for common equilibrium in *per capita* income (Barro and Sala-i-Martin 1990; Barro et al. 1991, 1995). In other words, countries tend to converge toward a 'steady-state' equilibrium, but they experience gradual decreases in their rate of growth (Kangasharju 1999), however, under the rigid assumption of identical cross-country growth rates. In other words, the convergence process implies that initially poorer countries experience a relatively higher average annual growth rate, and thus catch up with the rich ones. The idea that poor countries tend to grow faster than rich ones is strictly attributed to Gerschenkron's[35] pioneering hypothesis of 'relative backwardness' (1960, 1962). Gerschenkron argues that backward economies take advantage of their economic underdevelopment[36] and by assimilating technology spillovers into high growth rates they catch up with the rich countries[37] (Verspagen 1994). Thus, the Veblen-Gerschenkron

---

[35] Although in many works Alexander Gerschenkron is cited as the first to introduce the idea of 'relative backwardness', the term was also used by Thorsten Veblen (1915) and Leibenstein (1957).

[36] Similarly, Findlay (1978a), Baumol (1986) and Romer (1993) consider relative backwardness to be a convergence facilitating factor.

[37] Gerschenkron's 'relative backwardness' idea (1962) was formalized in a model by Nelson-Phelps (1966), who argued that the growth of technology in an economically backward country is proportional to the gap between the backward country and the country using the most advanced technological solutions (located close to the Technology Frontier Area) (Gomulka 2006).

hypothesis links economic convergence[38] with the initial size of the gap with world technology frontiers (Stokke 2004)[39]; while Abramovitz (1986) points out that backward countries have a potential for rapid advances, but he also stresses the importance of social capabilities which can enhance or hinder the catching-up process (Abramovitz 1989).

Technically speaking, convergence occurs if average annual growth rates are inversely correlated with initial *per capita* income. A straightforward implication of undisturbed convergence is that—in a long-term perspective—cross-country disparities should inevitably be eradicated. If this is not the case, countries instead experience divergence and the gap between 'rich' and 'poor' enlarges. Empirically, the convergence can be tested using two standard approaches, namely sigma (σ)-convergence and beta (β)-convergence. σ-convergence is exhibited once disparity in *per capita* income decreases over time, which is measured by changes in the standard deviation (absolute approach) or the coefficient of variation (relative approach).[40] The standard deviation for country i in country set n and year t is as follows (Rodrik 2013; Thirlwall 2013):

$$\sigma_{i,t} = \left[\frac{1}{n} \sum_{i=1}^{n} \left(log\left(\frac{y_i}{y^x} \right) \right)^2 \right]^{1/2}, \tag{3.40}$$

if $y^x \equiv \frac{1}{n} \sum_{i=1}^{n} log(y_i)$, and y stands for *per capita* income. Over the period analyzed, the σ-convergence hypothesis is verified positively if $\sigma_{i,t} \to 0$ is satisfied.[41] This approach to convergence testing, although very interpretive and simple, has one main disadvantage: the standard deviation reveals a high sensitivity to the inclusion of outliers in the country set tested, and additionally it does not allow any causal mechanism provoking economic convergence among countries to be captured.[42]

---

[38] Productivity convergence.

[39] Findlay (1978a), however, argues that the gap to the world technology frontier cannot be *too* large, and countries located below a threshold value of the gap will not be able to catch-up economically.

[40] The coefficient of variation is highly useful in σ-convergence testing if two or more country groups are compared in terms of their internal convergence.

[41] If σ-convergence is tested with regard to the coefficient of variation, then the coefficient of variation is $\frac{\sigma_{i,t}}{\theta_{i,t}}$, where $\theta_{i,t}$ is the mean of the tested variable over the whole sample.

[42] The σ-convergence hypothesis was tested, *inter-alia*, in works by de la Fuente (2003), Canaleta et al. (2002), Rey and Dev (2006), Young et al. (2008), Egger and Pfaffermayr (2009), Garrido-Yserte and Mancha-Navarro (2010), Schmitt and Starke (2011), Smetkowski and Wójcik (2012), Delgado (2013) and Thirlwall (2013).

Fig. 3.10 Unconditional (absolute) convergence. Note: the figure refers to 'technology (ICT) convergence'; *P* poor country; *R* rich country

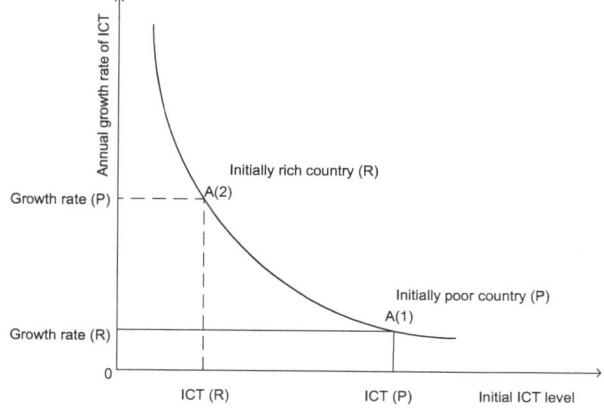

Following the neoclassical growth model (Sala-i-Martin 1995), the conditions for absolute (unconditional) β-convergence can be formulated as a regression equation[43]:

$$g_i = a + bv_{i,t_0} + \varepsilon_i, \tag{3.41}$$

where *i* denotes the country, t_0 is the initial year in the time span for the convergence test, v_{i,t_0} is the level of *per capita* income in t_0 expressed as its natural logarithm, and ε_i is a random error term. The coefficient b[44] in Eq. (3.41) stands for the convergence coefficient, indicating the speed of the process. Consider that, for example, $b = 1.5$, then one unit increase in ln (v_{i,t_0}) provokes an average annual growth in *per capita* income approximately 1.5 % higher in initially poorer countries. For economic interpretation, the sign—positive or negative—of b is crucial, since negative b indicates convergence (see Fig. 3.10), but positive b means divergence, yielding growing disparities between countries. Formally, if coefficient $b = 0$, then neither convergence nor divergence is reported and the gaps between countries are maintained over time.

Using the coefficient b from Eq. (3.41), the speed of convergence can be estimated. Assume that over the given time period $T = 0 \ldots \ldots t$, so that:

---

[43] Conventionally, Eq. (3.41) is estimated applying OLS. However, if we relax the assumption that the variables are normally distributed, the estimated coefficients might be biased and inefficient. Koenker and Bassett (1978) suggest the adoption of non-parametric quantile regression to avoid the problem. The quantile regression approach is highly useful when the original variable distribution is highly skewed (asymmetric). Standard β-convergence estimates allow for assessment of variable behaviour but are based on the conditional mean, while quantile regression (q-regression, q-convergence) introduces estimates in non-central locations (Koenker 2004; Hao and Naiman 2007). Using the quantile regression approach, it is possible to determine any number of quantiles for estimation, which allows modelling of variable behaviour in any pre-defined location of variable distribution.

[44] Also explaining the partial correlation between a variable growth rate and its initial level.

$$b = -\left(1 - e^{-\beta T}\right). \tag{3.42}$$

By extracting β from Eq. (3.42), we obtain:

$$\beta = -\ln(1 + b)/T, \tag{3.43}$$

where β indicates the rate at which convergence proceeds and countries head toward a steady state of *per capita* income. Consequently, we calculate the time span necessary for actual inter-country disparities to be halved:

$$HL_i = [-\ln(2)]/\beta. \tag{3.44}$$

Suppose that HL_i is 10 years. This implies that if the current convergence rate is maintained over the period the inter-country gaps will be halved within a 10-year period.

   The concept of unconditional convergence (both σ and β) is built on the rigid assumption that the process of convergence is 'automatic' and is not pre-conditioned by any country's individual characteristics. However, it is reasonable that the tendency of countries to converge (diverge) toward a steady state is conditioned by factors unobservable from their absolute convergence (Galor 1996; Quah 1996; Rodrik 2013). These can be technological development, social capital, institutional constraints, culture or many others.[45] The formalization of conditional convergence, however, requires that Eq. (3.41), needs to be modified by adding a vector (V_i) explaining a country's individual features. Thus, the regression is estimated as:

$$g_i = a + bv_{i,t_0} + \alpha V_i + \varepsilon_i, \tag{3.45}$$

with notation as in Eq. (3.41). The economic interpretation of b is analogous to that in the case of unconditional convergence.

   σ-, and β-convergence testing (both unconditional and conditional) is based on econometric procedures with cross-sectional data application (de la Fuente 2000). However, Bernard and Durlauf (1995) and Bernard and Jones (1996) argue that convergence is a dynamic process. They suggest an alternative approach to convergence analysis which is based on a time series.[46] They claim that economies should stochastically converge under the assumption that long-term growth forecasts for each country are close to equal.[47] Assume we have just two countries a and b, and

---

[45] The body of evidence on conditional convergence is *massive*. Seminal contributions in the field were made by, *inter alia*, Dowrick and Nguyen (1989), Barro and Sala-i-Martin (1990); Mankiw et al. (1992), Quah (1993, 1999), Pritchett (1997), Del Bo et al. (2010), Schmitt and Starke (2011), Barro (2012), and Yorucu and Mehmet (2014).

[46] Also labelled 'stochastic convergence' (see i.e. McGuinness and Sheehan 1998).

[47] The approach for convergence testing using a time-series has been applied in a multitude of studies, e.g. using empirical evidence on inter-regional stochastic convergence, by *inter alia*, Johnson (2000), Drennan et al. (2004), Alexiadis and Tomkins (2004), Herrerías and Monfort

for both the long-run GDP *per capita* forecasts are equal. Thus, the condition for absolute convergence can be expressed as (Bernard and Durlauf 1995):

$$\lim_{k \to \infty} E\left(\ln(GDPpc)_{a,t+k} - \ln(GDPpc)_{b,t+k} / \Pi_t \right) = 0, \qquad (3.46)$$

where t denotes time and Π is the stock of information which is available at a given point in time. The formula in Eq. (3.46), can be easily extended to any number of countries in the sample for which the absolute convergence hypothesis is to be tested.[48]

3.4.2 Convergence Clubs Hypothesis

Apart from the growing body of theoretical and empirical evidence on the convergence process, the concept of 'convergence clubs' has emerged. It was initially proposed and conceptualized by Baumol (1986) and consequently developed by Baumol and Wolff (1988), and Baumol et al. (1989). The 'convergence clubs' hypothesis assumes that a sub-set of countries (of the full sample) experience convergence[49] and head toward a common steady state (Alexiadis and Tomkins 2004; Alexiadis 2013a), while the 'rest' of the countries are left outside the 'club' and gradually diverge. The general message is that convergence occurs only for a subset of countries, while the 'rest' are excluded from the 'very exclusive organization' (Baumol 1986). Alexiadis and Alexandrakis (2008) argue that convergence clubs arise as some economically backward countries do not satisfy certain initial conditions and cannot fully realize their potential of catching-up with rich countries (Easterly et al. 1993; Ocampo et al. 2007). Thus, a group of initially poor countries grows at lower rates than rich countries, and the gap between the two increases.[50]

In the literature, there exist two main approaches providing a theoretical framework for the detection of convergence clubs. The first one, proposed by Baumol (1986), derives from the absolute convergence to 'steady state' approach (see, e.g., Barro and Sala-i-Martin 1990; Barro et al. 1991, 1995), and the second—developed

---

(2013), Lin et al. (2013); or inter-country stochastic convergence as in the works of Datta (2003), Bentzen (2005), and Canarella et al. (2010).

[48] The possibility of applying the formula in Eq. (3.46) to use it for absolute convergence testing, however, is determined by specific econometric tests. The most commonly used for this purpose is the Augmented Dickey Fuller test (1979, 1981), which introduces cointegration and unit root procedures to the empirical analysis of time-series.

[49] Generally in terms of β-convergence.

[50] Quah (1997, 1999) argues that countries may form 'coalitions' and behave non-linearly in their convergence patterns for three main reasons: countries' behaviour along their development paths are heavily preconditioned by other counties (e.g. by trade flows, human labour flows); countries tend to specialize to boost economies of scale; and human capital, culture, social and absorptive capabilities matter for development (see also Abramovitz 1989).

by Chatterji (1992)—is based on 'convergence in gaps'.[51] Formally, the test for the existence of convergence clubs in a set of countries consists in augmenting the standard procedure for β-convergence testing (see Eq. (3.41)) by introducing the square term of the explanatory variable. Inserting these square terms into Eq. (3.41), generates the possibility of identifying multiple *equilibria* (Alexiadis 2013a) as the convergence path exhibits non-linearities (Desdoigts 1999; Quah 1997, 1999; Fiaschi and Lavezzi 2007; Artelaris et al. 2011). Following the theoretical specification developed by Baumol (1986) and Baumol and Wolff (1988), the basic condition for convergence club emergence is expressed in a quadratic model:

$$g_i = a + b_1 v_{i,t_0} + b_2 v_{i,t_0}^2 + \varepsilon_i. \tag{3.47}$$

Equation (3.47) is an augmented version of the standard regression (see Eq. (3.41)) applied for β-convergence testing. The hypothesis on convergence clubs is supported only in the case that the coefficients b_1 and b_2 emerge as negative and positive respectively. The model defined in Eq. (3.47) has several important implications. First, it shows that the convergence pattern with respect to a set of countries might not be linear, and the convergence as such is identified only in a subset of countries, while the rest are left behind (see Fig. 3.11). The function defined in Eq. (3.47) reaches its maximum when the first derivative of Eq. (3.48) reaches zero. Thus:

$$f'_{g_i, v_{i,t_0}} = \frac{dg_i}{dv_{i,t_0}} = 0. \tag{3.48}$$

Extracting v_{i,t_0} from Eq. (3.48), gives the level of *per capita* income corresponding to the maximum of the function in Eq. (3.47), which can be calculated as:

$$v_{threshold} = \frac{-b_1}{2b_2}. \tag{3.49}$$

The *per capita* income in (t_0) calculated applying Eq. (3.49) stands for the '*threshold value*' ('*threshold condition*') (Alexiadis 2013a) and enables the identification of convergence club members.

Thus, in the case of countries that initially exceed the '*threshold value*' of *per capita* income $(v_{i,t_0} - v_{threshold} > 0)$ the relationship between the average annual growth rate and the level of *per capita* income in (t_0) is negative. Hence β-convergence is confirmed, and they form a convergence club. However, for

---

[51] The evidence on convergence club identification, mainly with respect to *per capita* income, can be found in works by, *inter alia*, Ben-David (1994, 1998), Armstrong (1995, 2002), Dewhurst and Mutis-Gaitan (1995), Fagerberg and Verspagen (1996), Verspagen (1997), Desdoigts (1999), Baumont et al. (2003), Durlauf (2003), Su (2003), Canova (2004), Fischer and Stirböck (2006), Le Gallo and Dall'Erba (2006), Alexiadis (2013b), Lechman (2012c), Song et al. (2013), Brida et al. (2014) and Fischer and LeSage (2014)

Fig. 3.11 Convergence clubs—theoretical specification. Source: own elaboration based on Baumol (1986) and Alexiadis (2013a) concepts. Note: the figure refers to 'technology (ICT) convergence clubs'

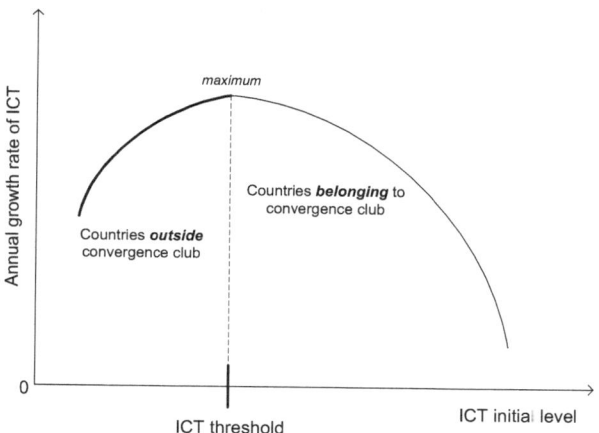

countries that were initially located below the *'threshold value'* of *per capita* income ($v_{i,t_0} - v_{threshold} < 0$), the relationship between the average annual growth rate and the level of *per capita* income in (t_0) is positive and β-convergence is not reported.[52] Hence, these are left outside the club.

Following the original concept of technology gaps developed, *inter alia*, by Gomulka (1971, 1986), Chatterji (1992) proposed a different approach to convergence club detection. He argues that positive verification of the β-convergence hypothesis is not sufficient for the reduction of gaps among countries. Cross-country disparities may grow over time and thus convergence 'in gaps' is not reported. The procedure proposed by Chatterji (1992) for the detection of convergence clubs may be treated as an enriched and more sophisticated version of σ-convergence. Its theoretical specification is the following. Assume we have a set of countries over the period (t_0 ....T) with a leading economy (L). We follow the rigid assumption that both in the initial and the terminal year the leading economies remain unchanged. The gap (divide) between the L-economy and any other country in the set is defined as (Chatterji and Dewhurst 1996; Kangasharju 1999):

$$G_{i,t_0} = ln\left(\frac{V_{leader,t_0}}{V_{i,t_0}}\right),$$ (3.50)

in the initial year (t_0), and in the terminal year (T) it is:

$$G_{i,T} = ln\left(\frac{V_{leader,T}}{V_{i,T}}\right).$$ (3.51)

---

[52] It is important to note that Baumol (1986) approach to convergence club identification is heavily pre-conditioned by the initial level of *per capita* income.

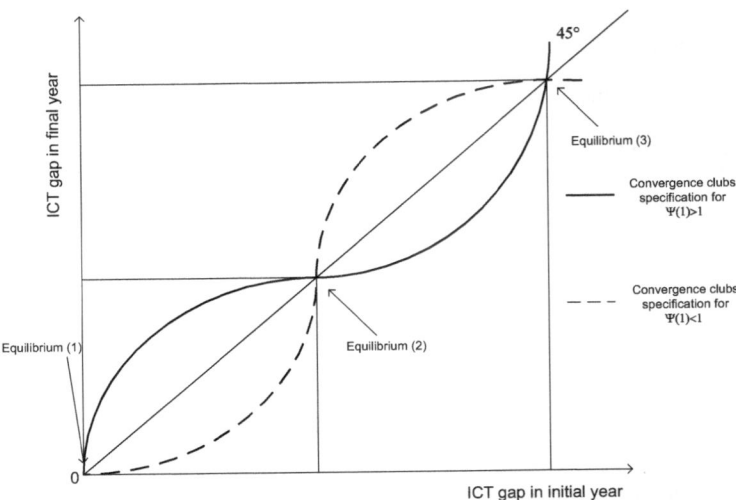

Fig. 3.12 Convergence clubs—theoretical specification. Source: own elaboration based on concepts by Chatterji (1992), Kangasharju (1999) and Alexiadis (2013a). Note: the 45° line indicates whether $G_{i,t_0} = G_{i,T}$, $G_{i,t_0} > G_{i,T}$ or $G_{i,t_0} < G_{i,T}$. Points below 45° indicate convergence and above 45° divergence. The figure refers to 'technology (ICT) convergence clubs'

Following Kangasharju (1999), the condition for the identification of convergence clubs in a given country set is defined as a three-equilibria model:

$$G_{i,T} = \Psi_1\left(G_{i,0}\right) + \Psi_2(G_{i,0})^2 + \Psi_3\left(G_{i,0}\right)^3. \tag{3.52}$$

The third degree polynomial[53] (Eq. 3.52) yields the existence of three different equilibrium (see Fig. 3.12) satisfying $G_{i,0} = G_{i,T}$. This also suggests that convergence follows a cubic behaviour, and at every equilibrium the gap between economy i and economy L is constant.

Following Chatterji (1992) and Alexiadis (2013a), the steady-state values ($G_{1,t_0} \rightarrow$ Equilibrium (1), $G_{2,t_0} \rightarrow$ Equilibrium (2), and $G_{3,t_0} \rightarrow$ Equilibrium (3)) that determine club membership are defined as:

$$G_{2,t_0} = \frac{-\Psi_2 - \sqrt{(\Psi_2)^2 - 4\Psi_3(\Psi_1 - 1)}}{-2\Psi_3}$$

$$G_{3,t_0} = \frac{-\Psi_2 + \sqrt{(\Psi_2)^2 - 4\Psi_3(\Psi_1 - 1)}}{-2\Psi_3} \tag{3.53}$$

and G_{1,t_0} is zero by definition.

---

[53] Cubic specification.

Convergence behaviour and convergence club formation strictly depend on the value Ψ_1. If $\Psi_1 < 1$, then countries with an initial gap lower than G_{2,t_0} exhibit convergence. Thus for a convergence club the gap between country i and economy L is gradually decreasing. Conversely, countries with an initial gap between G_{2,t_0} and G_{3,t_0} are instead diverging from economy L and are excluded from the club, increasing their distance to economy L. The most backward economies with an initial gap above G_{3,t_0} may converge, but only toward the third equilibrium point. If $\Psi_1 > 1$, the situation is just the opposite. Countries exhibiting convergence (forming convergence clubs) are those with an initial gap varying from G_{2,t_0} to G_{3,t_0}, while countries with an initial gap below G_{2,t_0} instead tend to diverge from economy L. Again, the poorest countries, with initial gaps above G_{3,t_0}, converge, but toward the 'lower' equilibrium.

References

Abramovitz, M. (1986). Catching up, forging ahead, and falling behind. *The Journal of Economic History, 46*(02), 385–406.

Abramovitz, M. (1989). *Thinking about growth: And other essays on economic growth and welfare*. Cambridge: Cambridge University Press.

Aghion, P., Akcigit, U., & Howitt, P. (2013). *What do we learn from Schumpeterian growth theory?* (No. w18824). National Bureau of Economic Research.

Alexiadis, S. (2013a). *Convergence clubs and spatial externalities: Models and applications of regional convergence in Europe*. Berlin: Springer.

Alexiadis, S. (2013b). EU-27 regions: Absolute or club convergence?. In *Convergence clubs and spatial externalities* (pp. 119–139). Berlin: Springer.

Alexiadis, S., & Alexandrakis, A. (2008). Threshold conditions' and regional convergence in European agriculture. *International Journal of Economic Sciences and Applied Research, 1*(2), 13–37.

Alexiadis, S., & Tomkins, J. (2004). Convergence clubs in the regions of Greece. *Applied Economics Letters, 11*(6), 387–391.

Allen, P. (1988a). Evolution, innovation and economics. In G. Dosi et al. (Eds.), *Technical change and economic theory* (pp. 95–119). London: Pinter.

Allen, D. (1988b). New telecommunications services: Network externalities and *critical mass*. *Telecommunications Policy, 12*(3), 257–271.

Antonelli, C. (1986). The international diffusion of new information technologies. *Research Policy, 15*(3), 139–147.

Antonelli, C. (1991). *The diffusion of advanced telecommunications in developing countries*. Paris: OECD.

Archibugi, D., & Coco, A. (2004a). A new indicator of technological capabilities for developed and developing countries (ArCo). *World Development, 32*(4), 629–654.

Archibugi, D., & Coco, A. (2004b). Measuring technological capabilities at the country level: A survey and a menu for choice. *Research Policy, 34*(2), 175–194.

Archibugi, D., & Coco, A. (2005). Measuring technological capabilities at the country level: A survey and a menu for choice. *Research Policy, 34*(2), 175–194.

Armstrong, H. W. (1995). Convergence among regions of the European Union, 1950–1990. *Papers in Regional Science, 74*(2), 143–152.

Armstrong, H. W. (2002). European Union regional policy: Reconciling the convergence and evaluation evidence. In J. R. Cuadrado & M. Parellada (Eds.), *Regional convergence in the European Union* (pp. 231–272). Berlin: Springer.

Arroyo-Barrigüete, J. L., Ernst, R., López-Sánchez, J. I., & Orero-Giménez, A. (2010). On the identification of *critical mass* in Internet-based services subject to network effects. *The Service Industries Journal, 30*(5), 643–654.

Artelaris, P., Arvanitidis, P. A., & Petrakos, G. (2011). Convergence patterns in the world economy: Exploring the nonlinearity hypothesis. *Journal of Economic Studies, 38*(3), 236–252.

Azariadis, C., & Drazen, A. (1990). Threshold externalities in economic development. *The Quarterly Journal of Economics, 105*, 501–526.

Baldwin, R. E., Martin, P., & Ottaviano, G. I. (2001). Global income divergence, trade, and industrialization: The geography of growth take-offs. *Journal of Economic Growth, 6*(1), 5–37.

Banks, R. B. (1994). *Growth and diffusion phenomena: Mathematical frameworks and applications* (Vol. 14). Berlin: Springer.

Baraldi, A. L. (2004). Equilibrium size in network with indirect network externalities. *Rivista italiana degli economisti, 9*(3), 475–494.

Baraldi, A. L. (2012). The size of the *critical mass* as a function of the strength of network externalities: A mobile telephone estimation. *Economics of Innovation and New Technology, 21*(4), 373–396.

Barro, R. J. (2012). *Convergence and modernization revisited* (No. w18295). National Bureau of Economic Research.

Barro, R. J., Mankiw, N. G., & Sala-i-Martin, X. (1995). *Capital mobility in neoclassical models of growth* (No. w4206). National Bureau of Economic Research.

Barro, R. J., & Sala-i-Martin, X. (1990). *Economic growth and convergence across the United States* (No. w3419). National Bureau of Economic Research.

Barro, R. J., Sala-i-Martin, X., Blanchard, O. J., & Hall, R. E. (1991). Convergence across states and regions. *Brookings Papers on Economic Activity, 1*(4), 107–182.

Bass, F. M. (1969). A new product growth model for consumer durables. *Management Science, 15*, 215–227.

Bass, F. M. (1974). The theory of stochastic preference and brand switching. *Journal of Marketing Research, 11*(1), 1–20.

Bass, F. M. (1980). The relationship between diffusion rates, experience curves, and demand elasticities for consumer durable technological innovations. *The Journal of Business, 53*(3), S51–S67.

Bass, F. M. (2004). Comments on "a new product growth for model consumer durables the bass model". *Management Science, 50*(12_suppl), 1833–1840.

Bass, F. M., & Parsons, L. J. (1969). Simultaneous-equation regression analysis of sales and advertising. *Applied Economics, 1*(2), 103–124.

Bassanini, A., Scarpetta, S., & Visco, I. (2000). *Knowledge, technology and economic growth: Recent evidence from OECD countries* (No. 259). Paris: OECD.

Battisti, G. (2008). Innovations and the economics of new technology spreading within and across users: Gaps and way forward. *Journal of Cleaner Production, 16*(1), S22–S31.

Baumol, W. J. (1986). Productivity growth, convergence, and welfare: What the long-run data show. *The American Economic Review, 76*, 1072–1085.

Baumol, W. J., Blackman, S. A. B., & Wolff, E. N. (1989). *Productivity and American leadership: The long view* (pp. 225–250). Cambridge, MA: MIT Press.

Baumol, W. J., & Wolff, E. N. (1988). Productivity growth, convergence, and welfare: Reply. *The American Economic Review, 78*, 1155–1159.

Baumont, C., Ertur, C., & Le Gallo, J. (2003). Spatial convergence clubs and the European regional growth process, 1980–1995. In B. Fingleton (Ed.), *European regional growth* (pp. 131–158). Berlin: Springer.

Becker, G. S., Murphy, K. M., & Tamura, R. (1994). Human capital, fertility, and economic growth. In G. S. Becker (Ed.), *Human capital: A theoretical and empirical analysis with special reference to education* (3rd ed., pp. 323–350). Chicago: University of Chicago Press.

Bell, M., & Pavitt, K. (1995). The development of technological capabilities. *Trade, Technology and International Competitiveness, 22,* 69–101.

Bell, M., & Pavitt, K. (1997). Technological accumulation and industrial growth: Contrasts between developed and developing countries. In D. Archibugi & J. Michie (Eds.), *Technology, globalization and economic performance* (pp. 83–137). Cambridge: Cambridge University Press.

Ben-David, D. (1993). Equalizing exchange: Trade liberalization and income convergence. *The Quarterly Journal of Economics, 108*(3), 653–679.

Ben-David, D. (1994). *Convergence clubs and diverging economies* (No. 922). London: CEPR.

Ben-David, D. (1998). Convergence clubs and subsistence economies. *Journal of Development Economics, 55*(1), 155–171.

Bentzen, J. (2005). Testing for catching-up periods in time-series convergence. *Economics Letters, 88*(3), 323–328.

Bernard, A. B., & Durlauf, S. N. (1995). Convergence in international output. *Journal of Applied Econometrics, 10*(2), 97–108.

Bernard, A. B., & Jones, C. I. (1996). Technology and convergence. *Economic Journal, 106*(437), 1037–1044.

Bertram, G. W. (1963). Economic growth in Canadian industry, 1870-1915: The staple model and the take-off hypothesis. *Canadian Journal of Economics and Political Science, 29,* 159–184.

Bhargava, S. C. (1995). A generalized form of the Fisher-Pry model of technological substitution. *Technological Forecasting and Social Change, 49*(1), 27–33.

Blackman, A. W., Jr. (1971). The rate of innovation in the commercial aircraft jet engine market. *Technological Forecasting and Social Change, 2*(3), 269–276.

Bonacich, P., Shure, G. H., Kahan, J. P., & Meeker, R. J. (1976). Cooperation and group size in the n-person prisoners' dilemma. *Journal of Conflict Resolution, 20*(4), 687–706.

Brida, J. G., Garrido, N., & Mureddu, F. (2014). Italian economic dualism and convergence clubs at regional level. *Quality & Quantity, 48*(1), 439–456.

Cabral, L. (1990). On the adoption of innovations with 'network' externalities. *Mathematical Social Sciences, 19*(3), 299–308.

Cabral, L. M. (2006). Equilibrium, epidemic and catastrophe: Diffusion of innovations with network effects. In C. Antonelli, D. Foray, B. H. Hall, & W. D. Steinmueller (Eds.), *New frontiers in the economics of innovation and new technology: Essays in honour of Paul A. David.* Cheltenham, UK: Edward Elgar.

Canaleta, C. G., Arzoz, P. P., & Gárate, M. R. (2002). Structural change, infrastructure and convergence in the regions of the European Union. *European Urban and Regional Studies, 9*(2), 115–135.

Canarella, G., Miller, S. M., & Pollard, S. K. (2010). *Stochastic convergence in the Euro area* (No. 2010-32).

Canova, F. (2004). Testing for convergence clubs in income per capita: A predictive density approach. *International Economic Review, 45*(1), 49–77.

Carrington, P. J., Scott, J., & Wasserman, S. (Eds.). (2005). *Models and methods in social network analysis.* New York: Cambridge University Press.

Castellacci, F. (2002). Technology gap and cumulative growth: Models and outcomes. *International Review of Applied Economics, 16*(3), 333–346.

Castellacci, F. (2006a). Innovation, diffusion and catching up in the fifth long wave. *Futures, 38*(7), 841–863.

Castellacci, F. (2006b). Convergence and divergence among technology clubs. In *DRUID conference, Copenhagen,* June, Vol. 30, No. 07. http://www.druid.dk/wp/pdf_files.org/06-21.pdf

Castellacci, F. (2007). Technological regimes and sectoral differences in productivity growth. *Industrial and Corporate Change, 16*(6), 1105–1145.

Castellacci, F. (2008). Technology clubs, technology gaps and growth trajectories. *Structural Change and Economic Dynamics, 19*(4), 301–314.

Castellacci, F. (2011). Closing the technology gap? *Review of Development Economics, 15*(1), 180–197.

Castellacci, F., & Archibugi, D. (2008). The technology clubs: The distribution of knowledge across nations. *Research Policy, 37*(10), 1659–1673.

Castellacci, F., & Natera, J. M. (2013). The dynamics of national innovation systems: A panel cointegration analysis of the coevolution between innovative capability and absorptive capacity. *Research Policy, 42*(3), 579–594.

Chatterji, M. (1992). Convergence clubs and endogenous growth. *Oxford Review of Economic Policy, 8*(4), 57–69.

Chatterji, M., & Dewhurst, J. L. (1996). Convergence clubs and relative economic performance in Great Britain: 1977–1991. *Regional Studies, 30*(1), 31–39.

Coe, D. T., & Helpman, E. (1995). International R&D spillovers. *European Economic Review, 39*(5), 859–887.

Cohen, W. M., & Levinthal, D. A. (1990). Absorptive capacity: A new perspective on learning and innovation. *Administrative Science Quarterly, 35*(1), 128–152.

Comin, D., & Hobijn, B. (2004). Cross-country technology adoption: Making the theories face the facts. *Journal of Monetary Economics, 51*(1), 39–83.

Comin, D., & Hobijn, B. (2006). *An exploration of technology diffusion* (No. w12314). National Bureau of Economic Research.

Comin, D., & Hobijn, B. (2011). Technology diffusion and postwar growth. In *NBER macroeconomics annual 2010* (Vol. 25, pp. 209–246). University of Chicago Press.

Comin, D., Hobijn, B., & Rovito, E. (2006). *Five facts you need to know about technology diffusion* (No. w11928). National Bureau of Economic Research.

Cool, K. O., Dierickx, I., & Szulanski, G. (1997). Diffusion of innovations within organizations: Electronic switching in the Bell System, 1971-1982. *Organization Science, 8*(5), 543–559.

Coontz, S. H. (2013). *Population theories and their economic interpretation* (Vol. 8). London: Routledge.

Cramer, J. S. (2003). The origins and development of the logit model. In J. S. Cramer (Ed.), *Logit models from economics and other fields* (pp. 149–158). Cambridge: Cambridge University Press.

Darwin, C. (1968). *On the origin of species by means of natural selection.* London: Murray. 1859.

Datta, A. (2003). Time-series tests of convergence and transitional dynamics. *Economics Letters, 81*(2), 233–240.

David, P. A. (1969). *A contribution to the theory of diffusion.* Research Center in Economic Growth Stanford University.

David, P. A. (1986). Technology diffusion, public policy, and industrial competitiveness. In R. Landau & N. Rosenberg (Eds.), *The positive sum strategy: Harnessing technology for economic growth* (pp. 373–391). Washington, DC: National Academy Press.

Davies, S. (1979). The diffusion of process innovations. CUP Archive.

De la Fuente, A. (2000). *Mathematical methods and models for economists.* Cambridge: Cambridge University Press.

De la Fuente, A. (2003). Convergence equations and income dynamics: The sources of OECD convergence, 1970-1995. *Economica, 70*, 655–671.

Del Bo, C., Florio, M., & Manzi, G. (2010). Regional infrastructure and convergence: Growth implications in a spatial framework. *Transition Studies Review, 17*(3), 475–493.

Delgado, F. J. (2013). Are taxes converging in Europe? Trends and some insights into the effect of economic crisis. *Journal of Global Economics, 1*(1), 102.

Desdoigts, A. (1999). Patterns of economic development and the formation of clubs. *Journal of Economic Growth, 4*(3), 305–330.

Dewhurst, J. H., & Mutis-Gaitan, H. (1995). Varying speeds of regional GDP per capita convergence in the European Union, 1981-91. In H. W. Armstrong & R. W. Vickerman (Eds.), *Convergence and divergence among European regions* (pp. 22–39). London: Pion.

Dickey, D. A., & Fuller, W. A. (1979). Distribution of the estimators for autoregressive time series with a unit root. *Journal of the American Statistical Association, 74*(366a), 427–431.

Dickey, D. A., & Fuller, W. A. (1981). Likelihood ratio statistics for autoregressive time series with a unit root. *Econometrica: Journal of the Econometric Society, 49*, 1057–1072.

Dosi, G. (1991). The research on innovation diffusion: An assessment. In N. Nakićenović & A. Grübler (Eds.), *Diffusion of technologies and social behavior* (pp. 179–208). Berlin: Springer.

Dosi, G., & Nelson, R. R. (1994). An introduction to evolutionary theories in economics. *Journal of Evolutionary Economics, 4*(3), 153–172.

Dowrick, S. (1992). Technological catch up and diverging incomes: Patterns of economic growth 1960-88. *Economic Journal, 102*(412), 600–610.

Dowrick, S., & Nguyen, D. T. (1989). OECD comparative economic growth 1950-85: Catch-up and convergence. *American Economic Review, 79*(5), 1010–1030.

Dowrick, S., & Rogers, M. (2002). Classical and technological convergence: Beyond the Solow-Swan growth model. *Oxford Economic Papers, 54*(3), 369–385.

Drennan, M. P., Lobo, J., & Strumsky, D. (2004). Unit root tests of sigma income convergence across US metropolitan areas. *Journal of Economic Geography, 4*(5), 583–595.

Durlauf, S. N. (2003). *The convergence hypothesis after 10 years*. Social Systems Research Institute, University of Wisconsin.

Easterly, W. (2006). Reliving the 1950s: The big push, poverty traps, and takeoffs in economic development. *Journal of Economic Growth, 11*(4), 289–318.

Easterly, W., Kremer, M., Pritchett, L., & Summers, L. H. (1993). Good policy or good luck? *Journal of Monetary Economics, 32*(3), 459–483.

Economides, N., & Himmelberg, C. (1995a). *Critical mass* and network size with application to the US fax market.

Economides, N., & Himmelberg, C. (1995b). *Critical mass* and network evolution in telecommunications. In *Toward a competitive telecommunications industry: Selected papers from the 1994 telecommunications policy research conference* (pp. 47–63). College Park, MD: University of Maryland.

Egger, P., & Pfaffermayr, M. (2009). On testing conditional sigma—Convergence. *Oxford Bulletin of Economics and Statistics, 71*(4), 453–473.

Ehrnberg, E. (1995). On the definition and measurement of technological discontinuities. *Technovation, 15*(7), 437–452.

Evans, P. B. (1995). *Embedded autonomy: States and industrial transformation* (pp. 3–21). Princeton: Princeton University Press.

Evans, D. S., & Schmalensee, R. (2010). Failure to launch: *Critical mass* in platform businesses. *Review of Network Economics, 9*(4), 1–26.

Fagerberg, J. (1987). A technology gap approach to why growth rates differ. *Research Policy, 16* (2), 87–99.

Fagerberg, J. (1994). Technology and international differences in growth rates. *Journal of Economic Literature, 32*, 1147–1175.

Fagerberg, J., & Verspagen, B. (1996). Heading for divergence? Regional growth in Europe reconsidered. *JCMS: Journal of Common Market Studies, 34*(3), 431–448.

Fagerberg, J., & Verspagen, B. (2002). Technology-gaps, innovation-diffusion and transformation: An evolutionary interpretation. *Research Policy, 31*(8), 1291–1304.

Fiaschi, D., & Lavezzi, A. M. (2007). Nonlinear economic growth: Some theory and cross-country evidence. *Journal of Development Economics, 84*(1), 271–290.

Findlay, R. (1978a). Relative backwardness, direct foreign investment, and the transfer of technology: A simple dynamic model. *The Quarterly Journal of Economics, 92*(1), 1–16.

Findlay, R. (1978b). Some aspects of technology transfer and direct foreign investment. *The American Economic Review, 68*(2), 275–279.

Fischer, M. M., & LeSage, J. P. (2014), A Bayesian space-time approach to identifying and interpreting regional convergence clubs in Europe. *Papers in Regional Science*. doi:10.1111/pirs.12104

Fischer, M. M., & Stirböck, C. (2006). Pan-European regional income growth and club-convergence. *The Annals of Regional Science, 40*(4), 693–721.

Fisher, R. A. (1930). *The genetical theory of natural selection*. Oxford, UK: Clarendon Press.

Fisher, J. C., & Pry, R. H. (1972). A simple substitution model of technological change. *Technological Forecasting and Social Change, 3*, 75–88.

Foley, D. K. (2009). The history of economic thought and the political economic education of Duncan Foley.

Frohlich, N., Oppenheimer, J. A., & Young, O. R. (1971). *Political leadership and collective goods* (p. 13). Princeton, NJ: Princeton University Press.

Fudenberg, D., & Tirole, J. (1985). Pre-emption and rent equalization in the adoption of new technology. *Review of Economic Studies, 52*, 383–401.

Galor, O. (1996). *Convergence? Inferences from theoretical models* (No. 1350). CEPR Discussion Papers.

Garrido-Yserte, R., & Mancha-Navarro, T. (2010). The Spanish regional puzzle: Convergence, divergence and structural change. In J. R. Cuadrado Roura (Ed.), *Regional policy, economic growth and convergence* (pp. 103–124). Berlin: Springer.

Geroski, P. A. (1990). Innovation, technological opportunity, and market structure. *Oxford Economic Papers, 42*(3), 586–602.

Geroski, P. A. (2000). Models of technology diffusion. *Research Policy, 29*(4), 603–625.

Gerschenkron, A. (1962). *Economic backwardness in historical perspective*. Cambridge, MA: Harvard University Press.

Gompertz, B. (1825). On the nature of the function expressive of the law of human mortality, and on a new mode of determining the value of life contingencies. *Royal Society of London Philosophical Transactions Series I, 115*, 513–583.

Gomulka, S. (1971). *Inventive activity, diffusion, and the stages of economic growth* (Vol. 24). Aarhus: Aarhus University, Institute of Economics.

Gomulka, S. (1986). *Growth, innovation and reform in Eastern Europe*. Brighton: Wheatsheaf Books.

Gomulka, S. (2006). *The theory of technological change and economic growth*. New York: Routledge.

Gottschalk, P. G., & Dunn, J. R. (2005). The five-parameter logistic: A characterization and comparison with the four-parameter logistic. *Analytical Biochemistry, 343*(1), 54–65.

Grajek, M. (2003). *Estimating network effects and compatibility in mobile telecommunications* (No. SP II 2003-26). Discussion papers//WZB, Wissenschaftszentrum Berlin für Sozialforschung, Forschungsschwerpunkt Markt und Politische Ökonomie, Abteilung Wettbewerbsfähigkeit und Industrieller Wandel.

Grajek, M. (2010). Estimating network effects and compatibility: Evidence from the Polish mobile market. *Information Economics and Policy, 22*(2), 130–143.

Grajek, M., & Kretschmer, T. (2012). Identifying critical mass in the global cellular telephony market. *International Journal of Industrial Organization, 30*(6), 496–507.

Gray, V. (1973). Innovation in the states: A diffusion study. *The American Political Science Review, 67*(4), 1174–1185.

Griliches, Z. (1957). Hybrid corn: An exploration in the economics of technological change. *Econometrica, Journal of the Econometric Society, 25*(4), 501–522.

Gruber, H., & Verboven, F. (2001). The evolution of markets under entry and standards regulation—The case of global mobile telecommunications. *International Journal of Industrial Organization, 19*(7), 1189–1212.

Hall, B. H., & Khan, B. (2003). *Adoption of new technology* (No. w9730). National Bureau of Economic Research.

Hao, L., & Naiman, D. Q. (2007). *Quantile regression. Quantitative applications in the social sciences* (Vol. 149). Thousand Oaks, CA: Sage.

Hardin, R. (1982). *Collective action* (pp. 38–49). Baltimore: Resources for the Future.

Helpman, E. (Ed.). (1998). *General purpose technologies and economic growth.* Cambridge, MA: MIT Press.

Herrerías, M. J., & Monfort, J. O. (2013). Testing stochastic convergence across Chinese provinces, 1952–2008. *Regional Studies.* doi:10.1080/00343404.2013.786825 (ahead-of-print).

Hoselitz, B. F. (1957). Noneconomic factors in economic development. *The American Economic Review, 47*(2), 28–41.

Ireland, N., & Stoneman, P. (1986). Technological diffusion, expectations and welfare. *Oxford Economic Papers, 38,* 283–304.

Jaber, M. Y. (Ed.). (2011). *Learning curves: Theory, models, and applications.* Boca Raton, FL: CRC Press.

James, J. (2003). *Bridging the global digital divide.* Cheltenham: Edward Elgar.

James, J. (2011). Are changes in the digital divide consistent with global equality or inequality? *The Information Society, 27*(2), 121–128.

Johnson, P. A. (2000). A nonparametric analysis of income convergence across the US states. *Economics Letters, 69*(2), 219–223.

Kangasharju, A. (1999). Relative economic performance in Finland: Regional convergence, 1934-1993. *Regional Studies, 33*(3), 207–217.

Kapur, S. (1995). Technological diffusion with social learning. *Journal of Industrial Economics, 43*(2), 173–195.

Kapur, D. (2001). Diasporas and technology transfer. *Journal of Human Development, 2*(2), 265–286.

Karshenas, M., & Stoneman, P. L. (1993). Rank, stock, order, and epidemic effects in the diffusion of new process technologies: An empirical model. *The RAND Journal of Economics, 24*(4), 503–528.

Karshenas, M., & Stoneman, P. (1995). Technological diffusion. In P. Stoneman (Ed.), *Handbook of the economics of innovation and technological change* (pp. 265–297). Oxford: Blackwell.

Katz, M. L., & Shapiro, C. (1985). Network externalities, competition, and compatibility. *The American Economic Review, 75*(3), 424–440.

Katz, M. L., & Shapiro, C. (1986). Technology adoption in the presence of network externalities. *The Journal of Political Economy, 94*(4), 822–841.

Katz, M. L., & Shapiro, C. (1992). Product introduction with network externalities. *The Journal of Industrial Economics, 40,* 55–83.

Keller, W. (1996). Absorptive capacity: On the creation and acquisition of technology in development. *Journal of Development Economics, 49*(1), 199–227.

Keller, W. (2004). International technology diffusion. *Journal of Economic Literature, 42*(3), 752–782.

Kim, M. S., & Kim, H. (2007). Is there early take-off phenomenon in diffusion of IP-based telecommunications services? *Omega, 35*(6), 727–739.

Kindleberger, C. P. (1995). Technological diffusion: European experience to 1850. *Journal of Evolutionary Economics, 5*(3), 229–242.

Kingsland, S. (1982). The refractory model: The logistic curve and the history of population ecology. *Quarterly Review of Biology, 57*(1), 29–52.

Koenker, R. (2004). Quantile regression for longitudinal data. *Journal of Multivariate Analysis, 91*(1), 74–89.

Koenker, R., & Bassett, G., Jr. (1978). Regression quantiles. *Econometrica: Journal of the Econometric Society, 84,* 33–50.

Koski, H., & Kretschmer, T. (2005). Entry, standards and competition: Firm strategies and the diffusion of mobile telephony. *Review of Industrial Organization, 26*(1), 89–113.

Kubielas, S. (2009). *Technology gap approach to industrial dynamics and sectoral systems of innovation in transforming CEE economies* (Working paper). Warsaw University.

Kucharavy, D., & De Guio, R. (2011). Logistic substitution model and technological forecasting. *Procedia Engineering, 9*, 402–416.

Kudryashov, N. A. (2013). Polynomials in logistic function and solitary waves of nonlinear differential equations. *Applied Mathematics and Computation, 219*(17), 9245–9253.

Kumar, V., & Krishnan, T. V. (2002). Multinational diffusion models: An alternative framework. *Marketing Science, 21*(3), 318–330.

Kumar, U., & Kumar, V. (1992). Technological innovation diffusion: The proliferation of substitution models and easing the user's dilemma. *IEEE Transactions on Engineering Management, 39*(2), 158–168.

Kwasnicki, W. (1999). Technological substitution processes. An evolutionary model. Institute of Industrial Engineering and Management, Wroclaw University of Technology.

Kwasnicki, W. (2013). Logistic growth of the global economy and competitiveness of nations. *Technological Forecasting and Social Change, 80*(1), 50–76.

Lall, S. (1992). Technological capabilities and industrialization. *World Development, 20*(2), 165–186.

Le Gallo, J., & Dall'Erba, S. (2006). Evaluating the temporal and spatial heterogeneity of the European convergence process, 1980–1999. *Journal of Regional Science, 46*(2), 269–288.

Lechman, E. (2012a). Technology convergence and digital divides. A country-level evidence for the period 2000-2010. *Ekonomia, Rynek, Gospodarka, Społeczeństwo, No. 31.*

Lechman, E. (2012b). *Cross national technology convergence. An empirical study for the period 2000-2010*. Germany: University Library of Munich.

Lechman, E. (2012c). Catching-up and club convergence from cross-national perspective a statistical study for the period 1980-2010. *Equilibrium, 7*, 95–109.

Lee, M., Kim, K., & Cho, Y. (2010). A study on the relationship between technology diffusion and new product diffusion. *Technological Forecasting and Social Change, 77*(5), 796–802.

Leibenstein, H. (1957). *Economic backwardness and economic growth*. New York: Wiley.

Liebig, K. (2012). *Catching up through technology absorption: Possibilities for developing countries* (Vol. 540). Munich: GRIN.

Lim, B. L., Choi, M., & Park, M. C. (2003). The late take-off phenomenon in the diffusion of telecommunication services: Network effect and the *critical mass*. *Information Economics and Policy, 15*(4), 537–557.

Lin, P. C., Lin, C. H., & Ho, I. L. (2013). Regional convergence or divergence in China? Evidence from unit root tests with breaks. *The Annals of Regional Science, 50*(1), 223–243.

Loch, C. H., & Huberman, B. A. (1999). A punctuated-equilibrium model of technology diffusion. *Management Science, 45*(2), 160–177.

Lotka, A. J. (1920). Undamped oscillations derived from the law of mass action. *Journal of the American Chemical Society, 42*(8), 1595–1599.

Mahajan, V., & Peterson, R. A. (Eds.). (1985). *Models for innovation diffusion* (Vol. 48). Beverly Hills, CA: Sage.

Mahler, A., & Rogers, E. M. (1999). The diffusion of interactive communication innovations and the *critical mass*: The adoption of telecommunications services by German banks. *Telecommunications Policy, 23*(10), 719–740.

Mankiw, N. G., Romer, D., & Weil, D. N. (1992). A contribution to the empirics of economic growth. *The Quarterly Journal of Economics, 107*(2), 407–437.

Mansfield, E. (1961). Technical change and the rate of imitation. *Econometrica: Journal of the Econometric Society, 29*(4), 741–766.

Mansfield, E. (1968). *The economies of technological change*. New York: WW Norton.

Mansfield, E. (1971). *Technological change: An introduction to a vital area of modern economics*. New York: WW Norton.

Mansfield, E. (1986). Microeconomics of technological innovation. In R. Landau & N. Rosenberg (Eds.), *The positive sum strategy* (pp. 307–325). Washington, DC: National Academies Press.

Marchetti, C., & Nakicenovic, N. (1980). *The dynamics of energy systems and the logistic substitution model.* International Institute for Applied Systems Analysis.

Markus, M. L. (1987). Toward a *"critical mass"* theory of interactive media universal access, interdependence and diffusion. *Communication Research, 14*(5), 491–511.

Marwell, G., & Oliver, P. (1993). *The critical mass in collective action.* New York: Cambridge University Press.

Mcguinness, S., & Sheehan, M. (1998). Regional convergence in the UK, 1970-1995. *Applied Economics Letters, 5*(10), 653–658.

Metcalfe, J. S. (1987). *The diffusion of innovation: An interpretive survey.* University of Manchester, Department of Economics.

Metcalfe, J. S. (1997). On diffusion and the process of technological change. In G. Antonelli & N. De Liso (Eds.), *Economics of structural and technological change* (pp. 123–144). London: Routledge.

Metcalfe, J. S. (2004). Ed Mansfield and the diffusion of innovation: An evolutionary connection. *The Journal of Technology Transfer, 30*(1-2), 171–181.

Meyer, P. (1994). Bi-logistic growth. *Technological Forecasting and Social Change, 47*(1), 89–102.

Meyer, P. S., Yung, J. W., & Ausubel, J. H. (1999). A primer on logistic growth and substitution: The mathematics of the Loglet Lab software. *Technological Forecasting and Social Change, 61*(3), 247–271.

Miranda, L., & Lima, C. A. (2013). Technology substitution and innovation adoption: The cases of imaging and mobile communication markets. *Technological Forecasting and Social Change, 80*(6), 1179–1193.

Modis, T. (2003). A scientific approach to managing competition. *Industrial Physicist, 9*(1), 25–27.

Modis, T. (2007). Strengths and weaknesses of S-curves. *Technological Forecasting and Social Change, 74*(6), 866–872.

Molina, A., Bremer, C. F., & Eversheim, W. (2001). Achieving *critical mass*: A global research network in systems engineering. *Foresight, 3*(1), 59–65.

Morris, S. A., & Pratt, D. (2003). Analysis of the Lotka–Volterra competition equations as a technological substitution model. *Technological Forecasting and Social Change, 70*(2), 103–133.

Nakicenovic, N. (1987). Technological substitution and long waves in the USA. In T. Vasko (Ed.), *The long-wave debate* (pp. 76–103). Berlin: Springer.

Nakicenovic, N. (Ed.). (1991). *Diffusion of technologies and social behavior.* Berlin: Springer.

Nelson, R. R. (1982). *An evolutionary theory of economic change.* Cambridge, MA: Harvard University Press.

Nelson, R. R., & Phelps, E. S. (1966). Investment in humans, technological diffusion and economic growth. *The American Economic Review, 56*, 69–75.

Nurkse, R. (1955). *Problems of capital formation in lesser-developed areas.* New York: Oxford University Press.

Ocampo, J. A., Jomo, K. S., & Vos, R. (Eds.). (2007). *Growth divergences: Explaining differences in economic performance.* London: Orient Longman.

Oliver, P., Marwell, G., & Teixeira, R. (1985). A theory of the critical mass. I. Interdependence, group heterogeneity, and the production of collective action. *American Journal of Sociology, 91*, 522–556.

Olson, M. (1965). *The logic of collective action.* Cambridge, MA: Harvard University Press

Oren, S. S., Smith, S. A., & Wilson, R. B. (1982). Nonlinear pricing in markets with inter-dependent demand. *Marketing Science, 1*(3), 287–313.

Pearl, R., & Reed, L. J. (1922). A further note on the mathematical theory of population growth. *Proceedings of the National Academy of Sciences of the United States of America, 8*(12), 365.

Perez, C., & Soete, L. (1988). Catching up in technology: Entry barriers and windows of opportunity. In G. Dosi, C. Freeman, R. Nelson, G. Silverberg, & L. Soete (Eds.), *Technical change and economic theory* (pp. 458–479). London: Pinter.

Pritchett, L. (1997). Divergence, big time. *Journal of Economic Perspectives, 11*, 3–18.

Puumalainen, K., Frank, L., Sundqvist, S., & Tappura, A. (2011). The *critical mass* of wireless communications: Differences between developing and developed economies. In *Mobile information communication technologies adoption in developing countries: Effects and implications* (pp. 1–17).

Quah, D. (1993). Galton's fallacy and tests of the convergence hypothesis. *The Scandinavian Journal of Economics, 95*(4), 427–443.

Quah, D. T. (1996). Empirics for economic growth and convergence. *European Economic Review, 40*(6), 1353–1375.

Quah, D. T. (1997). Empirics for growth and distribution: Stratification, polarization, and convergence clubs. *Journal of Economic Growth, 2*(1), 27–59.

Quah, D. (1999). *Ideas determining convergence clubs.* London School of Economics and Political Science.

Ranis, G., & Fei, J. C. (1961). A theory of economic development. *The American Economic Review, 51*(4), 533–565.

Rawls, J. (1999). *A theory of justice.* Cambridge, MA: Harvard University Press.

Reinganum, J. F. (1981a). Market structure and the diffusion of new technology. *Bell Journal of Economics, 12*(2), 618–624.

Reinganum, J. F. (1981b). On the diffusion of new technology: A game theoretic approach. *The Review of Economic Studies, 48*, 395–405.

Reinganum, J. F. (1989). The timing of innovation: Research, development, and diffusion. *Handbook of Industrial Organization, 1*, 849–908.

Rey, S. J., & Dev, B. (2006). σ-convergence in the presence of spatial effects. *Papers in Regional Science, 85*(2), 217–234.

Rodrik, D. (2013). Unconditional convergence in manufacturing. *The Quarterly Journal of Economics, 128*(1), 165–204.

Rogers, E. M. (1976). New product adoption and diffusion. *Journal of Consumer Research, 2*(4), 290–301.

Rogers, E. M. (2010). *Diffusion of innovations.* New York: Simon and Schuster.

Rogers, E. M., & Havens, A. E. (1962). Rejoinder to Griliches—Another false dichotomy. *Rural Sociology, 27*, 332–334.

Rogers, E. M., & Shoemaker, F. F. (1971). *Communication of innovations: A cross-cultural approach.* New York: Free Press.

Rohlfs, J. (1974). A theory of interdependent demand for a communications service. *The Bell Journal of Economics and Management Science, 5*(1), 16–37.

Romeo, A. A. (1977). The rate of imitation of a capital-embodied process innovation. *Economica, 44*, 63–69.

Romer, P. (1993). Idea gaps and object gaps in economic development. *Journal of Monetary Economics, 32*(3), 543–573.

Rosenberg, N. (1972). Factors affecting the diffusion of technology. *Explorations in Economic History, 10*(1), 3–33.

Rosenberg, N. (1982). *Inside the black box: Technology and economics.* Cambridge: Cambridge University Press.

Rosenstein-Rodan, P. N. (1943). Problems of industrialization of eastern and south-eastern Europe. *The Economic Journal, 53*, 202–211.

Rostow, W. W. (1956). The take-off into self-sustained growth. *The Economic Journal, 66*, 25–48.

Rostow, W. W. (1963). *The economics of take-off into sustained growth.* London: Macmillan.

Rostow, W. W. (1971). *Politics and the stages of growth.* CUP Archive.

Rostow, W. W. (1990). *The stages of economic growth: A non-communist manifesto.* Cambridge: Cambridge University Press.

Sala-i-Martin, X. (1995). *The classical approach to convergence analysis*. Centre for Economic Policy Research.

Sarkar, J. (1998). Technological diffusion: Alternative theories and historical evidence. *Journal of Economic Surveys, 12*(2), 131–176.

Satoh, D. (2001). A discrete bass model and its parameter estimation. *Journal of the Operations Research Society of Japan-Keiei Kagaku, 44*(1), 1–18.

Satoh, D., & Yamada, S. (2002). Parameter estimation of discrete logistic curve models for software reliability assessment. *Japan Journal of Industrial and Applied Mathematics, 19*(1), 39–53.

Saviotti, P. P. (2002). Black boxes and variety in the evolution of technologies. In G. Antonelli & N. De Liso (Eds.), *Economics of structural and technological change* (pp. 184–212). New York: Francis & Taylor.

Schmitt, C., & Starke, P. (2011). Explaining convergence of OECD welfare states: A conditional approach. *Journal of European Social Policy, 21*(2), 120–135.

Schoder, D. (2000). Forecasting the success of telecommunication services in the presence of network effects. *Information Economics and Policy, 12*(2), 181–200.

Schumpeter, J. (1984). *The theory of economic development*. Cambridge, MA: Harvard University Press.

Serranito, F. (2013). Heterogeneous technology and the technological catching-up hypothesis: Theory and assessment in the case of MENA countries. *Economic Modelling, 30*, 685–597.

Servon, L. J. (2008). *Bridging the digital divide: Technology, community and public policy*. Hoboken, NJ: Wiley.

Shapiro, C., & Varian, H. (1998). *Information rules*. Cambridge, MA: Harvard Business Press.

Shin, J. S. (2013). *The economics of the latecomers: Catching-up, technology transfer and institutions in Germany, Japan and South Korea*. London: Routledge.

Silverberg, G. (1994). *The economics of growth and technical change: Technologies, nations, agents*. Aldershot: Edward Elgar.

Silverberg, G., & Verspagen, B. (1995). *Evolutionary theorizing on economic growth*. Internat. Inst. for Applied Systems Analysis.

Simon, H. A. (1972). Theories of bounded rationality. *Decision and Organization, 1*, 161–176.

Smetkowski, M., & Wójcik, P. (2012). Regional convergence in central and eastern European countries: A multidimensional approach. *European Planning Studies, 20*(6), 923–939.

Soete, L., & Turner, R. (1984). Technology diffusion and the rate of technical change. *Economic Journal, 94*(375), 612–623.

Soete, L., & Verspagen, B. (1994). Competing for growth: The dynamics of technology gaps. In L. L. Pasinetti & R. M. Solow (Eds.), *Economic growth and the structure of long-term development: Proceedings of the IEA conference held in Varenna, Italy* (pp. 272–299). London: Macmillan Press.

Solow, R. M. (1956). A contribution to the theory of economic growth. *The Quarterly Journal of Economics, 70*, 65–94.

Song, P. C., Sek, S. K., & Har, W. M. (2013). Detecting the convergence clubs and catch-up in growth. *Asian Economic and Financial Review, 3*(1), 1–15.

Srinivasan, V., & Mason, C. H. (1986). Technical note-nonlinear least squares estimation of new product diffusion models. *Marketing Science, 5*(2), 169–178.

Srivastava, V. K. L., & Rao, B. B. (1990). *The econometrics of disequilibrium models* (Vol. 111). New York: Greenwood.

Stokey, N. L. (2012). *Catching up and falling behind* (No. w18654). National Bureau of Economic Research.

Stokke, H. E. (2004). Technology adoption and multiple growth paths: An intertemporal general equilibrium analysis of the catch-up process in Thailand. *Review of World Economics, 140*(1), 80–109.

Stone, R. (1980). Sigmoids. *Journal of Applied Statistics, 7*(1), 59–119.

Stoneman, P. (1995). *Handbook of the economics of innovation and technological change* (Blackwell handbooks in economics). Oxford: Blackwell.

Stoneman, P. (2001). *Technological diffusion and the financial environment* (No. 3). United Nations University, Institute for New Technologies.

Stoneman, P. (Ed.). (2002). *The economics of technological diffusion*. Oxford: Blackwell.

Stoneman, P., & Battisti, G. (2005). The intra-firm diffusion of new process technologies. *International Journal of Industrial Organization, 23*(1), 1–22.

Stoneman, P., & Battisti, G. (2010). The diffusion of new technology. *Handbook of the Economics of Innovation, 2*, 733–760.

Su, J. J. (2003). Convergence clubs among 15 OECD countries. *Applied Economics Letters, 10*(2), 113–118.

Thirlwall, A. P. (2013). *Regional disparities in per capita income in India: Convergence or divergence?* (No. 1313). Department of Economics, University of Kent.

Turk, T., & Trkman, P. (2012). Bass model estimates for broadband diffusion in European countries. *Technological Forecasting and Social Change, 79*(1), 85–96.

Valente, T. W. (1996). Social network thresholds in the diffusion of innovations. *Social Networks, 18*(1), 69–89.

Valente, T. W. (2005). Network models and methods for studying the diffusion of innovations. In P. Carrington, S. Wassermann, & J. Scott (Eds.), *Models and methods in social network analysis* (pp. 98–116). Cambridge: Cambridge University Press.

Van den Bulte, C., & Stremersch, S. (2004). Social contagion and income heterogeneity in new product diffusion: A meta-analytic test. *Marketing Science, 23*(4), 530–544.

Veblen, T. (1915). *Imperial Germany and the industrial revolution*. London: Macmillan.

Verhulst, P. F. (1838). Notice sur la loi que la population suit dans son accroissement. Correspondance Mathématique et Physique Publiée par A. *Quetelet, 10*, 113–121.

Verspagen, B. (1991). A new empirical approach to catching up or falling behind. *Structural Change and Economic Dynamics, 2*(2), 359–380.

Verspagen, B. (1994). Technology and growth: The complex dynamics of convergence and divergence. In G. Silverberg & L. Soete (Eds.), *The economics of growth and technical change: Technologies, nations, agents*. Aldershot: Edward Elgar.

Verspagen, B. (1997). European 'regional clubs': Do they exist, and where are they heading?; On economic and technological differences between European regions.

Vicente, M. R., & López, A. J. (2011). Assessing the regional digital divide across the European Union-27. *Telecommunications Policy, 35*(3), 220–237.

Villasis, G. (2008, November). The process of network effect. In *DEGIT conference papers* (No. c013_012). DEGIT, Dynamics, Economic Growth, and International Trade.

Volterra, V. (1926). Fluctuations in the abundance of a species considered mathematically. *Nature, 118*, 558–560.

Wang, M. Y., & Lan, W. T. (2007). Combined forecast process: Combining scenario analysis with the technological substitution model. *Technological Forecasting and Social Change, 74*(3), 357–378.

Ward, P. S., & Pede, V. (2013, June). *Spatial patterns of technology diffusion: The case of hybrid rice in Bangladesh*. Presentation at the Agricultural and Applied Economics Association. AAEA & CAES Joint Annual Meeting, Washington

Yorucu, V., & Mehmet, O. (2014). Absolute and conditional convergence in both zones of Cyprus: Statistical convergence and institutional divergence. *The World Economy, 37*(9), 315–1333.

Young, A. T., Higgins, M. J., & Levy, D. (2008). Sigma convergence versus beta convergence: Evidence from US county-level data. *Journal of Money, Credit and Banking, 40*(5), 1083–1093.

Information and Communication Technologies Diffusion Patterns in Developing Countries: Empirical Evidence

4

> *Diffusion is to be faster for simpler technologies where software knowledge is easily learned and transmitted, for population which are densely packed and where mixing is easily, where early users spread the word with enthusiasm, and in situations where the new technology is clearly superior to the old one and no major switching cost arise when moving from one to the other*
>
> Paul A. Geroski (2000)

Abstract

The chapter provides a detailed analysis of country-specific ICT diffusion patterns in 17 low-income and 29 lower-middle-income economies during the period 2000–2012. We propose using six ICT indicators extracted exclusively from the World Telecommunication/ICT Indicators database 2013 (17th Edition). These indicators include the following: Fixed telephone lines per 100 inhabitants, Mobile cellular telephone subscriptions per 100 inhabitants, Fixed Internet (Refers to narrowband networks.) subscriptions per 100 inhabitants, Fixed broadband Internet subscriptions per 100 inhabitants, Wireless-broadband subscriptions per 100 inhabitants, and number of Internet users. Additionally, the chapter examines technology substitution effects regarding fixed-telephone lines versus mobile cellular telephony, and fixed-internet networks versus wireless-broadband networks. The final parts report on technology convergence and trace technology club formation among developing and developed economies over the period 2000–2012.

Keywords

ICT diffusion • Technological substitution • Technology convergence • Technology convergence clubs • Developing countries

4.1   Introduction

Undeniably, over the last several decades, rapid diffusion of new information and communication technologies (ICTs) has been shaping and profoundly transforming the global landscape. Importantly, the ICT revolution has also been pervasive even in economically backward countries, where since the 1980s, the ICT penetration rates have been gradually increasing; by 2012, access to the ICT infrastructure had become ubiquitously available for the vast majority of people. According to ITU statistics (ITU 2013), the most tremendous changes have been witnessed regarding shifts in mobile cellular telephony penetration rates, which have resulted in connecting previously unconnected and underserved, geographically isolated regions (ITU 2011b). The progress with respect to backbone infrastructure that enables Internet access is far less spectacular, although a few developing countries are continuously progressing in this regard. Unfortunately, constrained access to the Internet services has resulted in negligible usage of the Internet network, especially in low-income countries.

The remainder of this chapter encompasses six logically structured sections. Section 4.2, explains data sources and rationale used in consecutive empirical analysis, while Sect. 4.3 demonstrates preliminary evidence on changes in ICT deployment. Section 4.4 provides insight into the ICT diffusion trajectories in selected 17 low-income and 29 lower-middle-income countries over the period 2000–2012. To this aim, we use logistic growth models and develop country-specific ICTs diffusion patterns, and that allows recognition of the dynamics of the process and its characteristic stages. Claiming that all of analyzed countries have been rapidly advancing in deployment of ICT over examined period; henceforth we target to discover whether fast diffusion of ICTs was followed by the technological substitution process resulting in switching from 'old' to 'new' technological solutions. Hence, Sect. 4.5 is fully dedicated to identification of the fixed-to-mobile telephony technological substitution and fixed-to-wireless Internet network technological substitutions. To follow the logic and to stay consistent with the main findings presented in Sects. 4.4 and 4.5, in Sect. 4.6 we examine the hypotheses on technology convergence and technology convergence clubs to discover whether the rapid expansion of ICTs in developing countries has enabled them to catch up with the developed economies with regard to access to and use of information and communication technologies. Finally, the last Sect. 4.7 concludes.

4.2   Data Explanation and Rationale

Our analysis concentrates on developing-country-specific ICT diffusion trajectories in low-income and lower-middle-income countries (hereafter labelled 'developing countries' or 'economically backward countries'), over the period 2000–2012. The time coverage is fully subjected to data availability, as for the years between 2000 and 2012, the balanced data set is acquirable for all countries included in the analysis. However, if longer time-series for particular countries are available, for

Table 4.1 List of selected low-, and lower-middle-income economies

Low-income economies ($ 1,035 or less)	Lower-middle-income economies ($1,036–4,085)	
Bangladesh	Armenia	Morocco
Benin	Bolivia	Nicaragua
Burkina Faso	Congo (Rep.)	Nigeria
Cambodia	Egypt	Pakistan
Comoros	El Salvador	Paraguay
Eritrea	Georgia	Philippines
Ethiopia	Ghana	Senegal
Kenya	Guyana	Sri Lanka
Madagascar	Honduras	Swaziland
Malawi	India	Syria
Myanmar	Indonesia	Ukraine
Nepal	Lao PDR	Viet Nam
Niger	Mauritania	Yemen
Rwanda	Moldova	Zambia
Togo	Mongolia	
Uganda		
Zimbabwe		

Source: Derived from World Bank country classification (accessed: March 2014)

those countries we extend the analysis time-span to demonstrate full ICT diffusion time-path. By convention, the empirical sample covers 17 low-income (out of 36 classified as such) economies, the low-income group, and 29[1] lower-middle-income economies (out of 48 classified as such), the lower-middle-income group.[2] All countries included in the study are listed in Table 4.1 below.

The low-income economies are those where annual GNI *per capita* is 1,035 or less[3]; the lower-middle-income economies are those where GNI per capita ranges from 1,036 to 4,085 in current US$ (adopting the Atlas Method). Analogously, in 2012, the average GNI *per capita* in the lower-middle-income economies was roughly US$2346.7, which corresponds to approximately 3,765.08 of GNI per capita in PPP 2005 constant international dollars). To achieve the empirical goals, we select a set of variables that approximate each country's individual achievements in information and telecommunication technologies access to and use of. Henceforth, we propose to use six ICT indicators, which are exclusively extracted from the World Telecommunication/ICT Indicators database 2013 (17th Edition). They are as follows (*i* denotes country and *y*, year):

---

[1] Regardless of data availability, we excluded from the sample small island states such as Cape Verde, Micronesia, Samoa, São Tomé and Principe, Solomon Islands, and Vanuatu.

[2] We excluded from our sample all countries for which data were incomplete or the existing time series were too short to ensure reliable estimates.

[3] According to the World Bank 2013 classification.

- Fixed telephone lines per 100 inhabitants—refers to the number of fixed telephone lines in a country for each 100 inhabitants (ITU 2010)[4]—$FTL_{i,y}$
- Mobile cellular telephone subscriptions per 100 inhabitants—refers to the number of mobile cellular subscriptions in a country for each 100 inhabitants (ITU 2010)—$MCS_{i,y}$
- Fixed Internet[5] subscriptions per 100 inhabitants—refers to the number of fixed Internet subscriptions in a country for each 100 inhabitants (ITU 2010)—$FIS_{i,y}$
- Fixed broadband Internet subscriptions per 100 inhabitants—refers to the number of fixed broadband Internet subscriptions in a country for each 100 inhabitants (ITU 2010)—$FBS_{i,y}$
- Wireless-broadband subscriptions per 100 inhabitants—refers to the number of wireless-broadband subscriptions in a country for each 100 inhabitants (ITU 2010)—$WBS_{i,y}$
- Internet users—refers to the 'proportion of individuals who used Internet from any location in the last three months' (ITU 2014b)—$IU_{i,y}$

The ICT indicators listed above are considered to be appropriate for the aims and scopes of our study because they provide broad information on newly emerged ICT infrastructure, access and usage in developing countries. The first four indicators are selected from the group of core indicators on ICT infrastructure and access', and the last one—Internet users—is derived from the group 'core indicators on access to, and use of, ICT by households and individuals' (ITU 2010). The term 'subscriptions' refers to entities that officially subscribe to telephone services and are obliged to pay for it. In the case of the indicators that explain Internet infrastructure ($FIS_{i,y}$, $FBS_{i,y}$ and $WBS_{i,y}$), the term 'subscriptions' refers to entities who officially subscribe to and pay for Internet access. The distinction between 'subscriptions' and 'users' will be clearly underlined—'users' are generally more numerous than 'subscriptions', especially in developing countries. 'Users' are those who use the Internet, not necessarily its legal owners. If the Internet is predominantly accessed and used by individuals in public places, the number of 'users' tends to be far higher than the number of 'subscriptions'. The two indicators fixed Internet and fixed broadband Internet subscriptions are closely related; the first covers access to both dial-up and total fixed broadband subscriptions.[6] Fixed broadband Internet defines access to a high-speed network[7] by cable modem, DSL, fibre or any other fixed broadband technology (ITU 2010).

By convention, our analysis covers the period between 2000 and 2012; however, as mentioned before, owing to better to data availability for particular countries, the evidence for some countries may be extended to a longer time span. In low-income

---

[4] A detailed explanation of core ICT indicators, including technical specifications, is presented in Appendix A.

[5] Refers to narrowband network.

[6] Accessed by cable modem, DSL or any other line.

[7] At least 256 kbit/s.

economies, the data on fixed telephone lines have been commonly available since 1975, on mobile cellular subscriptions since 1992, and on fixed Internet subscriptions since 1997. The data on $FBS_{i,\,y}$ and $WBS_{i,\,y}$ are more limited. Statistics on $FBS_{i,\,y}$ were firstly available in 2001 but only for Zimbabwe. In 2006, the data on $FBS_{i,\,y}$ were already available for 13 countries, and thus we take 2006 as the initial year for our analysis with regard to low-income economies. Wireless broadband was firstly introduced in low-income economies in 2009 (initially in 7 countries), so we analyse the $WBS_{i,\,y}$ changes for only a 4-year period (2009–2012). The data on Internet users have been broadly available since 1996 (in 12 out of 17 countries). Analogous to the low-income economies, in lower-middle-income countries, the data on fixed telephone line accessibility have been commonly available since 1975. The data on mobile cellular subscriptions have been officially available since 1984, the year data on $MCS_{i,\,y}$ were first gathered in Indonesia. From 1984 onward, data on $MCS_{i,\,y}$ appeared gradually for consecutive countries and by 2000 were finally available for all 29 lower-middle-income countries. In the case of fixed Internet subscriptions, the data were firstly available in 1994 (Sri Lanka) and for fixed broadband subscriptions, in 2000 (Indonesia, Nicaragua, Paraguay and Zambia). Similar to the low-income economies, data on $WBS_{i,\,y}$ generally appears in 2009; however, for India and Moldova, data were available in 2007 and for two more countries, Pakistan and Paraguay, in 2008. Data on Internet users for some countries (Egypt, Moldova, Nicaragua, Philippines, Sri Lanka, Ukraine, and Zambia) appear in 1993–1994, but for India and Syria, they were already available in 1992.

The consecutive sections demonstrate results of the empirical analysis, which covers preliminary descriptive statistics, development of country-specific ICT diffusion patterns, identification of technological substitution effects and finally in examination of technology convergence process along with technology convergence clubs formation.

4.3   Information and Communication Technologies in Developing Countries: Preliminary Evidence

Section 4.3 aims to explore key trends in the growing access to and use of ICTs in developing countries. It sheds light on the issues associated with the changing availability and usage of ICTs in low-income and lower-middle-income countries. It provides basic descriptive statistics on $FTL_{i,\,y}$, $MCS_{i,\,y}$, $FIS_{i,\,y}$, $FBS_{i,\,y}$, $WBS_{i,\,y}$ and $IU_{i,\,y}$ but also report on changes in their distribution. Table 4.2 summarises the ICT indicator descriptive statistics for low-income countries over the period

Table 4.2 Core ICT indicators. Summary statistics. 17 low-income economies. Period 2000–2012

Variable	# of observation	Mean	Min. value	Max. value	Lower percentile [10 %]	Upper percentile [90 %]	Gini[a] coeff.	Atkinson[b] coeff.
FTL$_{2000}$	17	0.065	0.18	1.99	0.209	1.28	0.36	0.10
FTL$_{2012}$	17	1.43	0.38	3.93	0.58	3.34	0.36	0.10
MCS$_{2000}$	17	0.45	0.00	2.13	0.00	1.06	0.57	0.34
MCS$_{2012}$	17	51.76	4.97	128.5	10.3	91.90	0.31	0.09
FIS$_{2000}$	17	0.058	0.001	0.23	0.003	0.127	0.48	0.20
FIS$_{2012}$	17	0.58	0.015	4.16	0.029	1.5	0.62	0.33
FBS$_{2006}$	17	0.007	0.00	0.08	0.00	0.015	0.806	0.62
FBS$_{2012}$	17	0.13	0.002	0.52	0.007	0.52	0.59	0.31
WBS$_{2009}$	17	0.03	0.00	0.29	0.00	0.18	0.83	0.72
WBS$_{2012}$	17	4.93	0.0009	28.13	0.003	23.78	0.70	0.45
IU$_{2000}$	17	0.18	0.0002	0.8	0.015	0.401	0.49	0.22
IU$_{2012}$	17	7.23	0.8	32.09	1.06	17.09	0.50	0.20

Source: Author's calculations based on the data derived from World Telecommunication/ICT Indicators database 2013 (17th Edition). Note: for FBS, y— estimates since 2006; for WBS, y—estimates since 2009

[a] In generic form, the Gini coefficient measures inequality of income distribution, while 0 corresponds to perfect income equality, and 1—total inequality. For given population attributed to values y_i, $i = 1 \ldots n$, and if $(y_i \leq y_{i+1})$, the general formula for Gini coefficient is defined as:

$$= \frac{1}{n} \left(n + 1 - 2 \frac{\sum_{i=1}^{n} (n+1-i)y_i}{\sum_{i=1}^{n} y_i} \right)$$

(Gini 1912). Graphically, value of Gini coefficient measures twice the surface between Lorenz curve, which shows the cumulative distribution of variable in the sample, and line presenting totally equal distribution. Here, instead of income, we deploy consecutive core ICT indicators

[b] Atkinson inequality measure (Atkinson 1970), with inequality aversion parameter ε included, has a general form: $A_{\varepsilon} = 1 - \left[\frac{1}{n} \sum_{i=1}^{n} \left[\frac{y_i}{\bar{y}} \right]^{1-\varepsilon} \right]^{\frac{1}{1-\varepsilon}}$, where $\bar{y}$ stands for average income of each individual in the sample, y_i—income of an individual, n is the number of individuals in the sample. Here, instead of income, we deploy consecutive core ICT indicators

2000–2012; while additional graphical evidence is provided in Fig. 4.1 (density curves[8]) and Fig. 4.2 (Lorenz curves[9]).

The results provided in Table 4.2 shed light on the disruptive changes in core ICT deployment in low-income countries. The changes in adoption of $MCS_{i, y}$ were extraordinary high; while the average levels of $MCS_{i, y}$ adoption were 0.45 and 51.8 in 2000 and 2012, respectively. In 2000, the Gini coefficient for $MCS_{i, y}$ was 0.57 and in 2012, it was 0.31, highlighting that fast adoption of mobile telephony resulted in a sharp reduction in the inequalities in access to mobile cellular services. A similar picture emerges from the elementary analysis of the density functions for $MCS_{i,2000}$ and $MCS_{i,2012}$ (see Fig. 4.1) and the Lorenz curves in analogous years (see Fig. 4.2). Regarding fixed narrowband ($FIS_{i, y}$) and fixed broadband Internet subscriptions ($FBS_{i, y}$), the changes are not as prominent as those for mobile cellular telephony. Although over the period 2000–2012, some positive changes are detectable, in 2012, the average adoption of $FIS_{i, y}$ and $FBS_{i, y}$ remained very low in low-income countries. The average achievements in the low-income countries in 2000 were $FIS_{i,2000} = 0.058$ and $FBS_{i,2006} = 0.007$. In 2012, the average $FIS_{i, y}$ and $FBS_{i, y}$ levels grew moderately, achieving 0.58 and 0.13, respectively. The picture emerging from the evidence on wireless broadband access and use is unlike those for $FIS_{i, y}$ and $FBS_{i, y}$. Although the period of wireless broadband technology adoption is short (2009–2012), it spread miraculously in a few low-income countries. Note that in 2009, the wireless broadband solutions were accessible exclusively in 7 low-income countries; however, by 2012, the situation had improved radically; the wireless-broadband technologies were available in all 17 - low-income economies under discussion. Surprisingly, despite the dynamic

---

[8] The density curves are plotted by adopting non-parametric estimation of the probability density function: $f(x) = \frac{d}{dx} F(x)$, where $F(x)$ explains the continuous distribution of random variable X. Kernel density estimator results were useful in this case they allowed for relaxing the restrictive assumptions on the shape that $f(x)$ should potentially hold; thus, it is flexible. The density curves that were generated by the kernel density estimator are continuous and show an "empirical' distribution of variables. To estimate density $f(x)$, we use its discrete derivative, a special case of the kernel estimator taking a general form: $f'(x) = \frac{1}{nh} \sum_{i=1}^{n} k \left(\frac{X_i - x}{h} \right)$, where $k(u)$ is a kernel function that satisfies $\int_{-\infty}^{\infty} k(u)du = 1$. $f'(x)$ shows the percentage of observations located near x. If many observations are located near x, then $f'(x)$ is large, and the opposite otherwise.

[9] By convention, the Lorenz curve is used for graphical explanations of distributions, e.g., of income or wealth. Formally, x stands for income and $F(x)$ is its distribution, which explains the proportion of individuals who have incomes less than or equal to x. The first moment distribution function may be defined as $F_1(x)$, and $F_1(x)$ explains the proportion of total income that was earned by individuals who have incomes less than or equal to x. If the previous is true, then the Lorenz curve expresses the relationship between $F(x)$ and $F_1(x)$. The area below the Lorenz curve is widely used to calculate the value of the Gini index (Gini $\rightarrow$ one minus twice the area below the Lorenz curve). The generalised Lorenz curves are commonly labelled 'concentration curves' and are broadly used as a tool to consider different aspects of distribution in economic analyses.

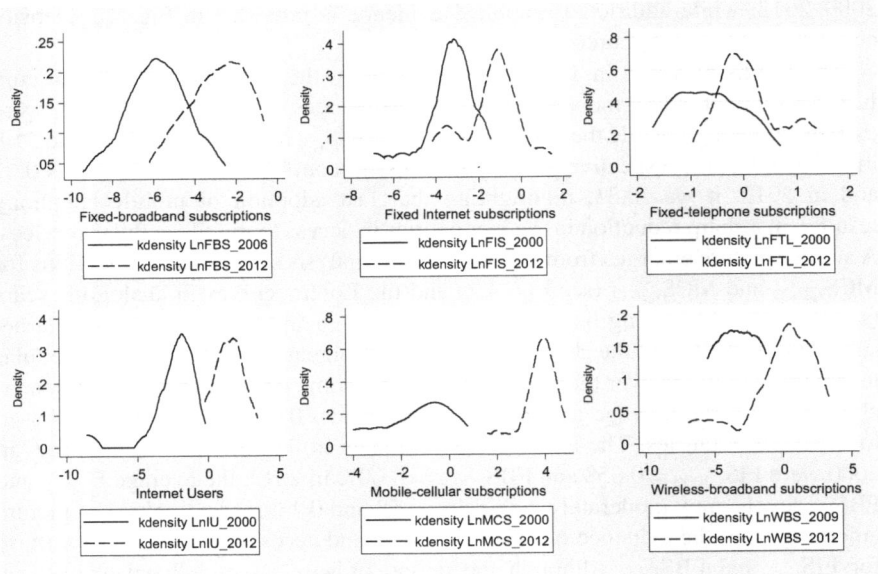

Fig. 4.1 Density representations for $FTL_{i,\,y}$, $MCS_{i,\,y}$, $FIS_{i,\,y}$, $FBS_{i,\,y}$, $WBS_{i,\,y}$ and $IU_{i,\,y}$ in low-income economies. *Solid lines* are for 2000 (or 2006 for $FBS_{i,\,y}$ (For $FBS_{i,\,y}$ variable)), *density curve* is estimated for 2006 instead of 2000. In the period 2000–2005 significant lacks of data disable adoption of kernel density estimator); and 2009 (In low-income economies, wireless-broadband network was firstly introduced in 2009) for $WBS_{i,\,y}$), *dashes* are *lines* for 2012. *Note* on x-axis—logged values of absolute data; kernel = epanechnikov (*Source*: Author's elaboration)

changes in Internet usage in low-income countries, related inequalities scarcely changed over the period 2000–2012 ($IU_{Gini,2000} = 0.49$; $IU_{Gini,2012} = 0.50$; for visual inspection, see Fig. 4.2). In 2000, the average number of Internet users (per 100 inhabitants) was $IU_{average,2000} = 0.18$, whereas in 2012, it was $IU_{average,2012} = 7.23$. In general, over the period 2000–2012, low-income countries experienced rapid growth in access to and use of basic ICTs. However, although the average annual ICT growth rates were high, in many of the analysed countries, ICT implementation was still low in the terminal year (2012).

Consideration of the cross-country distribution and inequalities with regard to core ICT indicators (in the years 2000 and 2012)[10] highlights significant changes in this regard. In Fig. 4.1, separate plotted density functions for each core ICT indicator support the evidence for the significant increase in access to and use of basic ICTs in low-income countries, although in the case of $FIS_{i,\,y}$, $FBS_{i,\,y}$ and $WBS_{i,\,y}$ cross-country inequalities are revealed to be persistent.

Analogously, Table 4.3 reports summary statistics for the core ICT indicators in the lower-middle-income countries over the period 2000–2012.

---

[10] For FBS: 2006 and 2012; for WBS: 2009 and 2012.

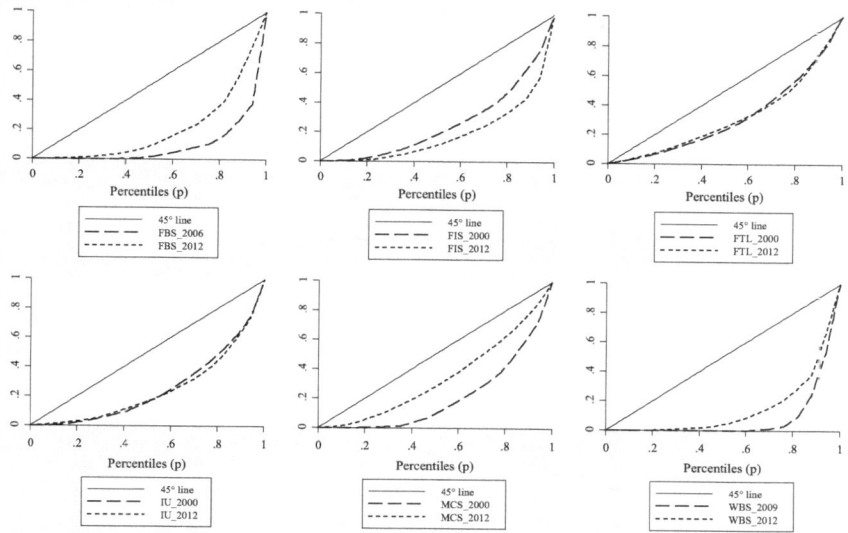

Fig. 4.2 *Lorenz curves* for $FTL_{i, y}$, $MCS_{i, y}$, $FIS_{i, y}$, $FBS_{i, y}$, $WBS_{i, y}$ and $IU_{i, y}$ in low-income economies. *Long-dash lines* are for 2000 (or 2006 for $FBS_{i, y}$; and 2009 for $WBS_{i, y}$), *short-dash lines* are for 2012 (*Source*: Author's elaboration)

Similar to what was observed in the low-income economies, the lower-middle-income countries experienced dynamic changes in the adoption and usage of ICTs. The average changes in mobile cellular telephony adoption are crucially different compared with those in $FTL_{i, y}$. Over 2000–2012, remarkable increases are observed; the average adoption of $MCS_{i, y}$ in 2000 was 3.28, whereas in 2012, it was 95.35 (*sic!*). The Gini coefficient decreased from 0.56 in 2000 to 0.14 in 2012; thus, cross-country disparities have nearly disappeared (see Fig. 4.4). The changes in fixed narrowband and fixed broadband Internet network access, are not as strong as those for $MCS_{i, y}$, although the penetration rates for both $FIS_{i, y}$ and $FBS_{i, y}$ climbed over the period 2000–2012, the average adoption of fixed narrowband and fixed broadband Internet connections remained low, achieving, respectively, $FIS_{average, 2012} = 2.56$ and $FBS_{average, 2012} = 2.45$.[11] Again, in both cases, the inequalities in access to fixed narrowband and fixed broadband Internet persisted over the examined period. Among the lower-middle-income countries in 2007, wireless broadband technologies were already available in India and Moldova; subsequently, and finally, in 2012, for each of the 29 examined countries, $WBS_{i, y}$ was available; however, its adoption levels varied significantly (see Figs. 4.3 and 4.4). Essential differences in $WBS_{i, y}$ adoption, in both 2009 and 2012, were mostly

---

[11] In 2012, the world averages for FIS and FBS adoption were, respectively, 13.0 and 11.0 (see ITU statistics).

Table 4.3 Core ICT indicators. Summary statistics. 29 lower-middle-income economies. Period 2000–2012

Variable	# of observation	Mean	Min. value	Max. value	Lower percentile [10 %]	Upper percentile [90 %]	Gini index	Atkinson index
FTL_{2000}	29	5.58	0.45	21.23	0.70	14.21	0.47	0.18
FTL_{2012}	29	10.0	0.24	34.30	0.58	26.7	0.49	0.21
MCS_{2000}	29	3.28	0.24	15.34	0.18	8.31	0.56	0.27
MCS_{2012}	29	95.35	58.27	147.67	64.70	130.33	0.14	0.016
FIS_{2000}	27	0.28	0.0037	1.74	0.028	0.54	0.55	0.26
FIS_{2012}	29	2.56	0.05	11.87	0.18	7.06	0.52	0.24
FBS_{2005}	26	0.15	0.0003	0.827	0.005	0.27	0.59	0.32
FBS_{2012}	29	2.45	0.008	11.87	0.105	8.002	0.58	0.30
WBS_{2009}	15	1.90	0.0062	11.69	0.065	6.55	0.69	0.43
WBS_{2012}	29	9.58	0.098	33.92	0.64	29.09	0.53	0.24
IU_{2000}	29	0.89	0.026	6.61	0.082	1.44	0.51	0.23
IU_{2012}	29	25.1	5.36	55.0	9.96	44.07	0.29	0.06

Source: Author's calculations based on data derived from World Telecommunication/ICT Indicators database 2013 (17th Edition). Note for FBS_i, y—estimates since 2005; for WBS_i, y—estimates since 2009

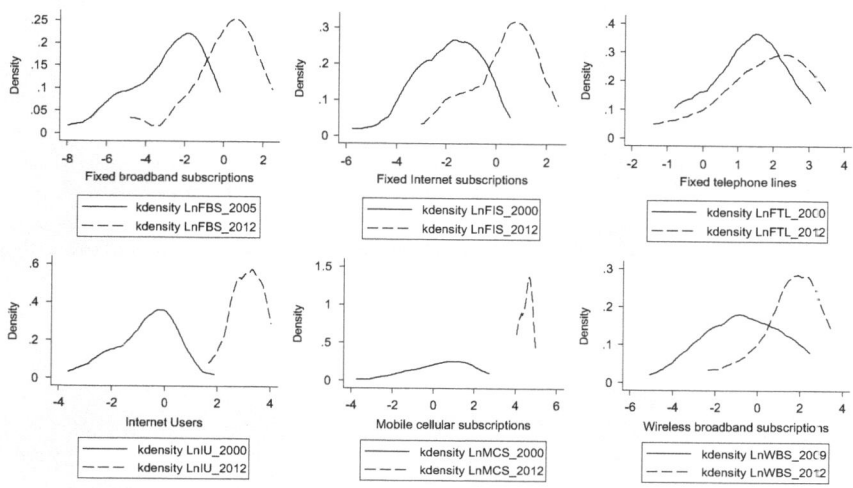

Fig. 4.3 *Density* representations for $FTL_{i, y}$, $MCS_{i, y}$, $FIS_{i, y}$, $FBS_{i, y}$, $WBS_{i, y}$ and $IU_{i, y}$ in lower-middle-income economies. *Solid lines* are for 2000 (or 2005 for $FBS_{i, y}$ (For $FBS_{i, y}$ variable, *density curve* is estimated for 2006 instead of 2000. In the period 2000–2005 significant lacks of data disable adoption of kernel density estimator.); and 2009 (In low-income economies, wireless-broadband network was firstly introduced in 2009.) for $WBS_{i, y}$), the *dashes lines* for 2012. Note: on x-axis—logged values of absolute data; kernel = epanechnikov (*Source*: Author's elaboration)

influenced by a few outliers that substantially outpaced other countries in this category. Considering Internet usage, the picture is slightly more promising. The average level of $IU_{i, y}$ rose from 0.89 % in 2000 to 25.1 % in 2012. Growing trends in the proportion of individuals who used the Internet over the period 2000–2012 were especially significant in Morocco and Georgia, where in 2012, the $IU_{i, y}$ levels were 55.0, 45.5 %, respectively.

Analogous to the changing penetration of mobile telephony, in the lower-middle-income countries over the period 2000–2012, there were significant decreases in cross-country inequalities in Internet use. In 2000, the Gini coefficient was 0.51, whereas in 2012, it fell to 0.29 (see the Lorenz curves in Fig. 4.4), reflecting significant improvements in diminishing cross-countries disparities in Internet use on one hand and on the other hand, the growing accessibility of Internet network infrastructures in the countries under study.

This preliminary evidence on changing ICT deployment, in both low-, and lower-middle-income countries, demonstrates continuous and rapid growth with respect to mobile telephony, fixed narrowband Internet, fixed broadband Internet and wireless broadband penetration rates over the period 2000–2012. Analogous trends are reported if the ratios of individuals who use the Internet are considered. The consecutive Sect. 4.3 coherently describes the ICT diffusion process in each of examined countries individually, shaping country-specific ICT diffusion trajectories. The evidence demonstrated throughout the next section provides the

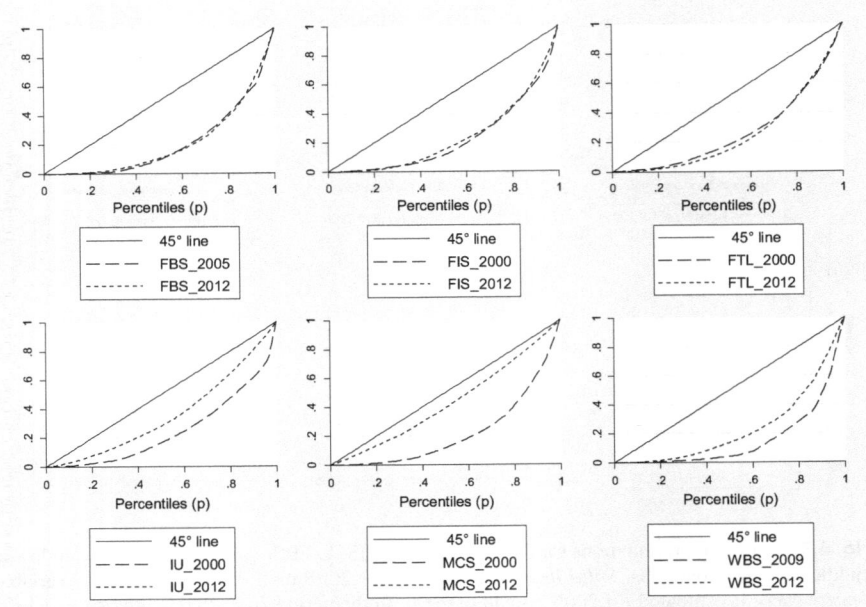

Fig. 4.4 *Lorenz curves* for $FTL_{i, y}$, $MCS_{i, y}$, $FIS_{i, y}$, $FBS_{i, y}$, $WBS_{i, y}$ and $IU_{i, y}$ in lower-middle-income economies. *Long-dash lines* are for 2000 (or 2005 for $FBS_{i, y}$; and 2009 for $WBS_{i, y}$), *short-dash lines* are for 2012 (*Source*: Author's)

deep in-sight into the dynamics of the ICT diffusion process in respective countries, unveiling its unique features and characteristics.

4.4   Shaping Country-Specific ICT Diffusion Trajectories

The main value of the following section is that it provides broad knowledge on how information and communication technologies were expanding in economically backward countries over the period 2000–2012. It presents a normative view of ICT diffusion in low- and lower-middle-income economies, and it provides a broad framework for evaluating country-specific ICT diffusion trajectories. The aim is to show the basic trends in ICT spread, specifically concentrating on demonstrating how countries' ICT profiles were transforming, which in turn gives an conclusive indication of the nature of ICTs and the process of their diffusion in economically backward economies. Section 4.4.1 is fully dedicated to explaining the mobile cellular telephony accessibility; while the consecutive Sect. 4.4.2 exclusively concentrates on elaboration of fixed-narrowband networks, fixed-broadband networks and wireless-broadband networks which enable the Internet network accessibility. The patterns of Internet usage are additionally demonstrated in this section.

The empirical evidence exhibited in Sect. 4.4 covers the following core ICT indicators: mobile cellular subscribers ($MCS_{i, y}$), fixed internet subscribers ($FIS_{i, y}$), fixed broadband subscribers ($FBS_{i, y}$), wireless broadband subscribers ($WBS_{i, y}$) and Internet users ($IU_{i, y}$). By convention, we develop technology-specific diffusion patterns (for each country separately) that give clear pictures of diffusion trajectories from when the indicator was first introduced until the year 2012. In this section, we exclude the fixed telephony indicator because it represents the 'old' technologies, which are not in the scope of this part of our analysis. To approximate country-specific ICT diffusion time paths (S-shaped time path) we use a logistic growth model formalised as:

$$N_{ICT,i,y}(t) = \frac{\kappa}{1 + e^{-\alpha t - \beta}}, \tag{4.1}$$

where $N_{ICT,i,y}(t)$ specifies the saturation of a given ICT in i country and y year.

The κ stands for the upper asymptote (growth limit), which explains the 'carrying capacity' of the system (country). The parameter α represents speed of diffusion, and β indicates the midpoint (T_m) when the logistic pattern reaches 0.5. The value of α yields poor economic interpretation; however, it facilitates determining the 'specific duration', defined as $\Delta t = \frac{\ln(81)}{\alpha}$. The value of Δt approximates time needed to pass from 10 %κ to 90 %κ (for broder discussion—see Chap. 3).

4.4.1 Mobile Cellular Telephony Diffusion

No other ICT service has ever had the same kind of impact in terms of subscriptions, particularly in the developing world, in so little time
Measuring Information Society 2011 (ITU 2011a)

(...) after the invention of the telephone, it took nearly 100 years for wired telephones to reach a population of one billion people around the world. With the invention of cellular communications, it took about 20 years to reach the same billion people
Gunasekaran and Harmantzis (2007)

Over the last two decades, in low-income and lower-middle-income countries, unprecedented speed and geographic coverage growth in mobile telephony was witnessed. Between 2000 and 2012, the average growth rate in the mobile sector in low-income countries was approximately 42.2 % per annum, and in lower-middle-income countries, it was 34.3 % per annum.[12] Both companies and individuals were fast adopting mobile cellular telephony, which was possible mainly owing to its relatively increasing availability (including rural areas), affordability and limited requirements for hard infrastructure (ITU 2006). From the 1990s onward, mobile

---

[12] Author's estimates.

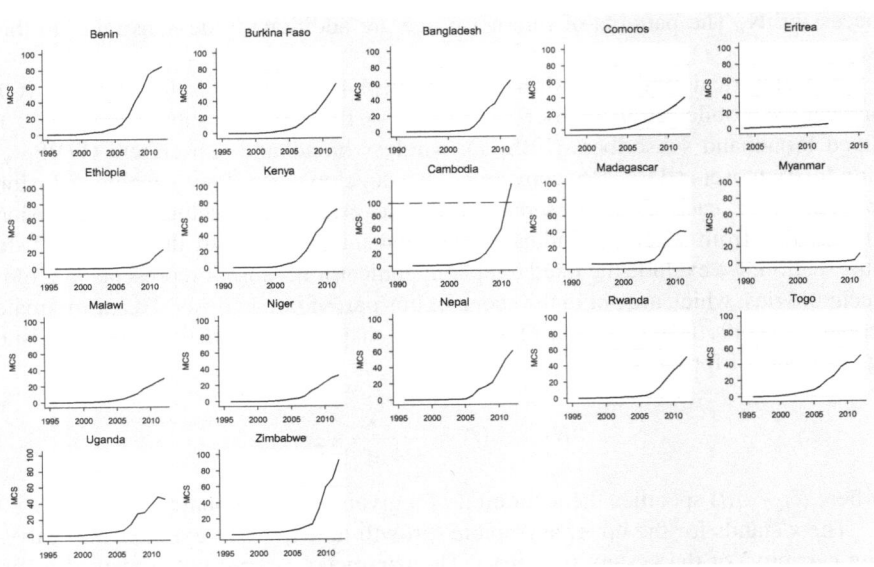

Fig. 4.5 MCS$_{i, y}$ diffusion paths. 17 low-income countries. Note: *horizontal long-dash line* corresponds to world average in MCs achievements in 2012 (*Source*: Author's elaboration)

telephony began to 'act globally', and from its very inception and first introduction to a worldwide audience, the market was characterised by growing competition. Continuous development of the prepaid card market contributed significantly to boosting the number of mobile telephony users because it allowed unqualified people to acquire their own mobile phones and, for mobile telephony operators, it reduced the risks of non-payment. The significant growth in mobile penetration was additionally eased by the gradual reduction in call costs and monthly charges for telephone usage. The latter was also perpetuated—at least partially—by mobile cellular market liberalisation in some developing countries, which did not take place in the case of fixed telephony services (Banerjee and Ros 2004; ITU 2011b). Figures 4.5 (low-income countries) and 4.6 (lower-middle-income countries) provide visual summaries of country-specific diffusion trajectories with respect to mobile cellular telephony. The value of this graphical evidence is enhanced by the MCS$_{i, y}$ logistic growth estimates reported in Tables 4.4 (low-income countries) and 4.6 (lower-middle-income countries). Considering exclusively low-income countries, the first observation is that the shifts in mobile cellular telephony access and adoption are disruptive. The most prominent example is Cambodia, where in 1993, MCS$_{KHM,1993} = 0.04$, but because of an average annual growth rate of approximately 40 %, in 2012, the country reached MCS$_{KHM,2012} = 128.5$ per 100 inhabitants. Another three countries, namely, Zimbabwe, Benin and Kenya, managed to grow dramatically in terms of mobile teledensity,[13] and in

---

[13] Described as the number of mobile telephony subscribers per 100 inhabitants.

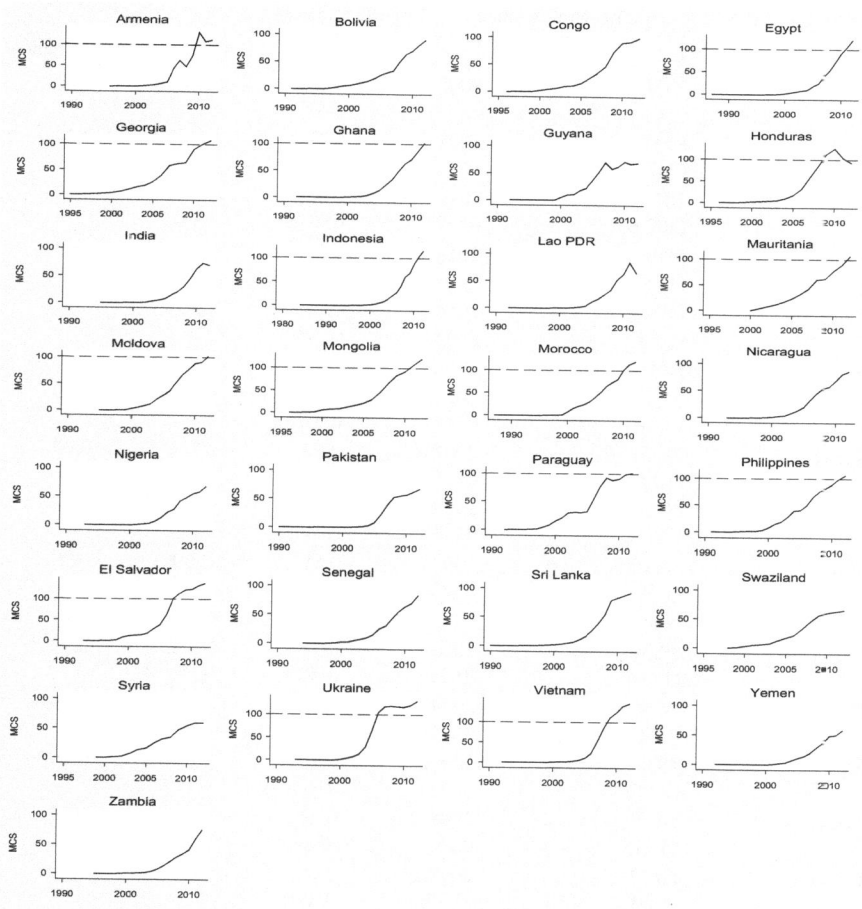

Fig. 4.6 $MCS_{i, y}$ diffusion paths. 29 lower-middle-income countries. Note: *long-dash horizontal line* corresponds to world average in 2012 (*Source*: Author's elaboration)

2012, the $MCS_{i, y}$ penetration rates reached, respectively, $MCS_{Zwe,2012} = 91.9$, $MCS_{Ben,2012} = 83.7$ and $MCS_{Ken,2012} = 71.2$. The Kenyan and Cambodian examples deserve special attention and may be cited as a 'best practice' in the early liberalisation of the mobile telephony market, which promoted the rapid spread of these services. In Kenya in 1992, once mobile telephony services were introduced to the domestic market, the prices of mobile telecommunication services was extraordinarily high, and only a small part of the population could afford to buy them. However, in 1999, the Kenyan government established the Communications Commission of Kenya,[14] which strongly affected the future mobile communication

---

[14] See website: http://www.cck.go.ke/ (accessed May 2014).

Table 4.4 $MCS_{i, y}$ logistic growth estimates. 17 low-income countries

Country	κ	T_m	α	deltaT	R-squared
Bangladesh	73.6 (5.6)	2008.95 (0.42)	0.53 (0.057)	8.24	0.99
Benin	89.99 (2.02)	2008.3 (0.12)	0.76 (0.051)	5.78	0.99
Burkina Faso	102.73 (102.6)	2011.2 (0.27)	0.46 (0.016)	9.40	0.99
Cambodia	909.9[a] (1151.3)	2016.4 (4.0)	0.4 (0.03)	10.81	0.99
Comoros	60.9[b] (6.34)	2010.8 (0.48)	0.48 (0.039)	9.38	0.99
Eritrea	Not applicable[c]				
Ethiopia	41.2[b] (9.4)	2011.7 (0.64)	0.74 (0.078)	5.87	0.99
Kenya	78.24 (1.55)	2007.8 (0.15)	0.56 (0.025)	7.82	0.99
Madagascar	40.55[b] (0.74)	2000.7 (0.11)	1.01 (0.062)	4.33	0.99
Malawi	36.18[b] (1.39)	2009.4 (0.22)	0.55 (0.02)	7.99	0.99
Myanmar	Not applicable[c]				
Nepal	89.39 (11.0)	2010.7 (0.48)	0.58 (0.07)	7.35	0.99
Niger	36.03[b] (0.80)	2009.9 (0.11)	0.67 (0.029)	6.5	0.99
Rwanda	58.49[b] (2.69)	2009.7 (0.18)	0.71 (0.044)	6.18	0.99
Togo	54.5[b] (3.31)	2008.7 0.28	0.52 (0.03)	8.33	0.99
Uganda	50.73[b] (4.00)	2008.2 (0.39)	0.68 (0.08)	6.44	0.99
Zimbabwe	114.04 (14.8)	2010.2 (0.51)	0.75 (0.12)	5.79	0.99

Source: Author's estimates
[a]Overestimates, statistically not significant. Estimation method—NLS. In parenthesis (robust SE)
[b]Countries that in 2012 were located in exponential growth phase
[c]Not applicable, countries in initial phase of growth

market in the country, which was a turning point because since then onward, the telecommunication market has been privatised, and the free competition between the two biggest market players—Safaricom Limited and Celtel Kenya—facilitated essential price drops.[15] In 2013 in Kenya, there were 4 large prepaid and post-paid telecom market operators: Safaricom Limited, Airtel Networks Limited

---

[15] Compare ITU World Telecommunication/ICT Indicators database 2013.

Subscriptions, Essar Telecom Limited and Telkom Kenya (Orange) (CCK 2013). The latter increased access to and use of mobile telephony in Kenya, which as one of the world's poorest countries, in 2012 enjoyed an $MCS_{i, y}$ penetration rate much higher than the world average. In Cambodia, extensive national action was undertaken to break down the state monopoly in telecommunication services. As result, until 2011, many internationally supported (e.g., by ASEAN or IDRC) projects that promoted broader access to ICTs were successfully completed. Special emphasis was put on broadening access to wireless solutions, which potentially could have resulted in rural-urban inequalities in access to telecommunication services. Because of having established advantageous legal conditions and market regulations that enhanced competition in the telecommunication sector, in 2011, Cambodia had 4 fixed-telephone[16] and 10 mobile telephony[17] operators.

The country-specific charts in Fig. 4.5 suggest that the 'laggard countries', in 2012, were still in the early exponential growth phase in $MCS_{i, y}$ diffusion. Hence, the future diffusion of mobile telephony is marked by high uncertainty, and the estimated respective κ_i parameters for Comoros, Ethiopia, Madagascar, Malawi, Niger, Rwanda, Togo and Uganda may suffer from heavy under-, or overestimates. Eritrea and Myanmar are examples of countries that in 2012 still suffered from very low access to and usage of mobile cellular telephony. In Myanmar, the first data on $MCS_{i, y}$ were available for 1993, but over the period 1993–2011, $MCS_{i, y}$ penetration barely climbed, from $MCS_{Mmr,1993} = 0.001$ to $MCS_{Mmr,2011} = 2.37$; finally, in 2012, it increased to $MCS_{Mmr,2012} = 10.3$ subscriptions per 100 inhabitants. In Eritrea, mobile penetration remained very low over the period 2004–2012,[18] and the country barely improved access to and use of mobile telephony. In 2004, mobile penetration was $MCS_{Eri,2004} = 0.42$, and in 2012, it was $MCS_{Eri,2012} = 5.0$. Even keeping in mind that Eritrea and Myanmar, in 2012, were still in the very initial phase of growth with respect to MCS, the logistic growth estimates are not applied because the generated parameters might be misleading and inconclusive. The empirical analysis of diffusion trajectories may be enriched by providing additional specification of the predicted of mobile telephony diffusion paths.[19] Table 4.5 summarises the predicted development of mobile telephony in low-income countries. The reported predictions are considered under two distinct assumptions: first, a country approaches $\kappa = 100$[20] (per. 100 inhab.), which corresponds to the world average in MCS penetration in 2012; and second, a country approaches $\kappa = 51$ (per 100 inhab.), which corresponds to the low-income group average in

---

[16] Telecom Cambodia, Camintel, MFone, Viettel (Cambodia).

[17] CamGSM, MFone, CADCOMMS, HELLO, Applifone, Viettel (Cambodia), GT-TELL, Latelz, Sotelco, KTC.

[18] Eritrea became an independent country in 2001.

[19] All predictions are calculated under rigid assumptions that κ is fixed, the speed of diffusion (α) remains unchanged and countries follow the theoretical S-time trajectory. Assuming the previous, the T_m(midpoint), deltaT and 5-year interval—when a country should hypothetically reach approximately 100 % of assumed κ—are predicted.

[20] In here, $\kappa = 100$ (per 100 inhab.) may be interpreted $\kappa = 100$ %.

Table. 4.5 Predicted MCS$_{i, y}$ diffusion paths. 17 low-income countries

Country	$\kappa = 100$ (per 100 inhab.) (fixed)	Tm	deltaT	5-year interval to achieve ≅ 100 % of fixed κ ($\kappa = 100$ per 100 inhab.)	$\kappa = 51$ (per 100 inhab.) (fixed)	Tm	deltaT	5-year interval to achieve ≅ 100 % of fixed κ ($\kappa = 51$ per 100 inhab.)
Bangladesh	100	2010.4	10.53	2015–2020	In 2012 exceeded group average			
Benin	100	2008.7	6.89	2013–2018	In 2012 exceeded group average			
Burkina Faso	100	2011.13	9.27	2016–2021	In 2012 exceeded group average			
Cambodia	In 2012 exceeded world average				In 2012 exceeded group average			
Comoros	100	2013	11.36	2020–2025	51	2010	8.29	2013–2018
Eritrea	100	2022.7	16.29	2030–2035	51	2019.8	15.8	2025–2030
Ethiopia	100	2014	7.56	2020–2025	51	2012.3	6.44	2015–2020
Kenya	100	2009.1	10.4	2015–2020	In 2012 exceeded group average			
Madagascar	100	2012.2	12.7	2020–2025	51	2008.6	7.14	2010–2015
Malawi	100	2014.3	13.11	2020–2025	51	2011	10.36	2015–2020
Myanmar	100	2013.5	3.2[a]	2015–2020	51	2012.9	3.00[a]	2010–2015
Nepal	100	2011.1	8.05	2015–2020	In 2012 exceeded group average			
Niger	100	2013.7	12.07	2020–2025	51	2010.59	9.1	2015–2020
Rwanda	100	2011.8	9.15	2015–2020	51	2009.2	5.18	2010–2015
Togo	100	2011.59	13.68	2020–2025	51	2007.7	7.57	2010–2015
Uganda	100	2011.7	12.09	2020–2025	51	2008.3	6.49	2010–2015
Zimbabwe	100	2009.8	4.92	2010–2015	In 2012 exceeded group average			

Source: Author's estimates. Note: $\kappa = 100$ (per 100 inhab.)—corresponding to world average in 2012; $\kappa = 51$ (per 100 inhab.)—corresponding to low-income-group average in 2012

[a]Underestimates due abrupt growth over 2011/2012

$MCS_{i, y}$ penetration in 2012. Not surprisingly, it is demonstrated that countries[21] such as Bangladesh, Benin, Burkina Faso, Kenya, Nepal and Rwanda should potentially achieve 100 (per 100 inhab.) saturation before the year 2020. Reported in Table 4.4, parameters T_m and Δt deserve special attention.[22] In countries[23] such as Bangladesh, Benin, Burkina Faso, Kenya, Nepal and Zimbabwe, the midpoint varies from 2007.8 years in Kenya to 2011.2—in Burkina Faso, suggesting that over the period 2007–2011, each analysed country passed the inflection point and reached 50 % of predicted saturation (κ). The shortest specific duration— 5.7 years—was reported for Benin and Zimbabwe; these countries exhibited the highest α parameters—(0.76) and (0.75), respectively. This demonstrates that Benin and Zimbabwe were the two best-performing countries (in the low-income group) in terms of dynamic mobile cellular telephony diffusion.

In the lower-middle-income countries, the spread of mobile cellular telephony was as rapid as it was in the low-income economies. Figure 4.6 and Table 4.6 summarise the empirical evidence on country-specific diffusion curves with respect to mobile telephony. Preliminary graphical inspection of the mobile telephony diffusion trajectories in the examined countries leads to the conclusion that most countries followed a diffusion path that resembled an S-shaped curve. Many of these countries, adopting various strategies and policies, made tremendous progress in promoting the deployment of mobile telephony at the national level. The estimated midpoint (T_m) for the listed countries ranges from 2005 (Ukraine, Paraguay) to 2009 (Egypt) reflecting that lower-middle-income countries[24] managed to achieve the mobile telephony midpoint, on average, 2 years earlier compared with the best-performing low-income countries.[25] Special attention should also be paid to El Salvador and Viet Nam, which did extraordinary well in terms of rapid mobile telephony diffusion; in 2012, each achieved the highest MCS penetration rates among lower-middle-income countries, $MCS_{Slv,2012} = 137.3$ and $MCS_{Vnm,2012} = 147.6$, respectively. If mobile telephony diffusion, in both El Salvador and Viet Nam, follows the S-shaped trajectory, the predicted 'full' saturation is 140.3 for El Salvador and 148.5 for Viet Nam (see the estimates summarised in Table 4.6). In Viet Nam, the telecom market has experienced pervasive changes since 2002–2003 because the national government has allowed for competition among telecom operators and Internet providers (Wireless Broadband Masterplan Until 2020 for the Socialist Republic of Viet Nam, ITU 2012). Owing to increasing affordability, the national mobile telephony market steadily

---

[21] For Zimbabwe, the estimates might be slightly misleading because in 2012, the country had already reached a penetration rate of $MCS_{Zwe,2012} = 91.9$ per 100 inhabitants.

[22] As already mentioned, the α parameter has little economic meaning, and thus its values are not interpreted directly but are used to calculate the 'specific duration' (deltaT).

[23] The remaining 11 low-income countries are not considered for β and detlaT interpretation because the latter may be unreliable → countries in the early exponential or initial phase of growth.

[24] Refers exclusively to the best-performing countries.

[25] To remind: refers to the period 2007–2011.

Table 4.6 $MCS_{i, y}$ logistic growth estimates. 29 lower-middle-income countries

Country	κ	T_m	α	delta T	R-squared
Armenia	125.9 (13.1)	2007.7 (0.76)	0.64 (0.18)	6.78	0.96
Bolivia	132.1 (13.2)	2009.5 (0.76)	0.32 (0.02)	13.5	0.99
Congo	112.8 (7.02)	2008.1 (0.39)	0.57 (0.06)	7.63	0.99
Egypt	151.16 (2.66)	2009.2 (0.08)	0.47 (0.01)	9.29	0.99
El Salvador	140.3 (3.53)	2006.03 (0.24)	0.61 (0.058)	7.12	0.99
Georgia	134.8 (13.06)	2008.3 (0.78)	0.38 (0.04)	11.3	0.99
Ghana	115.06 (8.17)	2008.7 (0.35)	0.52 (0.038)	8.46	0.99
Guyana	69.2[a] (1.63)	2004.5 (0.21)	0.71 (0.11)	6.14	0.99
Honduras	110.02 (9.18)	2006.7 (0.22)	1.13 (0.2)	3.94	0.98
India	82.01[a] (7.61)	2008.6 (0.36)	0.67 (0.07)	6.49	0.99
Indonesia	142.08 (5.62)	2008.9 (0.26)	0.45 (0.017)	9.58	0.99
Lao P.D.R.	79.7[a] (13.6)	2008.1 (0.6)	0.7 (0.12)	6.19	0.98
Mauritania	139.3 (19.1)	2009.2 (0.9)	0.37 (0.03)	11.75	0.99
Moldova	110.02 (3.65)	2007.2 (0.17)	0.48 (0.016)	9.09	0.99
Mongolia	145.04 (7.26)	2008.5 (0.33)	0.44 (0.03)	9.76	0.99
Morocco	151.6 (9.22)	2008.03 (0.48)	0.33 (0.026)	13.07	0.99
Nicaragua	94.3 (4.24)	2007.5 (0.32)	0.47 (0.036)	9.19	0.99
Nigeria	67.7[a] (3.73)	2007.4 (0.26)	0.60 (0.046)	7.28	0.99
Pakistan	62.5[a] (2.2)	2006.6 (0.09)	1.07 (0.08)	4.09	0.99
Paraguay	114.1 (7.69)	2005.8 (0.73)	0.38 (0.054)	11.4	0.98
Philippines	119.6 (4.15)	2006.6 (0.27)	0.35 (0.017)	12.22	0.99
Senegal	99.2 (5.87)	2008.6 (0.29)	0.44 (0.019)	9.83	0.99
Sri Lanka	97.8 (3.00)	2007.4 (0.23)	0.65 (0.06)	6.72	0.99

(continued)

Table 4.6 (continued)

Country	κ	T_m	α	delta T	R-squared
Swaziland	71.8[a] (2.43)	2007.04 (0.24)	0.53 (0.039)	8.2	0.99
Syria	66.3[a] (3.46)	2007.2 (0.36)	0.50 (0.05)	8.73	0.99
Ukraine	122.8[b] (2.27)	2004.8 (0.07)	1.31 (0.11)	3.34	0.99
Viet Nam	148.5 (2.4)	2007.7 (0.069)	0.88 (0.046)	4.98	0.99
Yemen	67.3[a] (2.59)	2008.4 (0.2)	0.5 (0.03)	8.73	0.99
Zambia	162.4[a, c] (53.2)	2012.4 (1.62)	0.37 (0.04)	11.63	0.99

Source: Author's estimates
[a]Countries that in 2012 were located in exponential growth phase
[b]Underestimates. Estimation method—NLS. In parenthesis (robust SE)
[c]Potential overestimates

followed the growth pattern, and in 2012, the country achieved impressive MCS penetration rates,[26] far exceeding world averages. In El Salvador, from 2000 onward, mobile telephony spread at exponential rates, reaching the midpoint in 2006, and finally saturating in 2012. The unprecedented success of mobile telephony in El Salvador was highly facilitated by liberal telecom market legislation, ensuring a competitive environment, which encouraged global telecom companies to invest in the country.[27] The market competition provoked significant declines in the prices of mobile services and the rapid development of prepaid services, which motivated people to acquire mobile telephony despite their relatively poor economic conditions. In contrast with the previously discussed examples of countries that did extraordinarily well in terms of increasing mobile cellular telephony penetration rates, across the group of lower-middle-income economies, there exists a subset of countries that performed relatively worse in this category (15 out of 29 lower-middle-income economies), and these are: Bolivia, Yemen, Syria, Lao P. D.R., Swaziland, Nigeria, Pakistan, Guyana, India, Zambia, Senegal, Nicaragua, Sri Lanka, Honduras and Congo. Despite the fact that these economies are *relatively* lagging in terms of mobile telephony diffusion, it should be noted that in

---

[26] In the case of Viet Nam, there is however a risk that the reported statistics on ICT penetration rates are—to a point—significantly overestimated (Wireless Broadband Masterplan Until 2020 for the Socialist Republic of Viet Nam, ITU 2012). In 2011, some 'corrections' were introduced by the government to the national ICT statistics. However, despite the previous, mobile telephony penetration rates are still reported as remarkably high compared with world averages.

[27] In 2012 in El Salvador, there were five global telecom operators: Tigo El Salvador, Movistar El Salvador, CTE Telecom Personal (Claro), Digicel El Salvador, and Intelfon S.A. (RED).

2012, reached comparable levels of $MCS_{i, y}$ penetration are, which are still high. The other six countries—Bolivia, Senegal, Nicaragua, Sri Lanka, Honduras and Congo—are rapidly approaching *full* penetration, $\kappa = 100$ % ($MCS_{i, y} = 100$ per 100 inhabitants). Relying on the predicted further diffusion of mobile telephony (see Table 4.7), countries such as Bolivia and Honduras should achieve $MCS_{i, y} = 100$ per 100 inhab. before the year 2015; Congo and Sri Lanka should achieve this goal before 2018.[28] Prospects for the remaining 11 economies are less promising; however, enormous under-, or overestimates may be the case here. It is critical to understand that from the long-term perspective, the future development of telecom markets in developing economies is heavily preconditioned by country-specific features, policies and regulations, and even social norms. All of the previous may strongly promote or hinder investments in mobile telephony infrastructure, shaping access to and use of mobile telephony at national levels.[29]

4.4.1.1 Final Remarks

The analysis of mobile telephony diffusion patterns in low-, and lower-middle-income countries, demonstrated dramatic growth with respect to mobile cellular penetration rates in the great majority of the examined economies. However, the analysis also revealed the existence of significant differences both among particular countries and between income groups. Notwithstanding, the success of mobile telephony's broad and rapid diffusion in most of the economically backward countries was preconditioned by multiple factors. First, over the analysed period, mobile telephony became relatively accessible and affordable for a majority of inhabitants in low- and lower-middle-income economies (ITU 2011c). The increased affordability and access were mainly determined by the establishing of prepaid systems, which eliminates all of the constraints associated with post-paid systems, e.g., opening a bank account and building the whole infrastructure that would enable bank payment methods. The broad implementation of prepaid systems, which are well-tailored for low-income economies, is of critical importance among low-skilled and relatively economically poor societies (UNCTAD 2007). Additionally, the broad adoption of mobile telephony with well-developed prepaid systems overcomes one of the basic barriers associated with insufficient and inadequate hard infrastructure, the slow pace of acquiring a fixed telephone line. Moreover, mobile cellular services were available to people living in remote and underserved regions (UNCTAD 2007). Rapid infrastructure development was also critically affected by growing market liberalisation that enhanced competition, which contributed to lowering the prices of mobile telephony (World Bank

---

[28] Under the rigid assumption that the diffusion path is approximated by a theoretical sigmoid curve.

[29] The country facing the most significant constraints in the broad diffusion of mobile telephony is Yemen. Classified as one of the most economically backward countries, Yemen must address severe internal military conflicts, which cause permanent instability and constrain the broad diffusion of mobile telephony. Although the telecom market was liberalised in 2001, the fixed-telephony market is still a monopoly (ITU 2014a)

Table 4.7 Predicted MCS$_{i,y}$ diffusion paths. 29 lower-middle-income countries

Country	κ = 100 (per 100 inhab.) (fixed)	T$_m$	Delta T	5–10 year interval to achieve 100 % of κ (κ = 100 per 100 inhab.)	κ = 95 (per 100 inhab.) (fixed)	T$_m$	deltaT	5–10 year interval to achieve 100 % of κ (κ = 95 per 100 inhab.)
Armenia	In 2012 exceeded world average				In 2012 exceeded group average			
Bolivia	100	2007.9	10.7	2010–2015	95	2007.4	10.2	2010–2015
Congo	100	2007.7	6.3	2013–2018	In 2012 exceeded group average			
Egypt	In 2012 exceeded world average				In 2012 exceeded group average			
El Salvador	In 2012 exceeded world average				In 2012 exceeded group average			
Georgia	In 2012 exceeded world average				In 2012 exceeded group average			
Ghana	In 2012 exceeded world average				In 2012 exceeded group average			
Guyana	100	2007.07	13.8	2015–2020	95	2006.6	12.8	2015–2020
Honduras	100	2006.5	3.45	2010–2015	95	2006.4	3.2	2010–2015
India	100	2009.4	8.4	2015–2020	95	2009.2	7.9	2015–2020
Indonesia	In 2012 exceeded world average				In 2012 exceeded group average			
Lao P.D.R.	100	2009.13	8.6	2015–2020	95	2008.8	8.03	2015–2020
Mauritania	In 2012 exceeded world average				In 2012 exceeded group average			
Moldova	In 2012 exceeded world average				In 2012 exceeded group average			
Mongolia	In 2012 exceeded world average				In 2012 exceeded group average			
Morocco	In 2012 exceeded world average				In 2012 exceeded group average			
Nicaragua	100	2007.8	9.89	2015–2020	95	2007.5	9.3	2015–2020
Nigeria	100	2009.6	11.5	2020–2025	95	2009.3	11.03	2018–2023
Pakistan	100	2009.08	10.9	2020–2025	95	2008.7	10.3	2015–2020
Paraguay	In 2012 exceeded world average				In 2012 exceeded group average			
Philippines	In 2012 exceeded world average				In 2012 exceeded group average			
Senegal	100	2009.08	10.9	2020–2025	95	2008.7	10.3	2015–2020
Sri Lanka	100	2007.5	6.9	2013–2018	95	2007.3	6.4	2010–2015

(continued)

Table 4.7 (continued)

Country	$\kappa = 100$ (per 100 inhab.) (fixed)	T_m	Delta T	5–10 year interval to achieve 100 % of $\kappa = 100$ per 100 inhab.	$\kappa = 95$ (per 100 inhab.) (fixed)	T_m	deltaT	5–10 year interval to achieve 100 % of $\kappa = 95$ per 100 inhab.
Swaziland	100	2009.05	12.5	2020–2025	95	2008.7	11.9	2015–2020
Syria	100	2009.8	13.4	2020–2025	95	2009.5	12.8	2018–2023
Ukraine	In 2012 exceeded world average				In 2012 exceeded group average			
Viet Nam	In 2012 exceeded world average				In 2012 exceeded group average			
Yemen	100	2010.7	12.1	2020–2025	95	2010.4	11.8	2015–2020
Zambia	100	2010.2	9.4	2020–2025	95	2009.9	9.05	2015–2020

Source: Author's estimates. Note: $\kappa = 100$ %— corresponds to world average in 2012; $\kappa = 95$ %—corresponds to lower-middle-income group average in 2012

2006). The latter was possible because certain pro-telecommunication policies were implemented; thus, the gradual opening of telecommunication markets was enabled.

4.4.2 Internet Networks Diffusion and Internet Usage

Internet networks that facilitate unrestricted access to knowledge and information are important prerequisites for overcoming the socio-economic development obstacles that low-income and lower-middle-income countries are still facing. Internet access promotes all types of economic activities, enables access to education, brings people to new opportunities and broadens their horizons (ITU, UNESCO 2011). Internet access makes major inroads (ITU 2011a) in broad knowledge acquisition, providing basic infrastructure for communication and information exchange. In developing countries, the significant constraint on rapid and broad spread of the Internet is poorly developed infrastructure, which impedes the diffusion of wired (fixed) solutions that offer access to data transfer. One solution that may help to overcome the barrier of poor infrastructure is the extensive implementation of wireless technologies, which connect unconnected areas. The trends observed in the diffusion of fixed versus wireless connections clearly demonstrate that the latter are the best solutions for many countries that routinely face low penetrations rates with regard to wired connections; universalising wireless broadband networks is especially challenging in economically backward countries.

Over the period 2000–2012, in many low- and lower-middle-income countries, much effort was to foster Internet access, facilitated by both fixed and wireless networks. In addition, access to broadband technologies has been intensively promoted because high-speed solutions offer more efficient (compared with low-speed[30]) channels for data acquisition. As such, the shifts away from analogue or primitive digital data transmission channels (e.g., ISDN)[31] were substituted by 'fibre optics', which offer high-speed connections. However, despite the relatively high average annual growth rates (over 1997–2012), with respect to $FIS_{i, y}$, $FBS_{i, y}$ and $WBS_{i, y}$, in 2012, the average penetration rates remained low in both low-, and lower-middle-income countries. Sheds light on and explores major trends in the diffusion of $FIS_{i, y}$, $FBS_{i, y}$ and $WBS_{i, y}$ in respective countries. Additionally, the section provides evidence on the changes in the number of Internet users ($IU_{i, y}$) over the analogous period. Toward these aims, we develop diffusion patterns for fixed (wired) narrowband ($FIS_{i, y}$), fixed broadband Internet ($FBS_{i, y}$) and wireless broadband ($WBS_{i, y}$) connections.

Figures 4.7 and 4.8 plot country-specific diffusion patterns with respect to examined indicators in 17 low-income economies. The principal observation of note is that among the low-income countries, the average penetration of fixed

---

[30] Below 256 kbit/s.

[31] Mainly fixed.

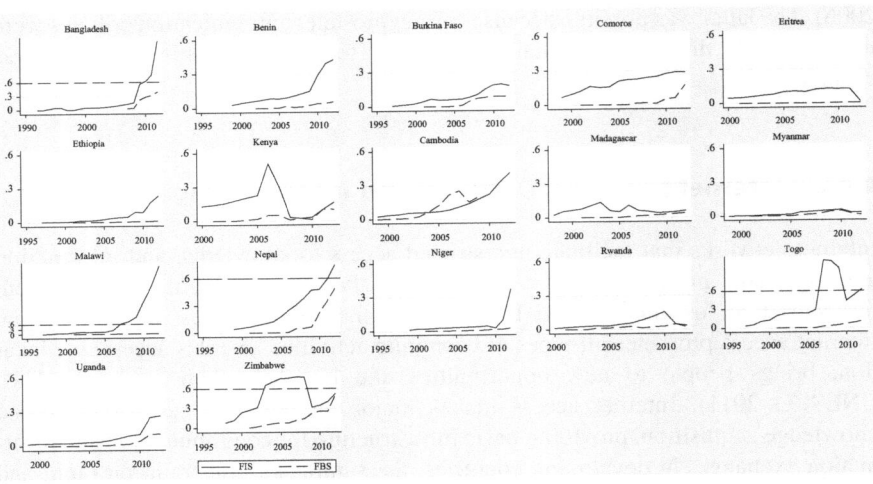

Fig. 4.7 $FIS_{i, y}/FBS_{i, y}$ diffusion paths. 17 low-income countries. Note: *horizontal long-dash-line* → average achievements in $FIS_{i, y}$ in low-income countries in 2012. On *vertical axis*—$FIS_{i, y}/FBS_{i, y}$ (*Source*: Author's elaboration)

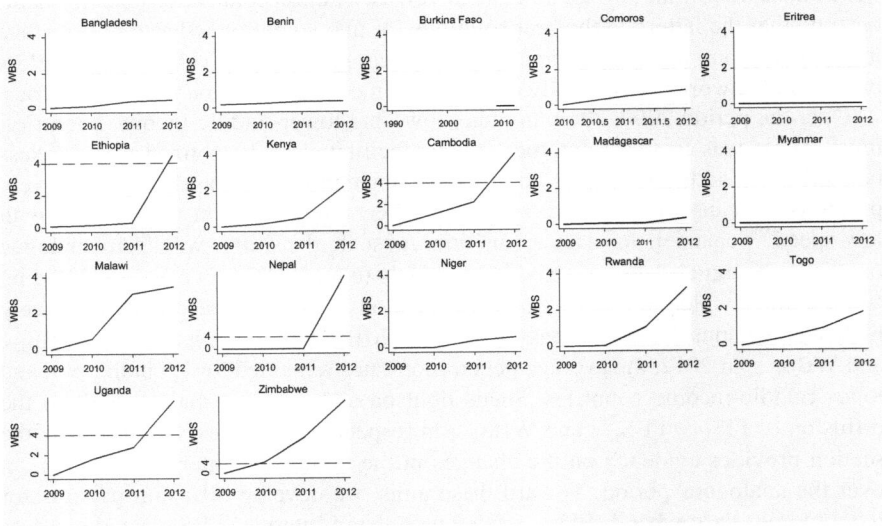

Fig. 4.8 $WBS_{i, y}$ diffusion paths. 17 low-income countries. Note: *horizontal long-dash-line* → average achievements in $WBS_{i, y}$ in low-income countries in 2012 (*Source*: Author's elaboration)

Internet ($FIS_{i, y}$) and fixed broadband Internet ($FBS_{i, y}$) changed very little over examined period and in 2012 remained significantly below 1 per 100 inhabitants. For $FIS_{i, y}$ and $FBS_{i, y}$, the respective average values in 2012 were reported as

$FIS_{average,2012} = (0.59)$ and $FBS_{average,2012} = (0.137)$. In five countries, Malawi, Bangladesh, Nepal, Togo and Zimbabwe, the growth of fixed Internet subscribers importantly differed compared with the remaining 14 economies. The remaining analysed low-income countries did not demonstrate considerable growth in terms of fixed Internet (narrowband) subscriptions. The state of fixed broadband network adoption in low-income countries was similarly poor over the analysed period. Fixed broadband connections were first available in 2001, in Zimbabwe ($FBS_{Zwe,2001} = 0.006$ per 100 inhabitants), and in consecutive years, it was gradually implemented in another 16 of the considered countries. However, despite efforts in this area, in 2012, the average fixed broadband penetration rate was approximately $FBS_{average,2012} = 0.137$. Since 2009, wireless broadband solutions emerged (see Fig. 4.8). Growth in wireless broadband connections was predominantly driven by rising awareness of the potential benefits of wireless broadband, engaging the privet sector, and legal regulations that created 'friendly' environments for investments in this type of communication infrastructure. This is of special importance in economically backward economies because wireless communication constitutes an important alternative in countries with fairly poorly developed wired infrastructures. In the absence of a fixed infrastructure, wireless point-to-point connections may be used effectively to connect people (especially beneficial for people living in geographically remote areas) if they are affordable. In 2009, wireless broadband connections were initially available in 7 countries (Bangladesh, Benin, Burkina Faso, Ethiopia, Kenya, Myanmar and Zimbabwe), but according to ITU statistics (ITU 2013), 4 years later in 2012, wireless broadband was observed in each of the considered low-income economies. The group average, in 2012, was approximately 4.9 per 100 inhabitants, but only four countries performed above this number: Zimbabwe, Nepal,[32] Uganda and Cambodia. However, still in 2012, the three bottom countries were Eritrea, Burkina Faso and Myanmar, where $WBS_{i,y}$ penetration rates remained considerably low, respectively, 0.001, 0.004 and 0.027. Enormous cross-country disparities in wireless-broadband deployment should not be ignored because they fundamentally shape the general picture of this topic, which suggests that a great majority of low-income economies still suffer from heavy digital deprivation and have only highly limited access to the Internet. As countries intensify their efforts toward higher $FIS_{i,y}$, $FBS_{i,y}$ and $WBS_{i,y}$ deployment, this will increase individual access to and use of the Internet. Over analyzed period, few low-income countries accelerated on the diffusion paths with regard to fixed and wireless connections, which should have promoted more universal use of Internet. Figure 4.9 provides visual representation of the changes in the numbers of individuals who had Internet access and used it. The principal observation is that limited access to basic ICT infrastructure in turn reduced the ability to increase the number of individuals who used the Internet.

---

[32] In the group of low-income countries, Zimbabwe and Nepal may be classified as outliers. Their $WBS_{i,y}$ coverage significantly exceeded the group average in 2012.

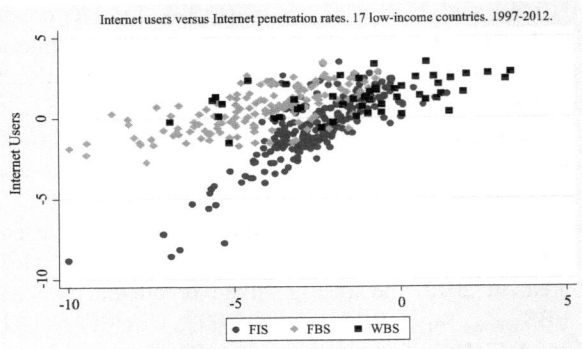

Fig. 4.9 $IU_{i, y}$ diffusion paths. 17 low-income countries. Note: *horizontal dash-line* refers to average achievements in $IU_{i, y}$ in low-income countries in 2012 (*Source*: Author's elaboration)

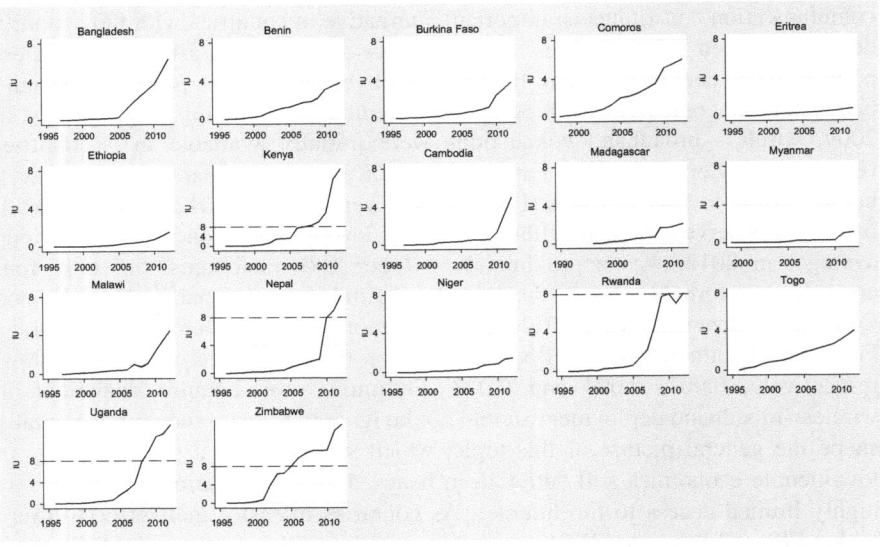

Fig. 4.10 Internet users vs. Internet subscription rates ($FIS_{i, y}$, $FBS_{i, y}$ and $WBS_{i, y}$). 17 - low-income countries (*Source*: Author's elaboration)

Figure 4.10 provides graphical explanation of the statistical relationship between $FIS_{i, y}$, $FBS_{i, y}$, and $WBS_{i, y}$ penetration rates and Internet users and suggests that the two covariates are strongly interrelated—increases in $FIS_{i, y}$, $FBS_{i, y}$ and $WBS_{i, y}$ increase the number of individuals who use the Internet. Although the first statistical data on number of Internet users trace back to 1996, in this year, the data were available for only a subset of countries (11 out of the 17 considered). Since 1999, a fully balanced data set on Internet usage in low-income countries has been available. Over the period 1999–2012, the average number of Internet users grew steadily, approximately 31.5 % per annum ($\text{IU}_{\text{average,1999}} = 0.11$ %, $\text{IU}_{\text{average,2012}} = 7.23$ %). However, over the analysed period, despite positive changes in increasing access to and usage of the Internet, visible inequalities emerged among the considered countries (compare evidence in Sect. 4.2). The latter allows concluding that a vast majority of low-income countries, in 2012, were still 'virtually locked', experiencing poor digitalisation. The important implication of the previous is that reliable estimates of logistics growth model parameters (for $\text{FIS}_{i,\,y}$, $\text{FBS}_{i,\,y}$, $\text{WBS}_{i,\,y}$ and $\text{IU}_{i,\,y}$) were disabled, and thus are not reported here.

Table 4.8 summarises two distinct predicted[33] scenarios of the future growth of Internet penetration rates ($\text{IU}_{i,\,y}$) in low-income countries. The first scenario forecasts $\text{IU}_{i,\,y}$ diffusion trajectories under the assumption that the ceiling is set at $\text{IU}_{i,\,y} = 42$ %, which corresponds to the world average Internet user rate in 2012[34]; the assumption of the second is $\kappa_{\text{IU}} = 8$ %,[35] which corresponds to the low-income group average in 2012. Table 4.8 shows the predicted midpoint (T_m), specific duration (deltaT) and estimated 5–10-year interval when the $\kappa_{\text{IU}} = 42$ % or $\kappa_{\text{IU}} = 8$ % may potentially be reached. Important to note is that the provided forecasts might be misleading because most of the examined countries, until 2012, were still in the early growth phase. What must also be remembered is that we presume that the future $\text{IU}_{i,\,y}$ diffusion scenarios are not purely random but rather are determined to follow the S-shaped trajectory. Thus, the forecasts are uncertain, and all predictions should be treated with caution, especially given that all predictions show high sensitivity to the historical data. Special attention should be focussed on predictions when the ceiling is fixed for $\kappa_{\text{IU}} = 42$ % because the accuracy of the forecast may be questionable and rather inconclusive, especially with respect to countries that in 2012 did not even surpass the low-income group average ($\text{IU}_{\text{average,2012}} \cong 8$ %). Such misleading predictions might be the case for Eritrea, Togo or Niger. These countries should potentially reach $\kappa = 42$ % in approximately 2050–2060 (Eritrea and Togo) or 2040–2050 (Niger). Relatively reasonable predictions (again if the fixed $\kappa_{\text{IU}} = 42$ % condition is satisfied) are reported for Kenya, Nepal, Uganda and Cambodia, and the error may be not as great as those for the previously discussed countries. The uncertainty of projections is clearly lower if $\kappa_{\text{IU}} = 8$ % is fixed. However, the given forecast (see Table 4.8) of a 5–10-year interval still may be heavily overestimated, especially with respect to Togo, Madagascar and Niger. These countries may achieve $\text{IU}_{i,\,y} = 8$ % more quickly if adequate policies that promote Internet infrastructure development are implemented, which would enable universal access to Internet connections and finally accelerate growth in the Internet penetration rate. The picture that emerges

---

[33] The forecasts were prepared under the rigid assumption that each country will follow the theoretical sigmoid diffusion curve; thus, the estimated equation based on which the forecast is provided follows $N_{IU}(t) = \frac{\kappa}{1+\,e^{-at-\beta}}$ (for details, see Chap. 3).

[34] Author's calculations.

[35] Author's calculations.

Table 4.8 Predicted $IU_{i,y}$ diffusion paths. 17 low-income countries

Country	κ = 41 % (fixed)	T_m	deltaT	5–10 year interval to achieve ≅100% of κ (κ = 41 %)	κ = 8 % (fixed)	T_m	deltaT	5–10 year interval to achieve ≅100% of κ (κ = %)
Bangladesh	42	2017.2	13.7	2020–2025	8	2009.8	8.56	2015–2020
Benin	42	2023.6	23.07	2040–2045	8	2012.2	17.9	2020–2025
Burkina Faso	42	2018.8	13.3	2025–2030	8	2012.3	10.3	2020–2025
Cambodia	42	2015.15	6.99	2020–2025	8	2011.5	4.5	2015–2020
Comoros	42	2019.7	20.2	2035–2040	8	2008.9	12.48	2015–2020
Eritrea	42	2037.4	28.8	2050–2060[a]	8	2025.5	27.6	2030–2035[a]
Ethiopia	42	2022.9	14.58	2040–2050[a]	8	2016.6	13.5	2030–2035[a]
Kenya	42	2010.2	8.09	2015–2020	In 2012 exceeded group average			
Madagascar	42	2026.9	23.2	2040–2050[a]	8	2016.2	20.1	2035–2040[a]
Malawi	42	2017.3	11.18	2025–2030	8	2011.7	8.05	2020–2025
Myanmar	42	2020.1	10.03	2030–2035[a]	8	2015.8	9.3	2025–2030[a]
Nepal	42	2013.9	9.53	2020–2025	In 2012 exceeded group average			
Niger	42	2025.0	17.45	2040–2050[a]	8	2017.4	16.07	2030–2035[a]
Rwanda	42	2016.3	15.9	2030–2040[a]	In 2012 exceeded group average			
Togo	42	2027.5	30.7	2050–2060[a]	8	2012.3	23.56	2030–2040[a]
Uganda	42	2013.2	13.2	2020–2030	In 2012 exceeded group average			
Zimbabwe	42	2013.8	21.9	2030–2035	In 2012 exceeded group average			

Source: Author's estimates. Note: κ = 42 %—corresponds to world average in 2012; κ = 8 %—corresponds to low-income-group average in 2012
[a]Estimates might be misleading—countries in very initial stage of diffusion paths

from the analogous evidence regarding changes in Internet access and use in lower-middle-income economies does not essentially differ compared with the low-income countries. The changes in fixed narrowband Internet ($FIS_{i, y}$), fixed broadband Internet ($FBS_{i, y}$) and wireless broadband ($WBS_{i, y}$) penetration rates are plotted in Figs. 4.11 and 4.13. Additionally, Fig. 4.14 provides visual representation of the spread of Internet users in lower-middle-income countries. Although the implementation of fixed narrowband and fixed broadband networks accelerated in the analysed countries, especially over the period[36] 2002–2012,[37] substantial cross-country differences were revealed to persist. In 2002, the average adoption of fixed narrowband Internet was roughly $FIS_{average,2002} = (0.49)$ per 100 inhabitants and grew approximately 16.5 % annually, reaching $FIS_{average,2012} = 2.56$ in 2012. Over the analogous period, the fixed broadband Internet network penetration rates grew approximately 56.06 % annually, which resulted in shifts in average $FBS_{i, y}$ from (0.009) in 2001 to 2.45 in 2012.

However, despite the noticeable growth in fixed narrowband and fixed broadband Internet penetration rates reported for a few lower-middle-income countries, by the end of 2012, many of the analysed economies still lagged heavily behind in this area. In, *inter alia*, Congo, Zambia, Mauritania or Nigeria, the evolvement of fixed narrowband and fixed broadband infrastructure was very slow over 2002–2012, and none of the listed countries reached the threshold of $FIS/FBS_{i,2012} = 1$ per 100 inhabitants. Hence, in the countries just mentioned, the fixed narrowband and fixed broadband Internet networks scarcely existed by the end of 2012. Visual inspection of the respective $FIS_{i, y}$ and $FBS_{i, y}$ diffusion curves (see Fig. 4.11) suggests that each of the lower-middle-income countries, by the end of the terminal year of analysis (2012) were still in the initial phase of growth. In Armenia, El Salvador, Georgia, Moldova, Mongolia, Philippines, Sri Lanka and Yemen, the trajectory of changes in $FIS_{i, y}$ and $FBS_{i, y}$ is relatively stable, whereas for the others, the growth trajectory is marked by random ups and downs. In Congo, Ghana, Lao P.D.R., Mauritania, Nigeria and Senegal, the growth path is mainly flat, which demonstrates extremely poor progress in fixed narrowband and fixed broadband infrastructure adoption. Similarly, as in the case of the low-income group, each of the considered lower-middle-income economies, by the end of 2012, was still 'locked' in the early stage of growth, which suggests the inability to provide reliable estimates of full diffusion curves approximated by logistic growth models[38]; thus, the latter are not reported. Figure 4.13 illustrates the diffusion process of wireless broadband technologies in lower-middle-income countries. Before 2007, no wireless broadband solutions are reported to exist in

---

[36] In lower-middle-income countries, the most appropriate time span for the analysis is 2002–2012 because before 2002, the availability of data on $FIS_{i,y}$ and $FBS_{i,y}$ indicators is limited.

[37] See: ITU ICT Eye – ICT Statistics, International Telecommunication Union 2014. Accessed: 20 May 2014.

[38] The theoretical specification of the logistic growth model, for FIS and FBS, would follow: $N_{FIS/FBS}(t) = \frac{1}{1+ e^{-at-\beta}}$ (for details see Chap. 3).

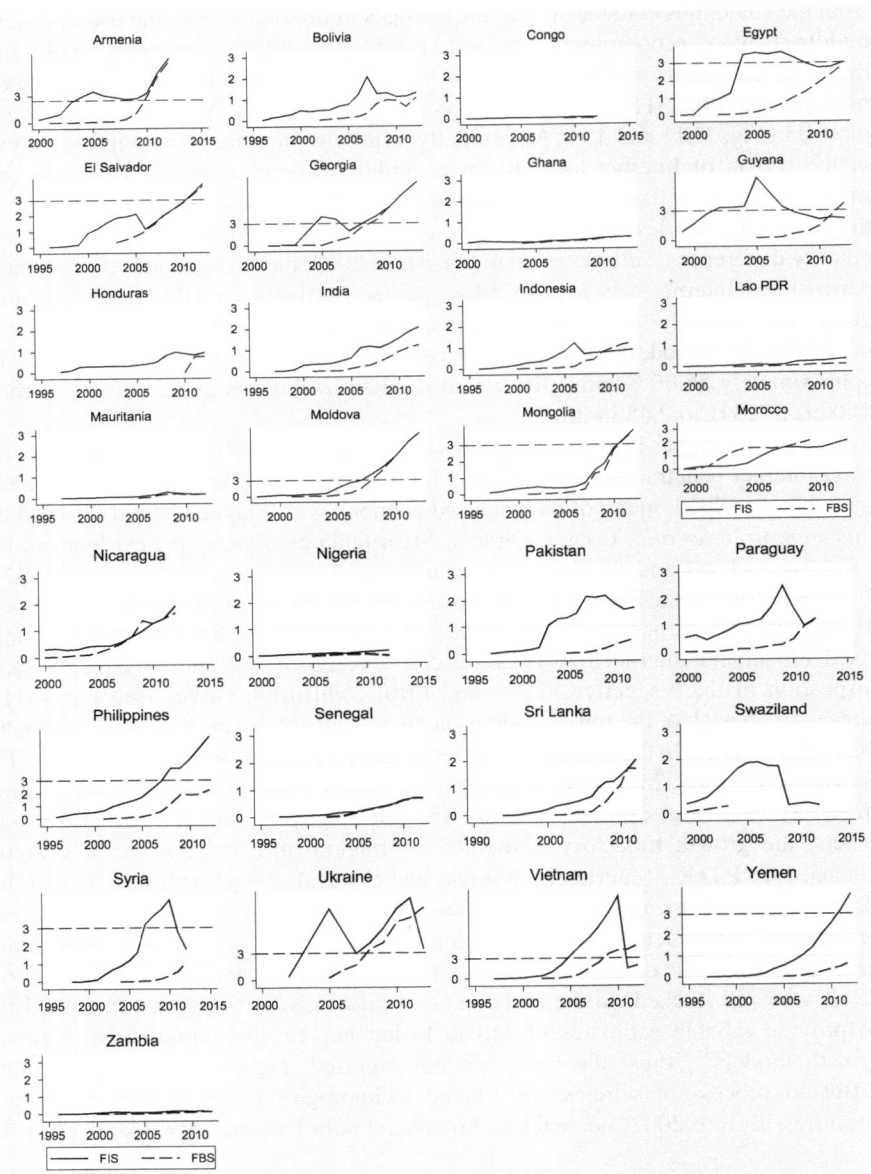

Fig. 4.11 $FIS_{i,y}/FBS_{i,y}$ diffusion paths. 29 lower-middle-income countries. Note: *horizontal long-dash-line* refers to average achievements in $FIS_{i,y}$ in lower-middle-income countries in 2012. On *vertical axis*—$FIS_{i,y}/FBS_{i,y}$ (*Source*: Author's elaboration)

Fig. 4.12 Fixed broadband penetration vs. wireless broadband penetration rates. 29 lower-middle-income countries. Year 2012 (*Source*: Author's elaboration)

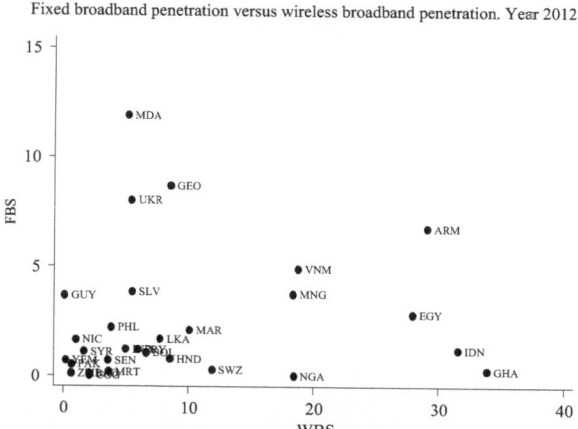

Fixed broadband penetration versus wireless broadband penetration. Year 2012.

any of the analysed countries; therefore, the analysis of $WBS_{i, y}$ spread is limited to the period 2007–2012. As was already highlighted, rapid development of wireless broadband communication channels is critical for economically backward countries. The physical infrastructure requirements that precondition broad wireless broadband diffusion are far less demanding compared with fixed narrowband or fixed broadband connection modes. Building networks based on wireless technologies seems to be of great relevance in countries that face permanent infrastructural shortages and in geographically remote regions with low population density and few opportunities to connect with the 'outside' world because no 'wired option' is available. Thus, special attention should be paid to encouraging the dynamic deployment of wireless broadband technologies that can serve as backup solutions for fixed lines. As shown in Fig. 4.12, in 2012, the correlation between fixed broadband and wireless broadband penetration rates was weak, which supports the supposition that in countries where infrastructure that supports landline broadband connections is weak, implementing wireless technologies may be a favourable alternative. The rapidly increasing demand for wireless broadband in countries such as Ghana, Indonesia, Egypt and Nigeria is predominantly driven by poorly developed fixed narrowband and/or fixed broadband infrastructures. Although some countries successfully implemented wireless broadband, the country-specific levels of $WBS_{i, y}$ still vary greatly in the analysed country set. Figure 4.13 explains wireless broadband technology diffusion in each country separately. The principal observation is that in only a few countries did wireless broadband deployment increase greatly over the analysed period. This was the case for, as mentioned, Ghana, Indonesia, Armenia and Egypt. However, and not surprisingly, despite the few examples of countries where the adoption and usage of wireless broadband technologies flourished, many lower-middle-income economies still demonstrate significantly low $WBS_{i, y}$ penetration rates. At the end of the spectrum, Guyana, Yemen, Zambia and Pakistan demonstrated $WBS_{i, y}$

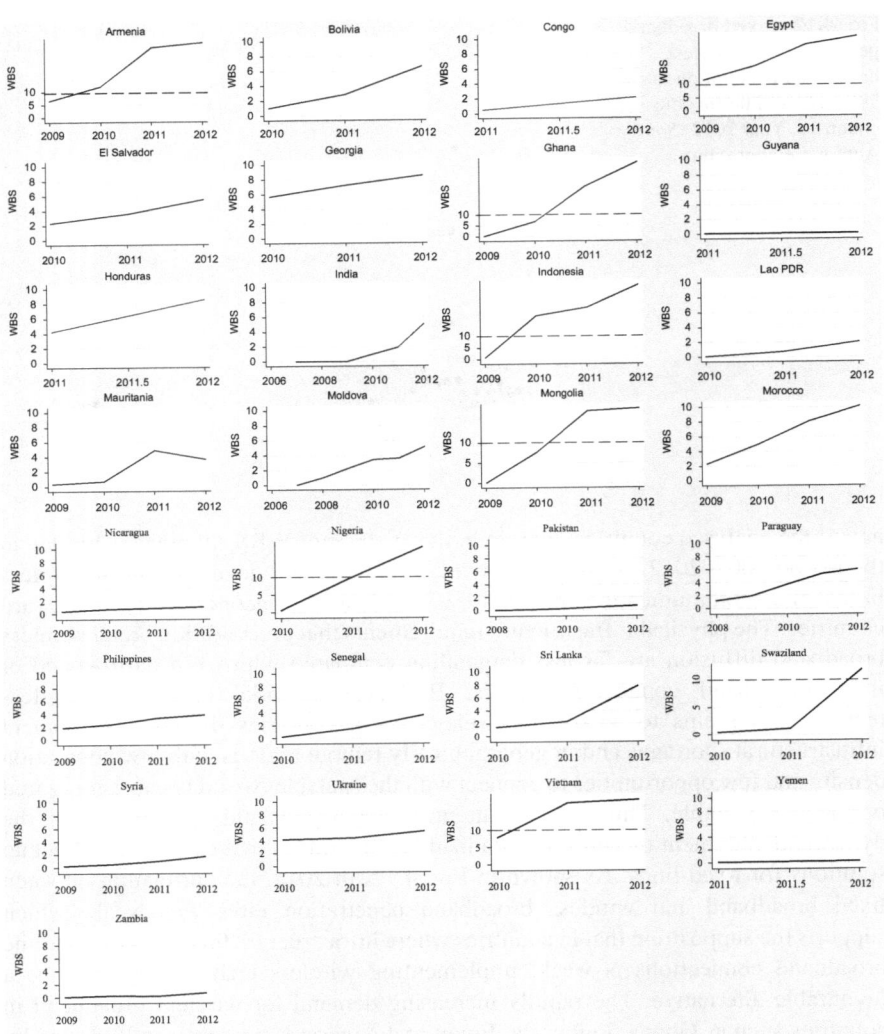

Fig. 4.13 $WBS_{i, y}$ diffusion paths. 29 lower-middle-income countries. Note: *horizontal long-dash-line* refers to average achievements in WBS development in lower-middle-income countries in 2012 (*Source*: Author's elaboration)

implementation far below 1 per 100 inhab. Aside from the numerical evidence just presented, the country-specific plots help to understand the differences in wireless broadband adoption dynamics. Whereas in some countries, the increase in wireless broadband implementation is evident (see, e.g., the prominent examples of diffusion in Armenia, Ghana, Indonesia or Moldova), for the remaining countries, the $WBS_{i, y}$ development time path is mostly flat.

Increasing access to fixed narrowband, fixed broadband and wireless broadband networks should increase the number of individuals who access and use the

Fig. 4.14 Internet users vs. Internet subscription rates (FIS$_{i, y}$, FBS$_{i, y}$ and WBS$_{i, y}$) (*Source*: Author's elaboration)

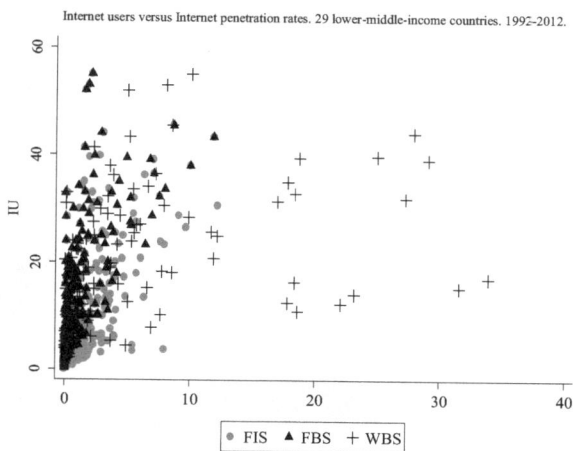

Internet users versus Internet penetration rates. 29 lower-middle-income countries. 1992-2012.

Internet. Figure 4.14 reports on the statistical relationships between the IU$_{i, y}$ indicator against FIS$_{i, y}$, FBS$_{i, y}$ and WBS$_{i, y}$ in 29 lower-middle-income countries over 2001–2012.[39] Combining the listed ICT indicators explains the interdependency between the universality of access to fixed or wireless technologies that enable data transfer and the number of individuals who use the Internet. Preliminary visual inspection suggests the existence of relatively strong and positive associations between consecutive pairs of variables (IU$_{i, y}$/FIS$_{i, y}$; IU$_{i, y}$/FBS$_{i, y}$; IU$_{i, y}$/WBS$_{i, y}$). Nevertheless, to confirm the previous, more formal statistical tests are reported. Finally, the calculated respective correlation coefficients[40] are as follows: r^2(IU$_{i, y}$/FIS$_{i, y}$) = 0.503, r^2(IU$_{i, y}$/FBS$_{i, y}$) = 0.58 and r^2(IU$_{i, y}$/WBS$_{i, y}$) = 0.36. Comparing the results, in contrast to what might have been expected, a relatively weaker relationship is reported between wireless broadband penetration rates and Internet users. However, these outcomes should be interpreted with caution. First, the period of analysis is significantly shorter[41] than that for IU$_{i, y}$/FIS$_{i, y}$ and IU$_{i, y}$/FBS$_{i, y}$; second, in the case of WBS$_{i, y}$ the data for 2009 are only available for 15 countries (out of 29), which heavily affected the final results.

Figure 4.15 and Tables 4.9 and 4.10 provide more detailed insight into the increases in Internet users in 29 lower-middle-income countries. Preliminary analysis of IU$_{i, y}$ development, which is graphically approximated by country-specific diffusion paths, suggests rather steady and consistent trends over time. The majority of countries, having passed the initial slow phase of growth, enter the exponential growth phase, and the number of individuals using Internet booms. Only a few

---

[39] For WBS the time period is limited to 2009–2012.

[40] Author's estimates.

[41] Additionally, in the case of WBS$_{i,y}$, the data for 2009 are only available for 15 countries (out of 29).

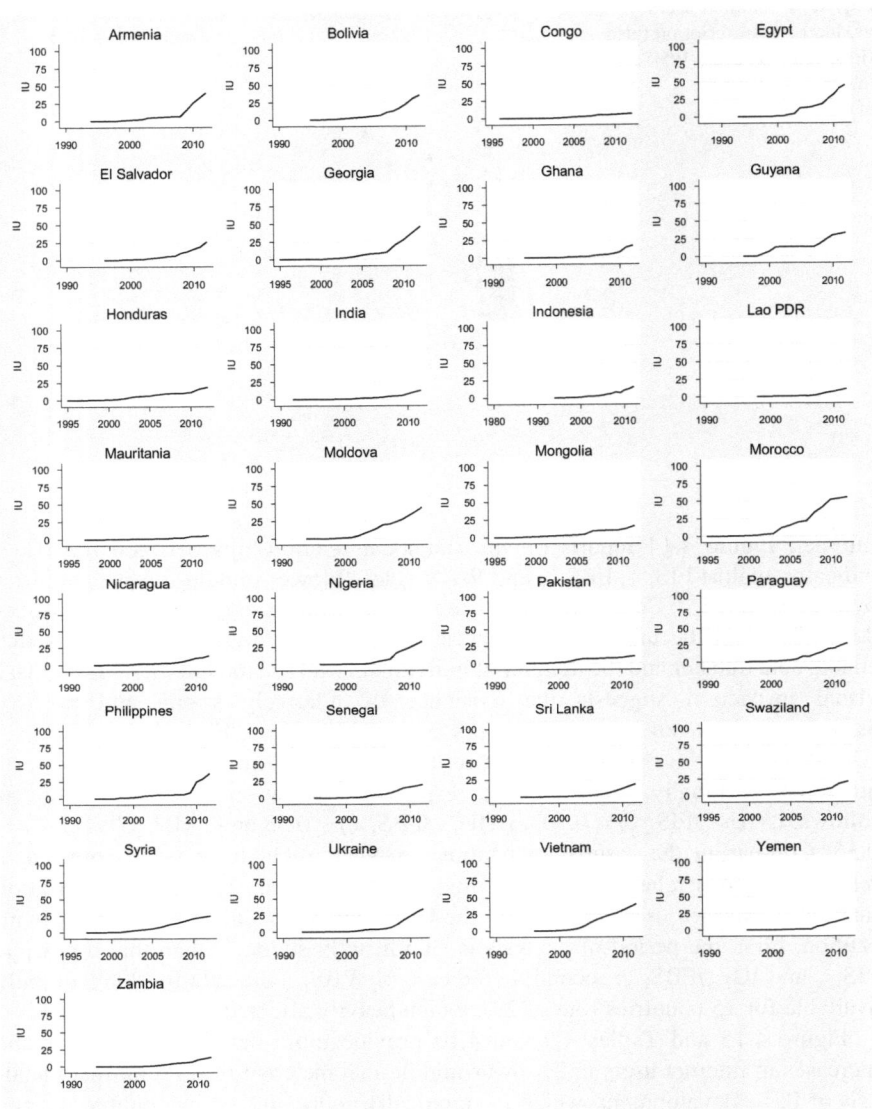

Fig. 4.15 $IU_{i, y}$ diffusion paths. 29 lower-middle-income countries (*Source*: Author's elaboration)

economies in 2012 were still 'locked' in the very early phase of growth: Congo, India, Mauritania, Lao P.D.R., Mongolia, Nicaragua, Pakistan and Zambia. The most dynamic expansion of Internet usage is reported for Morocco, Georgia, Egypt and Moldova, where the number of individuals using the Internet progressively grew, reaching $IU_{MAR,2012} = 55\%$, $IU_{GEO,2012} = 45.5\%$, $IU_{EGY,2012} = 44.07\%$ and

Table 4.9 $IU_{i, y}$ logistic growth estimates. 29 lower-middle-income countries

Country	κ	T_m	α	delta T	R-squared
Armenia	128.8[a, b] (117.4)	2014.00 (3.77)	0.39 (0.08)	11.23	0.98
Bolivia	95.0[a] (27.5)	2013.6 (1.49)	0.32 (0.02)	13.6	0.99
Congo	6.31[c] (0.23)	2007.2 (0.20)	0.56 (0.04)	7.7	0.99
Egypt	80.15[a] (18.7)	2011.3 (1.5)	0.31 (0.046)	13.9	0.99
El Salvador	174.4[a, b] (122.3)	2018.4 (3.26)	0.27 (0.014)	15.9	0.99
Georgia	117.4[a] (31.5)	2013.08 (1.15)	0.40 (0.037)	10.9	0.99
Ghana	Estimates unreliable[d]				
Guyana	83.3[a, b] (80.5)	2014.1 (9.7)	0.17 (0.06)	24.7	0.97
Honduras	30.76[a] (12.27)	2011.08 (3.23)	0.24 (0.048)	17.7	0.98
India	Estimates unreliable[d]				
Indonesia	53.0[a, b] (49.51)	2015.7 (5.85)	0.24 (0.03)	18.02	0.99
Lao P.D.R.	12.7[c] (1.02)	2009.5 (0.39)	0.63 (0.064)	6.9	0.99
Mauritania	9.87[c] (3.39)	2011.4 (1.78)	0.40 (0.059)	10.9	0.99
Moldova	53.6[a] (6.22)	2008.4 (0.81)	0.34 (0.037)	12.6	0.99
Mongolia	21.6[a] (6.46)	2009.3 (2.09)	0.32 (0.05)	13.6	0.98
Morocco	63.1[a] (3.45)	2007.7 (0.34)	0.50 (0.05)	8.6	0.99
Nicaragua	49.5[a, b] (34.4)	2015.6 (4.01)	0.26 (0.03)	16.3	0.99
Nigeria	35.9[a] (1.8)	2008.7 (0.26)	0.64 (0.05)	6.8	0.99
Pakistan	8.26[c] (0.55)	2003.0 (0.41)	0.64 (0.11)	6.7	0.98
Paraguay	32.7[a] (2.13)	2008.5 (0.40)	0.42 (0.04)	10.4	0.99
Philippines	Estimates unreliable[d]				
Senegal	22.8[a] (1.26)	2008.2 (0.37)	0.44 (0.04)	9.8	0.99
Sri Lanka	32.1[a] (3.76)	2011.3 (0.55)	0.43 (0.03)	10.1	0.99
Swaziland	Estimates unreliable[d]				
Syria	28.07[a] (0.72)	2007.9 (0.15)	0.45 (0.023)	9.5	0.99

(continued)

Table 4.9 (continued)

Country	κ	T_m	α	delta T	R-squared
Ukraine	49.8[a] (5.68)	2010.3 (0.54)	0.48 (0.049)	9.1	0.99
Viet Nam	41.5[a] (3.17)	2007.2 (0.46)	0.45 (0.049)	9.6	0.99
Yemen	19.6[a] (1.2)	2009.04 (0.22)	0.63 (0.07)	6.8	0.99
Zambia	23.8[a] (6.38)	2011.2 (1.6)	0.34 (0.046)	12.6	0.99

Source: Author's estimates based on data derived from World Telecommunication/ICT Indicators database 2013 (17th Edition)
[a]Countries that in 2012 were still located in exponential growth phase
[b]Over-, or underestimates, statistically not significant.
[c]Countries in initial phase of growth
[d]Definite overestimates due abrupt increase in $IU_{i, y}$. Estimation method—NLS. In parenthesis (robust SE)

$IU_{MDA,2012} = 43.3$ %, respectively,[42] in 2012. These best-performing economies were closely followed by Viet Nam, Armenia, The Philippines, Bolivia, Ukraine, Guyana and Nigeria, where the number of individuals using the Internet, on average, slightly exceeded 35.5 %. Tables 4.9 and 4.10 exhibit S-shaped trajectories and projected future diffusion pattern estimates regarding $IU_{i, y}$.[43] As previously concluded from visual inspection (see Fig. 4.15), in 2012, most of the lower-middle-income countries were still in the early exponential growth phase (along the S-shaped path), and the rest were in the very initial growth phase. The latter substantially impacted the quality of the estimated parameters of country-specific logistic growth trajectories, which these countries intend to follow.

In five countries, Armenia, El Salvador, Guyana, Indonesia and Nicaragua, the estimated parameters were statistically insignificant, and the obtained κ's were revealed to be substantially under- or overestimated. In another four countries, Ghana, India, The Philippines and Swaziland, the parameters are not reported because the estimates are heavily unreliable.[44] For the remaining twenty economies, the estimated parameters that explain the hypothetical saturation, mid-point and specific duration may be biased because the uncertainty is great regarding the future development scenarios for countries that are in either the initial or the early exponential phase of growth. Table 4.10 summarises two distinct predicted

---

[42] To compare, the world average in 2012 was approximately $IU_{WorldAverage,2012} = 42$ %.

[43] The forecasts were prepared under the rigid assumption that each country will follow the theoretical sigmoid diffusion curve; thus, the estimated equation based on which the forecast is provided follows $N_{IU}(t) = \frac{1}{1 + e^{-\alpha t - \beta}}$ (for details, see Chap. 3).

[44] The unreliability of the estimates for Ghana, India, The Philippines and Swaziland, is likely a consequence of abrupt growth in the $IU_{i,y}$ level.

Table 4.10 Predicted IU$_{i, y}$ diffusion paths. 29 lower-middle-income countries

Country	κ = 42 % (fixed)	T$_m$	deltaT	5–10 year interval to achieve 100 % of κ (κ = 42 %)	κ = 25 % (fixed)	T$_m$	deltaT	5–10 year interval to achieve 100 % of κ (κ = 25 %)
Armenia	42	2009.5	5.6	2013–2018	In 2012 exceeded group average			
Bolivia	42	2009.4	9.3	2015–2020	In 2012 exceeded group average			
Congo	42	2019.05	19.2	2030–2035	25	2015.9	17.8	2025–2030
Egypt	In 2012 exceeded world average				In 2012 exceeded group average			
El Salvador	42	2011.3	11.9	2020–2025	In 2012 exceeded group average			
Georgia	In 2012 exceeded world average				In 2012 exceeded group average			
Ghana	42	2012.9	10.2	2020–2025	25	2010.8	8.2	2015–2020
Guyana	42	2007.6	18.3	2020–2025	In 2012 exceeded group average			
Honduras	42	2013.5	19.7	2025–2030	25	2009.5	16.06	2020–2025
India	42	2015.05	15.0	2025–2030	25	2012.3	13.3	2020–2025
Indonesia	42	2014.3	17.2	2025–2030	25	2010.9	14.7	2020–2025
Lao P.D.R.	42	2014.7	12.2	2025–2030	25	2012.4	10.6	2020–2025
Mauritania	42	2018.3	14.8	2030–2035	25	2015.9	14.07	2020–2025
Moldova	In 2012 exceeded world average				In 2012 exceeded group average			
Mongolia	42	2013.9	18.02	2025–2030	25	2010.4	14.9	2020–2025
Morocco	In 2012 exceeded world average				In 2012 exceeded group average			
Nicaragua	42	2014.7	15.8	2025–2030	25	2011.7	13.6	2020–2025
Nigeria	42	2009.4	8.2	2015–2020	25	2009.4	13.6	2020–2025
Pakistan	42	2018.7	28.3	2035–2040	25	2013.6	24.9	2025–2030
Paraguay	42	2010.07	12.7	2020–2025	In 2012 exceeded group average			
Philippines	42	2009.9	6.05	2015–2020	In 2012 exceeded group average			
Senegal	42	2012.0	15.0	2025–2030	25	2008.7	10.8	2020–2025
Sri Lanka	42	2012.6	11.3	2020–2025	25	2010.2	8.6	2020–2025
Swaziland	42	2012.2	11.6	2020–2025	25	2009.7	8.8	2015–2020

(continued)

Table 4.10 (continued)

Country	κ = 42 % (fixed)	T_m	deltaT	5–10 year interval to achieve 100 % of κ (κ = 42 %)	κ = 25 % (fixed)	T_m	deltaT	5–10 year interval to achieve 100 % of κ (κ = 25 %)
Syria	42	2010.4	13.5	2020–2025	In 2012 exceeded group average			
Ukraine	42	2009.6	7.9	2015–2020	In 2012 exceeded group average			
Viet Nam	42	2007.3	9.7	2015–2020	In 2012 exceeded group average			
Yemen	42	2012.6	11.5	2020–2025	25	2010.1	8.7	2015–2020
Zambia	42	2014.4	15.08	2025–2030	25	2011.5	12.9	2020–2025

Source: Author's estimates. Note: κ = 42 %—corresponds to world average in 2012; κ = 25 %—corresponds to lower-middle-income group average in 2012

scenarios of future Internet usage growth in lower-middle-income countries. The first scenario is designed to forecast when $IU_{i, y}$ saturation will reach $\kappa_{IU} = 42$ %, which corresponds to the world average Internet user rate in 2012[45]; the second forecasts—$\kappa_{IU} = 25$ %,[46] which corresponds to the lower-middle-income group average in 2012. Table 4.10 also reports on predicted midpoint (T_m), specific duration (deltaT) and the estimated 5–10-year interval when the fixed $\kappa_{IU} = 42$ % or $\kappa_{IU} = 25$ % may be hypothetically achieved. Analogous to what was explained in accordance with the predictions for low-income economies, the forecasts might be unclear because most countries as of 2012 were still in the initial or early exponential growth phase. Special attention should be focused on predictions when the ceiling is fixed for $\kappa_{IU} = 42$ % because the accuracy of the forecasts may be in doubt, especially with respect to countries that in 2012 did not even surpass the low-income group average ($IU_{average,2012} \cong 25\%$). Misleading predictions might be the case for Pakistan, Honduras, Congo, Guyana and Mongolia. All of these countries should potentially reach $\kappa = 42$ % in approximately 2030–2040; thus, the error may be extremely large in this case. Relatively reasonable predictions are reported for Nigeria, The Philippines, Ukraine and Viet Nam; they should reach 42 % saturation in 2015–2020.

The uncertainty of projections is clearly lower for $\kappa_{IU} = 25$ %. However, owing to the rapid rise in wireless broadband technologies, 14 out of the 29 lower-middle-income economies had already surpassed the group average in terms of Internet user ratios by 2012. The projections for the remaining 15 countries are promising, especially for Ghana, Swaziland and Yemen because these countries are projected to reach $\kappa = 25$ % over the period 2015–2020. Forecasts are less promising for Pakistan, Congo, and Honduras because these countries seem to be stuck in the low-usage trap and are unable to take off.

4.4.2.1 Final Remarks

In 2012, the state of development of fixed narrowband, fixed broadband and wireless broadband architecture, in both low-income and lower-middle-income countries, reflected the tremendous efforts that had been made toward closing the digital divides. Beginning in 2001, the $FIS_{i, y}$, $FBS_{i, y}$ and $WBS_{i, y}$ penetration rates tended to increase at a rapid pace. Nevertheless, despite the reported spectacular achievements, especially in wireless broadband deployment, in 2012, many low-, and lower-middle-income countries were still greatly deprived of universal access to ICT devices that enable Internet connections. In the majority of low-, and lower-middle-income countries, access to broadband is still highly constrained because remarkably high prices for leasing lines impede its broad access and use; additionally, country-specific barriers such as unfavourable location, poorly developed

---

[45] Author's calculations based on data retrieved from the World Telecommunication/ICT Indicators database 2013 (17th Edition).

[46] Author's calculations based on data retrieved from the World Telecommunication/ICT Indicators database 2013 (17th Edition).

infrastructure, and permanent power supply problems significantly impede the broader introduction of fixed and wireless broadband technologies (for detailed discussion, see Chap. 5).

The evidence in Sect. 4.4 clearly reports that both low-income and lower-middle-income countries have made enormous progress in boosting ICT adoption. The latter might have potentially resulted in switching from 'old' to 'new' technologies; and the henceforth gave rise to the unique process of technological substitution. To stay in line with the previous, the consecutive part (Sect. 4.5) is fully dedicated to uncovering whether broad adoption of ICTs has induced the technological substitution regarding fixed-to-mobile telephony and fixed-to-wireless Internet connections.

4.5   Tracing the Technological Substitution

The analysis presented in Sect. 4.4 demonstrates that over the period 2000–2012, developing countries have experienced significant improvements in the access to and use of information and communication technologies. A broadly documented description of the diffusion patterns of mobile cellular telephony and wireless-broadband technologies raises questions about the extent to which the 'old' ICTs, such as fixed telephone lines, are being gradually substituted by the new mobile cellular communication. A similar question emerges when the fixed-narrowband Internet or fixed-broadband Internet lines are considered in relation to wireless-broadband technologies.

Thus far, the available literature concentrating on the empirical evidence for technological substitution with regard to fixed-to-mobile telephony and fixed-to-wireless Internet connections is scattered. Partial evidence on the technological substitution may be traced in the works of, for example, Gruber (2001), Sung and Lee (2002), Hamilton (2003), Rodini et al. (2003), Banerjee and Ros (2004), Garbacz and Thompson (2005, 2007), Gunasekaran and Harmantzis (2007), Narayana (2010), Briglauer et al. (2011), Caves (2011), Grzybowski (2011), Ward and Zheng (2012), Srinuan et al. (2012) and Wulf et al. (2013). Gruber (2001), in his study of 10 Central and Eastern European countries since the initial year of introduction of mobile telephony in 1998, finds that mobile telephony diffuses rapidly, surpassing wired telephony. His study confirmed the fixed-to-mobile substitution hypothesis, and he found that the process was mainly determined by the rapidly growing consumer demand for mobile telephony. Gruber also argues that mobile telephony is an adequate alternative for fixed mainlines in rural and remote areas lacking well-developed copper- or fibre-based telecommunication infrastructure. Sung and Lee (2002), using panel data at the provisional level, study the impact of mobile telephony diffusion on fixed telephony penetration rates in Korea over the period 1991–1998. They confirmed the substitutability between these two means of voice transmission, as the growing stock of mobile cellular telephony negatively impacted the wired telephony penetration rates, and the reverse was the case otherwise. Similar conclusions may be traced in the evidence provided by Rodini et al. (2003). They estimated that, due to the relatively higher prices of fixed as opposed to mobile connections, consumers are more likely to use mobile cellular telephony in place of wired telephony; thus, the fixed-to-mobile substitution occurs. Hamilton (2003), in the study for African countries covering the period 1985–1997, reports that fixed-to-mobile substitution occurs, although he found that the effects are relatively weak, as wired and mobile telecommunication services remain rather complementary goods, even in countries where the fixed telephony penetration rates are negligible. Banerjee and Ros (2004) compare patterns of diffusion of mobile cellular telephony alongside the wired patterns in 61 countries over the period 1995–2001. They claim that due to the rapid spread of modern mobile telephony, the fixed mainline penetration rates have dropped. Additionally, they observed that the process of substitution is dramatically more dynamic in relatively poorer countries. The evidence for fixed-to-mobile telephony substitution in low-income countries is formulated in the works of Waverman et al. (2005). Using the data of 19 low-income economies over the period 1996–2004, they found that although mobile cellular phones were perceived as 'luxury' goods (reported price elasticity was close to 2.0), consumers were more likely to buy mobile phones instead of installing fixed mainlines, which accounts for a strong fixed-to-mobile substitution effect. Gunasekaran and Harmantzis (2007) argue that growing worldwide availability of wireless-based technological solutions, allows broad promotion of such technologies in developing countries, enabling universal access to Internet networks and connectivity. They emphasize strongly that, compared to wired technologies, wireless networks may be adopted much faster, as they require far less initial infrastructure and investment and offer high flexibility and a similar quality of connectivity. In addition, Gunasekaran and Harmantzis claim that providing increased access to wireless networks is of special urgency for developing countries, where the lack of wired (both fibre-based and copper-based) infrastructure significantly impedes Internet penetration. Thus, for economically backward countries, wireless technologies are a good substitute for wired technologies. Garbacz and Thompson (2005, 2007), using the evidence from 53 developing and 32 developed countries over the period 1996–2003, demonstrate that the effect of the introduction of mobile telephony on fixed telephony penetration rates may be mixed. They suggest that in developing countries the fixed-to-mobile substitution is relatively strong; conversely, in developed countries, the process is relatively weaker and heavily affected by price competition among telecommunication operators. Narayana (2010), based on the empirical evidence of fixed-to-mobile telephony substitution in India exclusively, finds that since 1991, both fixed and mobile telephony are treated as substitutes. Interestingly, Narayana finds that the strongest fixed-to-mobile substitution is predominantly visible with regard to low-income people living in underdeveloped and underserved regions. The evidence suggests that the broad diffusion of mobile telephony reveals its unique importance in remote rural areas, where the wired-based infrastructure is hardly accessible. Briglauer et al. (2011), using time-series data for the Austrian telecommunication market over the period 2002–2007, report strong fixed-to-mobile substitutability, especially with respect to calls. With respect to data access, the substitution is still reported; however, the revealed effects are weaker and less robust. Caves (2011), adopting the econometric analysis for the cross-price elasticity of fixed and mobile telecommunication connections for the period 2001–2007 in the United States, reports on significant substitutability between fixed and mobile services. Grzybowski (2011) provides evidence on fixed-to-mobile substitution in the 27 European Union countries over the period 2005–2009. Grzybowski uses panel data on household choices with regard to telecommunication technologies (2011). He finds the fixed-to-mobile telephony substitution effect, which is mainly attributed to growing rates of the penetration of mobile telephony. A similar effect is reported with respect to shifting from fixed broadband to mobile broadband. Ward and Zheng (2012) address the fixed-to-mobile telephony substitution effect in 31 Chinese provinces for the period 1998–2007. Applying dynamic panel data analysis, they find that fixed-to-mobile telephony substitution is relatively weak, as wired connections stubbornly expand over the Chinese telecommunication market. Srinuan et al. (2012) investigate the substitutability between fixed broadband and mobile broadband networks. The analysis is based on 4,000 surveys of individuals in Sweden in 2009. They find that both fixed and mobile broadband networks are perceived as close substitutes; however, the consumer choices are predominantly driven by price relations, household type and location. Wulf et al. (2013), in their study on fixed-to-mobile broadband network substitution in 34 OECD countries over the period 2001–2009, use Voltera-Lotka competition equations and demonstrate that in developed countries, people rather perceive the mobile and fixed broadband as complementary goods. Hence, the fixed-to-mobile broadband substitution is relatively weak, which they explain as the performance of the broadband networks—at the time of analysis—being worse in terms of coverage, availability and data transfer capability than the fixed-networks.

The current state of development of telecommunication infrastructure in low-income and lower-middle-income countries that was discussed in Sect. 4.4 indicates the huge progress that developing countries have made towards higher penetration rates, especially with regard to mobile cellular telephony (ITU 2011b). According to International Telecommunication Union Statistics (ITU 2013), 2003 was the first year that the total number of mobile telephony subscriptions surpassed the total number of fixed mainline subscriptions. Since this '*threshold*' year, the number of mobile telephony subscriptions has continued to grow exponentially in many countries, while the increase in fixed mainline subscriptions has been significantly impeded. In 2005, the total number of $FTL_{i, y}$ subscriptions was approximately 1,243 million, and the number of $MCS_{i, y}$ was 2,205 millions, and the values in 2013 were 1,158 and 6,662 million, respectively (Key ICT indicators for developed and developing countries and the world; accessed: June 2014). The latter accounted for global substitution of fixed mainline to mobile telephony and crucially different development patterns with respect to both forms of telecommunication. Importantly, the explosion in mobile cellular telephony subscriptions was predominantly observed in developing countries, where the increase in 2005 of only 1,213 million subscriptions is documented, and in 2012 the number exceeded

5,171 million; in the developed world, it accounted for only a 'slight' jump from 992 in 2005 to 1,490 million in 2013. These tendencies in telecommunication services invite a detailed analysis of the substitutability of fixed and mobile services, as both of them are recognized as important means of voice and data transmission. Factors that contributed to the definite substitution of the 'old-fashioned' telephony (Fixed Telephone Lines—FTL) by the newly emerging means of communication (Hamilton 2003; Albon 2006; Vogelsang 2010; Grzybowski 2011) are numerous. First, in most economically backward countries (Banerjee and Ros 2004; ITU 2011a), people who never owned a fixed-telephone mainline, went straight to cellular technologies, which generated the boom in mobile networks, which started to develop and diffuse 'in place' of traditional fixed mainlines (Banerjee and Ros 2004), rather than alongside them. In the developing world, fixed and mobile services are not complementary, as is usually the case in high-income countries, but are rather purely substitutable. The relative ease of the countrywide deployment of the necessary infrastructure (compared to fixed infrastructure) for mobile services makes the mobile alternative highly attractive, and mobile phones are deployed instead of fixed ones (for broader discussion—see Sect. 4.4).

Moreover, in many low- and lower-middle-income countries, much effort has been made to foster access to the Internet, facilitated by both fixed (narrowband and/or broadband) and wireless connections. The analogue or primitive digital data transmission channels (i.e., ISDN)[47] have been gradually substituted by the fibre-based networks offering high-speed connections. It is broadly agreed that the fibre-optic networks are likely a backhaul network solution; however, in many economically backward economies, the geographical conditions heavily impede country-wide diffusion of such technologies. Hence, the fibre-optic-based connections should be intensively complimented by the advantageous wireless solutions, especially as the latter offer similar performance to wired connections. It is not surprising that in economically backward countries, the strong emphasis is put on wide deployment of wireless technologies, in the hope that the adoption of wireless solutions will provide direct access to the Internet and have the potential to replicate the successful diffusion of mobile cellular telephony. To move towards universal access to the international Internet connections, the question on backbone infrastructure availability arises. Many developing countries experience permanent shortages in the deployment of basic infrastructure, which preconditions widespread development of cable-based Internet connectivity. Thus, the 'unlimited' access to the Internet connections in low- and lower-middle-income countries is still perceived as a 'luxury good', not accessible and affordable for the wide audience. Thus, it prompts the question of how to accelerate the growth of high-speed Internet penetration. The special features of wireless broadband technologies and the uniquely broad spectrum of opportunities that are offered are worth mentioning, mainly because they help to overcome important infrastructural

---

[47] Mainly fixed.

barriers that heavily impede the growth of access to and use of the Internet in economically backward countries. If no fixed infrastructure is available, to enhance growth in the Internet network coverage and usage, assuring the ubiquitous access to wireless broadband networks is considered an important target to be achieved (Warren 2007). As indicated in Sect. 4.4, over the period 2007–2012, in a few low- and lower-middle-income countries, the fast expansion of wireless broadband networks was observed, whereas the fixed narrowband and fixed broadband penetration rates remained at extremely low levels. This suggests that wireless broadband technologies are an attractive alternative, becoming the main access channel to the Internet in the place of wired networks and offering connectivity to those previously unconnected (Gunasekaran and Harmantzis 2007; ITU 2011a). Wireless networks may be implemented faster, requiring less financial recourse compared to fixed-network deployment (i.e., installation of fibre or copper cables) (Gunasekaran and Harmantzis 2007), are more flexible and easily adjustable for local conditions and requirements. The advantages of shifting from fixed narrowband to wireless broadband is easily identifiable in remote, isolated and rural regions, as they are better suited for rough physical conditions and cost-effective solutions for low-income people (Gagnaire 1997; Galperin 2005; Proenza 2006; Warren 2007). Pentland et al. (2004) argue that wireless solutions are to be the first vital means of communication in undeserved and underdeveloped areas. In addition, similar to what was observed in the case of fixed-to-mobile substitution, in developing countries, people go straight to wireless broadband networks, rather than using both simultaneously or gradually switching from fixed to wireless technologies.

Adopting the theoretical framework explained in Chap. 3, the following section presents the empirical evidence on the dynamics and degree of fixed-to-mobile substitution[48] and fixed-to-wireless Internet connection substitution, which took place in low-income and lower-middle-income countries over the period 2000–2012 (as previously explained—in case of availability of longer time-series the analysis is extended). Figure 4.16 visualizes the technological substitution effects (both for fixed-to-mobile telephony and fixed-to-wireless Internet connections) encountered in low-income economies, while Tables 4.11 and 4.12 summarize the results of the estimated technological substitution models and changing relative market shares of competing technologies, respectively. The process of switching from fixed to mobile telephones and from fixed-narrowband Internet-to-wireless broadband connections can be easily traced in each of the 17 analysed countries (the exceptions are Eritrea and Nepal, where the data on wireless-broadband was available exclusively for one year; thus, the substitution is not reported). The process of fixed-to-mobile substitution is gradual, as 'prey' ($FTL_{i, y}$) and 'predator' ($MCS_{i, y}$) technologies fight to take the market over, and each technology passes three distinct phase of technological substitution—logistic growth, nonlogistic saturation and logistic decline. In each consecutive phase, the market share possessed by $FTL_{i, y}$ and $MCS_{i, y}$ or $FIS_{i, y}$ and $WBS_{i, y}$ technologies is different and

---

[48] The acronym for fixed-to-mobile substitution is 'FMS' (see, e.g., Grzybowski 2011).

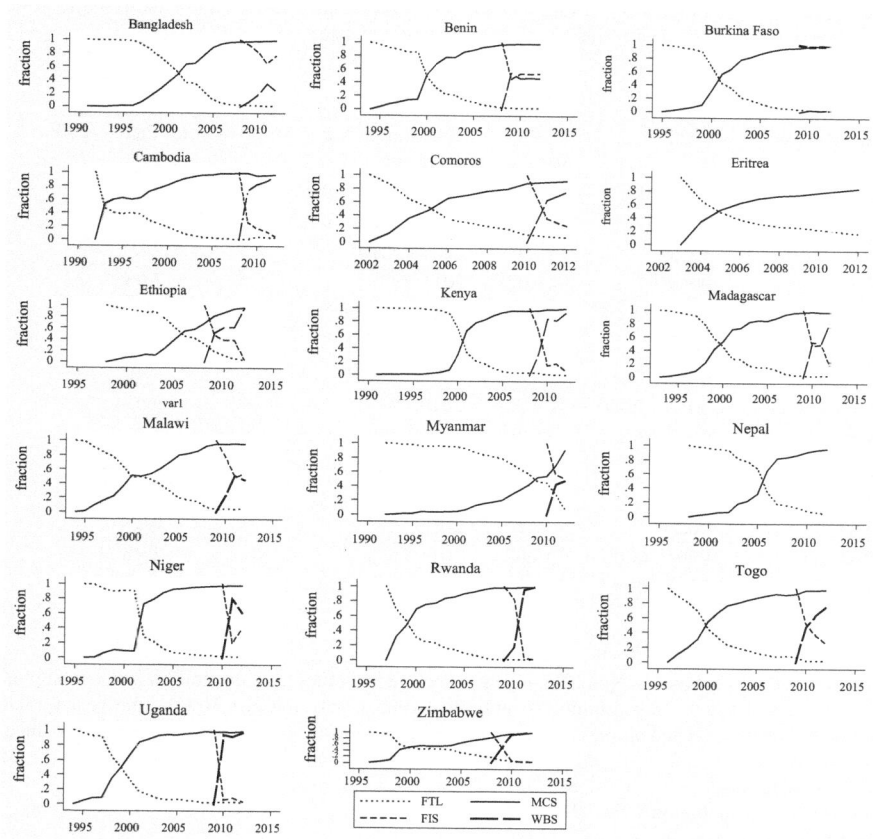

Fig. 4.16 Technological substitution: Fixed-to-mobile telephony and fixed-to-wireless Internet connections substitution patterns. 17 low-income economies. Note: in Nepal and Eritrea—WBS first available in 2012 → substitution not calculable (*Source*: Author's elaboration)

determined by the tempo of substitution. The rise of the $MCS_{i,\,y}$ and $WBS_{i,\,y}$ technologies are echoed by the falls in the $FTL_{i,\,y}$ and $FIS_{i,\,y}$ technologies, respectively. For the reasons explained previously, the process of invading the telecommunication market by mobile telephony was relatively easy and fast progressing. Through the year 2012, in each of the analysed low-income countries, the fixed mainline penetration rates achieved on average 1.43 persons per 100 inhabitants.[49] This indicates that since fixed telephony was first introduced in the group of low-income countries in 1960, the development trend has remained flat, and no significant periods of growth are reported. It supports the hypothesis that

---

[49] Author's estimates.

Table 4.11 Technological substitution estimates: Fixed-to-mobile telephony and fixed-to-wireless Internet connections substitution. 17 low-income countries

	FTL vs. MCS		FIS vs. WBS	
	T_m	dT	T_m	dT
Bangladesh	$2002.2^{T/E}$	7.44	Substitution not reported $\rightarrow$ since 2011—paths diverging	
Benin	$2001^T/2000^E$	9.69	Substitution not definite	
Burkina Faso	$2001^E/2002.2^T$	8.53	Substitution not reported	
Cambodia	$1993.0^{T/E}$	18.3	$2009^E/2010^T$	7.8
Comoros	$2005.6^{T/E}$	10.04	$2011^{T/E}$	7.95
Eritrea	$2004.8^T/2005^E$	17.6^a	Not calculable^b	
Ethiopia	$2006^{T/E}$	8.77	$2009.3^{T/E}$	5.06
Kenya	$2001.9^{T/E}$	7.57	$2009.1^T/2010^E$	4.49
Madagascar	$2001.2^T/2000^E$	9.72	$2010.2^{T/E}$	5.51
Malawi	$2000^E/2001.9^T$	11.49	Substitution not definite	
Myanmar	$2008.3^{T/E}$	13.96	Substitution not definite	
Nepal	$2006^{T/E}$	7.69	Not calculable^b	
Niger	$2002.6^{T/E}$	7.55	$2011^{T/E}$	0.3^c
Rwanda	$1999^T/2000^E$	10.83	$2011^{T/E}$	1.43
Togo	$2000^E/2005.5^T$	11.83^a	$2010^{T/E}$	7.62
Uganda	$2000^{T/E}$	10.66	$2000^{T/E}$	15.1
Zimbabwe	$2000^E/2002.6^T$	11.5^a	$2009^T/2010^E$	3.34

Source: Author's estimates. Note: Tm—year when the technological substitution is half-complete (^{T}—theoretical year of substitution; ^{E}—empirical year of substitution). If only one year for Tm reported—theoretical and empirical values are equal, or they differ less than 6 months. dT—'take-over' time
^{a}Misspecification
^{b}WBS value available only for 2012
^{c}Possible underestimates

due to multiple constraints impeding broad deployment of fixed telephony in economically backward countries, their inhabitants go straight to mobile cellular telephony instead of fixed.

Table 4.11 defines T_m as the period when the technological substitution process is half-complete; while dT indicates the 'take-over' time that designates the number of years necessary for the invading ('prey') technology to gain market share from 10 to 90 %. As the fixed-to-mobile substitution has been visually traced (see Fig. 4.16) in each of country in the low-income group, the parameters of the technological substitution model are fully reported. The average 'take-over' time is approximately $dT_{average} = 10.7^{50}$ years, which suggests that in the low-income countries it required nearly 11 years for the mobile telephony (as 'prey' technology) to achieve a 90 % share of the total telecommunication market. The process of

---

[50] Author's estimates. Parameters statistically insignificant—excluded from the average.

Table 4.12 Changing market shares: $FTL_{i,y}$ versus $MCS_{i,y}$ and $FIS_{i,y}$ versus $WBS_{i,y}$, 17 low-income countries

| | FTL versus MCS | | | | | | FIS versus WBS | | | |
| | 1992[a] | | 2000 | | 2012 | | 2009[b] | | 2012 | |
	FTL share	MCS share	FTL share	MCS share	FTL share	MCS share	FIS share	WBS share	FIS share	WBS share
Bangladesh	99.8	0.2	63.7	36.3	0.97	99.3	92.9	7.1	75.9	24.1
Benin	96.4 (1994)	3.6	48.2	51.8	1.8	98.2	45.3	54.7	53.4	46.6
Burkina Faso	98.4 (1996)	1.6	67.8	32.2	1.4	98.6	97.3	2.7	97.9	2.1
Cambodia	46.7 (1993)	53.3	19.2	80.8	2.9	97.1	26.03	73.97	5.7	94.3
Comoros	–	–	86.8 (2003)	13.2	7.8	92.2	36.5 (2011)	63.4	24.9	75.1
Eritrea	–	–	66.3 (2004)	33.7	16.4	83.6	–	–	94.1	5.9
Ethiopia	96.6 (1999)	3.4	92.8	7.2	3.7	96.3	48.2	51.8	4.8	95.2
Kenya	99.5	0.5	69.6	30.4	0.8	99.2	66.9	33.1	6.7	93.3
Madagascar	99.1 (1994)	0.9	46.6	53.4	2.7	97.3	47.5 (2010)	52.5	15.5	84.5
Malawi	98.8 (1995)	1.2	48.6	51.4	4.6	95.4	77.8 (2010)	22.2	54.6	45.4
Myanmar	99.4 (1993)	0.6	95.3	4.7	9.3	90.7	56.7 (2011)	43.3	52.3	47.7
Nepal	97.8 (1999)	2.2	96.3	3.7	4.8	95.2	–	–	3.1	96.9
Niger	99.4 (1997)	0.6	90.7	9.3	1.8	98.2	20.1 (2011)	79.9	39.3	60.7
Rwanda	68.4 (1998)	31.6	31.1	68.9	0.8	99.2	82.7 (2010)	17.3	1.1	98.9
Togo	89.4 (1997)	10.6	46.1	53.9	1.8	98.2	52.2 (2010)	47.7	25.7	74.3
Uganda	95.1 (1995)	4.9	32.7	67.3	1.9	98.1	5.9 (2010)	94.1	3.4	96.6
Zimbabwe	97.3 (1997)	2.7	48.3	51.7	2.3	97.7	51.9	48.1	1.8	98.2

Source: Author's calculations. Note: FTL share—share of FTL subscriptions in total number of both fixed and mobile phones subscriptions. MCS share—share of MCS subscriptions in total number of both fixed and mobile phones subscriptions. FIS share—share of FIS subscriptions in total number of both fixed Internet and wireless broadband subscribers. WBS share—shares of WBS subscriptions in total number of both fixed Internet and wireless broadband subscriptions

[a]Or year when data on $MCS_{i,y}$ was first available
[b]Or year when data on $WBS_{i,y}$ was first available

fixed-to-mobile telephony substitution proved to be the fastest in Bangladesh, Kenya, Nepal and Niger, where mobile telephony achieved a 90 % share of the market in only 7.5 years. The lack of favourable conditions for the development of fixed telephony infrastructure fostered the deployment of mobile solutions, as people are likely to deploy mobile cellular telephony as the only existing means of communication. It might also suggest that the extremely low fixed telephony penetration rates do not have a direct impact on economic underdevelopment or permanent lack of financial resources but are rather a negative indication of different noneconomic factors that hinder deployment of telecommunication infrastructure. The values of the T_m parameter vary significantly in the discussed country set. In Cambodia, the fixed-to-mobile telephony substitution was half-complete in 1993, which coincides with the year when mobile telephony was first introduced to the national telecommunication market. In the initial year of coexistence of fixed and mobile telephony in the Cambodian market, the total number of mobile cellular subscriptions already surpassed the total number of fixed mainlines subscriptions. However, afterwards, the dynamics of taking over the market by mobile telephony were relatively low compared to other countries in the group, which finally resulted in one of the highest dT, equalling 18.3 years. In the remaining countries, the T_m ranges from $T_m = 2000$ in Benin, Madagascar, Rwanda, and Uganda; to $T_m = 2008.3$ in Myanmar. Such disparities prove that the patterns of the fixed-to-mobile substitution process in the analysed countries are distinct in each case, although the mechanisms driving the technological substitution are analogous.

To a point, the value of T_m is predetermined by the year of the mobile telephony introduction to the market, as this is an initial period when the technological substitution process sets off. Second, the T_m is heavily subjected to the time when the invading technology takes off, entering the exponential growth phase. Thus, the speed of diffusion of the 'predator' technology is decisive in this case, as it determines the tempo of the logistic decline of the 'pray' technology. The higher the rate of diffusion of the 'predator' technology, the higher is the rate of decline of the 'prey' technology, and the T_m occurs earlier. In addition to these 'technical' conditions, the tempo of the diffusion of invading technologies is attributed to a wide array of factors, e.g., market regulations, risk aversion to acquiring innovations and price affordability, which are either identified or captured in the technological substitution modelling. However, all 'unobservable' factors may indirectly stimulate or impede the dynamics of process-shaping technological substitution patterns.

Calculated changing shares of the telecommunication market help to understand the dynamics and logic standing behind the technological substitution process. Table 4.12 explains the changing market shares that are simultaneously possessed by the 'prey' ($FTL_{i, y}$) and the 'predator' ($MCS_{i, y}$) technologies. A brief analysis of the calculations presented in Table 4.12 suggests that in each of the 17 low-income countries in 2012, the telecommunication market was totally dominated by mobile cellular telephony.

Compared to the process of shifts from fixed to mobile telephony, the examined technological substitution of fixed (narrowband) Internet-to-wireless broadband solutions are revealed as more dynamic. In case of the 'predator' technology ($WBS_{i,\,y}$), the reported logistic growth rates were extremely high, which resulted in the rapid fading away of the 'prey' technology ($FIS_{i,\,y}$). The average period of the market 'take over', denoted as dT, was approximately 5.8[51] years (two times lower than in the case of the fixed-to-mobile substitution process). The dT = 5.8 proves that it takes approximately 5.8 years for the $WBS_{i,\,y}$ technology to gain a fraction of the market at approximately 90 %. The shortest 'take over' time is reported for Rwanda—dT = 1.43 years; conversely, in Uganda the dT is 15.1 years, which makes the country an outlier in this respect. The phenomenal growth of the number of wireless broadband subscriptions resulted in a total substitution of fixed narrowband Internet connections, as in a very short time the wireless broadband technologies surpassed—in terms of number of subscribers—the 'old' fixed and narrowband solutions. In seven low-income countries, the fixed (narrowband) Internet-to-wireless broadband solution technological substitution was reported over the analysed period. In Burkina Faso, the wireless broadband technologies were introduced to the market in 2009, but in 2012, the share of the market this potentially invading technology possessed was only 2.1 % (see Table 4.12), which indicates extremely poor progress in the diffusion of wireless solutions. In Bangladesh, Benin, Malawi and Myanmar, the substitution is not definite, as the 'predator' technology still is not dominant in the market (see Table 4.12); however, the respective shares are gradually increasing. In two countries, namely, Eritrea and Nepal, the parameters of the technological substitution model are not reported, as the data on WBS have been available since 2012. Figure 4.17 indicates the technological substitution effects (both for fixed-to-mobile telephony and fixed-to-wireless Internet connections) encountered in lower-middle-income economies, while Tables 4.13 and 4.14 summarize the results of the estimated technological substitution models and the changing relative market shares of the competing technologies, respectively. As presented in Fig. 4.17, the country-specific technological substitution patterns demonstrate that all 29 lower-middle-income countries are fast moving towards high mobile telephony and wireless-broadband penetration rates.

The fixed-to-mobile substitution is reported in each case, while in some countries, the process of shifting from the 'old' technology to the 'new' technology is more dynamic and the total technological replacement occurs relatively earlier. As summarized in Table 4.13, the estimated parameters describing the technological substitution indicate significant variability in patterns of gradual replacement of the 'prey' ($FTL_{i,\,y}$ or $FIS_{i,\,y}$) by the 'predator' technology (MCS or WBS) that occurred in lower-middle-income countries over the period 1984–2012.

Surprisingly, the analysis of the fixed-to-mobile substitution process, the average 'take over' time, denoted by dT, equalled 10.9 years; hence, the dT in the low-income and lower-middle-income group hardly differs. This is an important and conclusive finding, suggesting that the nature of the mobile cellular

---

[51] Author's estimates. Parameters statistically insignificant—excluded from the average.

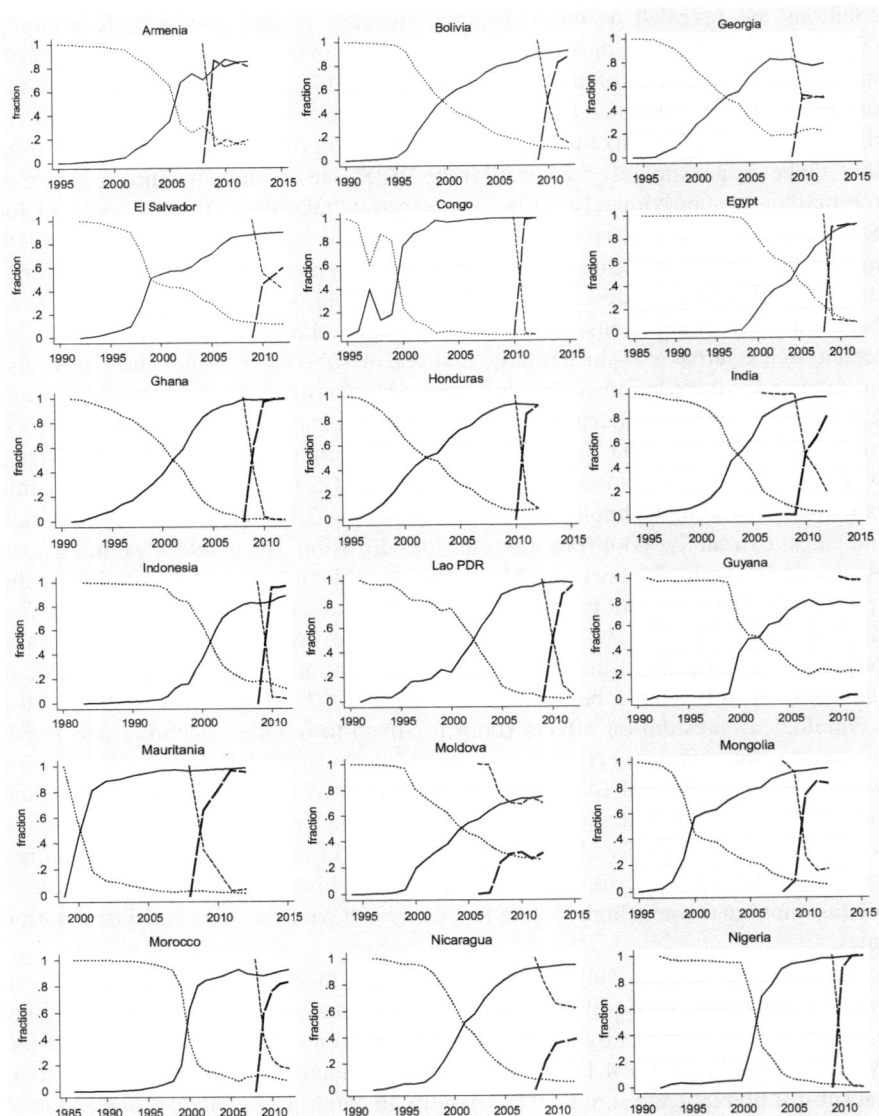

Fig. 4.17 (continued)

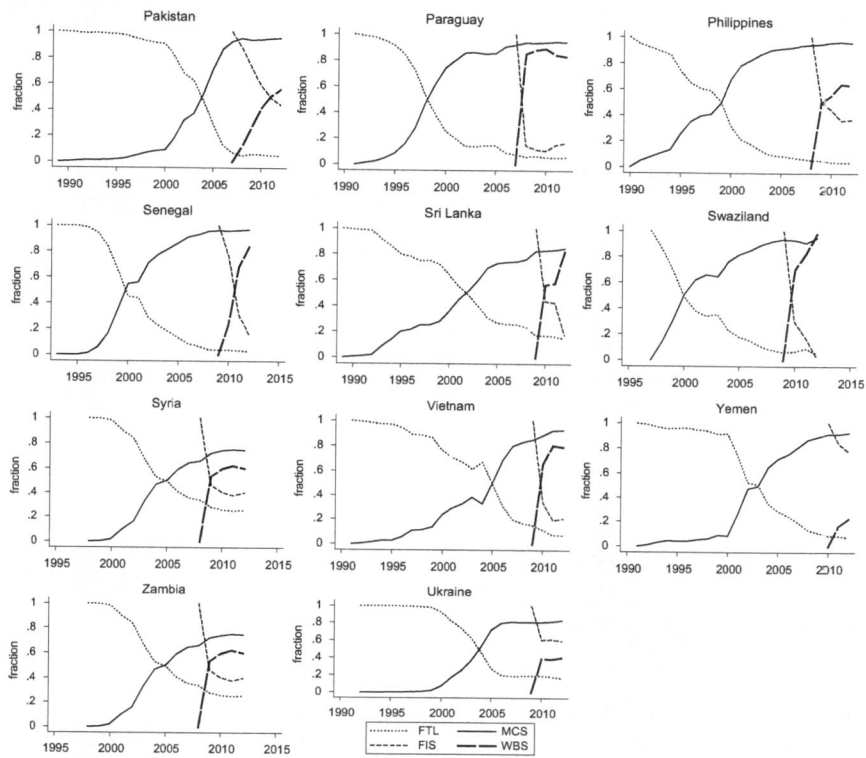

Fig. 4.17   Technological substitution: Fixed-to-mobile telephony and fixed-to-wireless Internet connections substitution patterns (*Source*: Author's elaboration)

technologies is so invading that a total market takeover is possible within approximately 10 years, regardless of the country and its unique characteristics. In addition, the latter may again support the supposition that, in developing countries, the mobile cellular solutions are perceived as an extremely attractive alternative to fixed telephony; thus, the spread of mobile phones is continuous and undisrupted by any external factor.

As indicated in Table 4.14, in 2012, in the majority of lower-middle-income countries, the national telecommunication markets were strongly dominated by mobile cellular telephony. It is easily recognized that in Congo, Ghana, India, Lao P.D.R., Mauritania, Nigeria, the Philippines, Senegal and Zambia, mobile cellular subscribers accounted for approximately 100 % of total telephony subscribers. The overwhelming diffusion of mobile telephony in the above-mentioned countries reconfirms the claim that in economies with poorly developed 'fixed' infrastructure and in which, in 2012, the $FTL_{i, y}$ penetration rates remained at extremely low levels, mobile telephony, as the exclusive communication

Table 4.13 Technological substitution estimates: Fixed-to-mobile telephony and fixed-to-wireless Internet connections substitution. 29 lower-middle-income economies

	FTL vs. MCS		FIS vs. WBS	
	T_m	dT	T_m	dT
Armenia	$2006.5^{T/E}$	8.2	$2009^{T/E}$	8.9
Bolivia	$2000^{E}/2002.6^{T}$	11.3	$2010^{T/E}$	5.03
Congo	$2000^{T/E}$	8.2	Not calculable—paths diverging[a]	
Egypt	$2005.2^{T/E}$	10.6	Not calculable—paths diverging[a]	
El Salvador	$2000^{E}/2002^{T}$	13.0	$2011^{T/E}$	15.8[b]
Georgia	$2003^{E}/2005^{T}$	10.2	Not applicable—shares close to equal	
Ghana	$2000.6^{T}/2002^{E}$	9.6	$2008^{T}/2009^{E}$	3.06
Guyana	$2003^{E}/2004.7^{T}$	13.3	Substitution not reported	
Honduras	$2002^{E}/2003.2^{T}$	11.3	Not calculable—paths diverging[a]	
India	$2004^{T/E}$	8.8	$2010^{T/E}$	3.09
Indonesia	$2002.8^{T/E}$	14.2	$2008^{T}/2009^{E}$	4.8
Lao P.D.R.	$2001.9^{T}/2003^{E}$	10.9	$2010^{T}/2011^{E}$	3.02
Mauritania	$1995^{T}/2001^{E}$	16.1[c]	$2008^{T}/2009^{E}$	4.8
Moldova	$2005^{E}/2006.5^{T}$	8.2	Substitution not reported	
Mongolia	$2000^{E}/2002.7^{T}$	10.7	$2010^{T/E}$	3.4
Morocco	$2000^{E}/2002.5^{T}$	9.4	$2007^{T}/2009^{E}$	11.4[b]
Nicaragua	$2001^{E}/2002.4^{T}$	10.4	Substitution not reported	
Nigeria	$2002^{T/E}$	7.8	$2010^{T/E}$	1.6
Pakistan	$2004^{T/E}$	9.9	$2011^{T/E}$	7.9
Paraguay	$1999^{E}/2000^{T}$	12.4	Not calculable—paths diverging[a]	
Philippines	$1998.7^{T/E}$	14.2	$2008^{T}/2010^{E}$	21.8[b]
Senegal	$2000^{E}/2002.5^{T}$	8.2	$2011^{T/E}$	3.2
Sri Lanka	$2002.8^{T/E}$	16.1	$2010^{T/E}$	6.8
Swaziland	$2000.7^{T/E}$	14.9	$2010^{T/E}$	3.04
Syria	$2007^{T/E}$	9.7	Not applicable—shares close to equal	
Ukraine	$2004^{E}/2005.7^{T}$	7.3	Substitution not reported	
Viet Nam	$2005^{T/E}$	11.8	$2007^{T/E}$	12.6[b]
Yemen	$2004^{T/E}$	11.8	Substitution not reported	
Zambia	$2001^{T/E}$	8.4	$2009^{T/E}$	8.9

Source: Author's estimates. Note: Tm—year when the technological substitution is half-complete ([T]—theoretical year of substitution; [E]—empirical year of substitution). If only one year for Tm reported—theoretical and empirical values are equal, or they differ less than 6 months. dT—'take over' time

[a]Since $WBS_{i, y}$ was introduced (in y-year)—$FIS_{i, y}$ and $WBS_{i, y}$ diffusion paths diverge $\rightarrow WBS_{i, y} > FIS_{i, y}$

[b]Possible overestimates

[c]Possible misspecification

Table 4.14 Changing market shares: $FTL_{i,y}$ versus $MCS_{i,y}$ and $FIS_{i,y}$ versus $WBS_{i,y}$, 29 lower-middle-income countries.

| | FTL vs. MCS | | | | | | FIS vs. WBS | | | |
| | Y_{MCS} | | 2000 | | 2012 | | Y_{WBS} | | 2012 | |
	FTL share	MCS share	FTL share	MCS share	FTL share	MCS share	FIS share	WBS share	FIS share	WBS share
Armenia	0.99 (1996)	0.01	0.96	0.04	0.15	0.85	0.13 (2009)	0.87	0.18	0.82
Bolivia	0.99 (1991)	0.01	0.46	0.54	0.08	0.92	0.47 (2010)	0.53	0.13	0.87
Congo	0.95 (1996)	0.05	0.23	0.77	0.01	0.99	0.02 (2010)	0.98	0.01	0.99
Egypt	0.99 (1987)	0.01	0.8	0.2	0.08	0.92	0.1 (2009)	0.90	0.09	0.91
El Salvador	0.99 (1993)	0.01	0.45	0.55	0.1	0.9	0.54 (2010)	0.46	0.41	0.59
Georgia	0.99 (1995)	0.01	0.72	0.28	0.21	0.79	0.48 (2010)	0.52	0.50	0.50
Ghana	0.99 (1992)	0.01	0.62	0.38	0.01	0.99	0.39 (2009)	0.61	0.01	0.99
Guyana	0.97 (1992)	0.03	0.63	0.37	0.21	0.79	0.97 (2011)	0.03	0.97	0.03
Honduras	0.98 (1996)	0.02	0.65	0.35	0.07	0.93	0.14 (2011)	0.86	0.08	0.92
India	0.99 (1995)	0.01	0.9	0.1	0.03	0.97	0.99 (2007)	0.01	0.19	0.81
Indonesia	0.99 (1984)	0.01	0.64	0.36	0.12	0.88	0.52 (2009)	0.48	0.03	0.97
Lao P.D.R.	0.96 (1996)	0.04	0.76	0.24	0.03	0.97	0.5 (2010)	0.5	0.05	0.95
Mauritania	–	–	0.55	0.45	0.01	0.99	0.34 (2009)	0.66	0.04	0.96
Moldova	0.99 (1995)	0.01	0.8	0.2	0.25	0.75	0.99 (2007)	0.01	0.69	0.31
Mongolia	0.98 (1996)	0.02	0.43	0.57	0.05	0.95	0.92 (2009)	0.08	0.17	0.83
Morocco	0.99 (1987)	0.01	0.37	0.63	0.07	0.93	0.4 (2009)	0.6	0.17	0.83
Nicaragua	0.99 (1993)	0.01	0.64	0.36	0.05	0.95	0.78 (2009)	0.22	0.62	0.38
Nigeria	0.97 (1993)	0.03	0.94	0.06	0.01	0.99	0.09 (2010)	0.91	0.01	0.99
Pakistan	0.99 (1997)	0.01	0.9	0.1	0.04	0.96	0.88 (2008)	0.12	0.44	0.56
Paraguay	0.98 (1992)	0.02	0.25	0.75	0.05	0.95	0.14 (2008)	0.86	0.16	0.84
Philippines	0.94 (1991)	0.06	0.32	0.68	0.03	0.97	0.5 (2009)	0.5	0.36	0.64
Senegal	0.99 (1994)	0.01	0.45	0.55	0.02	0.98	0.76 (2010)	0.24	0.16	0.84

(continued)

Table 4.14 (continued)

| | FTL vs. MCS | | | | | | FIS vs. WBS | | | |
| | Y_MCS | | 2000 | | 2012 | | Y_WBS | | 2012 | |
	FTL share	MCS share	FTL share	MCS share	FTL share	MCS share	FIS share	WBS share	FIS share	WBS share
Sri Lanka	0.99 (1990)	0.01	0.64	0.36	0.15	0.85	0.43 (2010)	0.57	0.17	0.83
Swaziland	0.86 (1998)	0.14	0.49	0.51	0.05	0.95	0.29 (2010)	0.71	0.02	0.98
Syria	0.99 (1999)	0.01	0.98	0.02	0.25	0.75	0.46 (2009)	0.54	0.4	0.6
Ukraine	0.99 (1993)	0.01	0.92	0.08	0.17	0.83	0.6 (2010)	0.4	0.59	0.41
Viet Nam	0.99 (1992)	0.01	0.76	0.23	0.07	0.93	0.34 (2010)	0.66	0.2	0.8
Yemen	0.98 (1992)	0.02	0.91	0.09	0.07	0.93	0.83 (2011)	0.17	0.77	0.23
Zambia	0.98 (1995)	0.02	0.45	0.55	0.01	0.99	0.49 (2009)	0.51	0.14	0.86

Note: Y_{MCS}—year when MCS was first introduced to the market; Y_{WBS}—year when $WBS_{i,\,y}$ was first introduced to the market

channel, is rapidly adopted by inhabitants, regardless of their material status. Conversely, in countries where the infrastructure allowing for fixed mainline installation is better developed, the pressure for rapid deployment of mobile cellular phones is weaker.

Over the period 2007–2012, in lower-middle-income economies, the fixed Internet-to-wireless broadband connection substitution process has been reported in only 18 countries (for visual inspection, see Fig. 4.17). In the remaining 11 countries, the respective technological substitution process was not reported for the cases of Guyana, Moldova, Nicaragua, Ukraine and Yemen; not calculable due to diverging $FIS_{i, y}$ and $WBS_{i, y}$ diffusion trajectories for Congo, Egypt, Honduras and Paraguay; and not applicable, as the shares of the market controlled by each 'technology' were approximately equal (substitution was not definite) in the cases of Georgia and Syria. In the group of 18 countries where the fixed-to-wireless broadband Internet connection substitution has been determined, the average 'take over' time (dT) was approximately 4.8 years,[52] i.e., more than two times shorter than the time reported for fixed-to-mobile substitution. The fastest substitution was demonstrated in Nigeria, where in barely 1.6 years (*sic!*), wireless broadband solutions took over 90 % of the respective market. For the sake of comparison, in Ghana, Lao P.D.R. and Swaziland, in approximately 3 years, 90 % of the market was controlled by wireless broadband providers. However, the fact that in 2012, the share of wireless broadband subscriptions (as the share of total number of fixed narrowband Internet subscriptions and wireless broadband subscriptions) was still below 90 % (to compare, see Table 4.14) is somewhat confusing. However, the returned dT parameters rather describe the *prospected* number of years, which are necessary for the 'predator' technology ($WBS_{i, y}$) to conquer the market.

It has been demonstrated that over the period 2007–2012, the wireless-broadband technologies have pervasively expanded in the national markets offering access to Internet. The process of switching from fixed to wireless Internet has been fast and explosive, especially in countries where the fixed (narrowband) Internet penetration rates in 2012 remained negligible. It suggests that, analogous to what was observed in the case of the fixed-to-mobile substitution process, significant shortages in fixed infrastructure foster the rapid spread of alternative solutions, in this case, wireless, which may be easily installed and adopted for use in unfavourable environments. The latter additionally reveals the unique effect, already identified in the case of fixed-to-mobile substitution that people go straight to wireless instead of fixed technologies if only the latter are offered to the market.

---

[52] Author's estimates. Parameters statistically insignificant/overestimates—excluded from the average.

4.5.1 Final Remarks

Our empirical evidence has demonstrated that both in low-income and lower-middle-income countries, the phenomenal process of rapid growth mobile telephony and wireless network (enabling data transmission) penetration rates have been demonstrated. The empirical analysis indicates that the rapid diffusion of mobile phones and Internet wireless networks has resulted in strong fixed-to-mobile telephony and fixed-to-wireless Internet connection substitution effects. In each of the analysed counties, the 'old' (prey) technology, i.e., fixed telephony has been totally replaced by the 'new' (predator) technology, i.e., mobile cellular telephony. Both in low-income and lower-middle-income countries, the average take over time has been estimated as approximately 10 years.[53] The analogous process of technological substitution has been revealed in terms of fixed-to-wireless Internet connections; however, in this case, the dynamics of switching from fixed to wireless connections is twice as high. Importantly, it shall be noted that the nature of the observed technological substitution effects in low-income and lower-middle-income countries warrants special attention. By convention, the technological substitution explains the process whereby 'old' technologies are gradually substituted with the 'new' technologies. Thus, the 'new' are used in place of the 'old' technologies. When the technological substitution effects are discussed with respect to low- and lower-middle-income countries, the *typical* technological substitution process is not observed. Through the year 2012, the state of development of the fixed-based solutions, such as fixed telephony, fixed narrowband and fixed broadband Internet networks, in the countries under consideration remained at extremely low levels. Thus, the technologies based on tangible grids did not diffuse widely over societies. However, once the mobile cellular telephony and wireless broadband networks emerged, the inhabitants in developing countries moved *straight* towards new technologies. The process of theoretical technological substitution, which occurred in low- and lower-middle-income countries, captures the broad idea of technological leapfrogging that accounts for the direct shift to mobile technologies without having previously deployed the wired (both copper- and fibre-based) telecommunication networks. We do not observe the process of quitting the 'fixed' and switching to the 'wireless' but rather the process of omitting the 'fixed stage' and going directly towards the 'wireless stage'. As a result of the previous, people are becoming completely dependent on mobile telephony. The latter accounts for the unique technological leapfrogging effect, revealed when countries do not follow the 'classical' technological development pattern but 'jump' directly from the initial stages of development to the advanced stages. The concept of technological leapfrogging originates from the supposition that heavily underdeveloped countries may *benefit* from relative backwardness and not follow the classical development path, thus omitting certain stages and directly moving towards the advanced stages of development. Steinmueller (2001) argues that

---

[53] Author's estimates; estimates statistically insignificant—excluded from the average.

technological leapfrogging consists of 'bypassing stages in capacity building or investment through which countries were previously required to pass during the process of economic development'. By definition, 'technology leapfrogging' explains the process of the successful adoption of the *superior* (more advanced) technological solutions, for cases in which the *prior* versions have never been adopted by the end users (Sharif 1989; Antonelli 1991; Ausubel 1991; James 2009, 2014). To rephrase the previous, technological leapfrogging consists of the rapid jump over several development stages, moving directly from the 'old' to the 'new' technologies. James (2009) argues that in underdeveloped countries where the fixed infrastructure is poorly developed, the direct transition to newly emerging technological solutions, especially those offering wireless broadband connections, technological leapfrogging may occur. The technologies that allow leapfrogging effects are predominantly those that do not rely on the development of tangible grids (Lee and Lim 2001; Galperin 2005; Kauffman and Techatassanasoontorn 2005); hence, wireless (especially broadband) technological solutions offer economically backward countries sustainable leapfrog-type technological development, which leads to breakthrough and revolutionary changes in countries where basic infrastructure is virtually nonexistent.

The last empirical part of Chap. 4 is designed to unveil, whether both—fast ICT diffusion (resulting in tremendously growing ICT penetration rates) and technological substitution (resulting in switching from the 'old' to 'new' technologies) have enabled the catching-up of developing countries with developed ones, regarding the level of ICT access to and use of. Henceforth, the forthcoming Sect. 4.6 is aimed to trace the technology convergence process and examine whether countries form specific technology convergence clubs. The latter is supposed to provide—at least partial—answer to the question if rapid advancements in ICTs adoption in developing countries have contributed to growing cross-country cohesion with this respect.

4.6 Technology Convergence, Divergence or Club Convergence? The Worldwide Evidence for the Period 2000–2012

The central target of Sect. 4.6 is to verify the hypothesis that impressive increases in ICT deployment in developing countries resulted in the emergence of the technology convergence process worldwide. Put another way, we aim to uncover whether countries that were initially 'technologically peripheral' economies[54] managed to *technologically* catch up with developed countries, so that the existing gaps were, at least partially, diminished. To this aim, we define 'technology convergence' as the processes whereby initially technologically poor countries tend to grow (in terms of average annual growth rates) faster compared to countries that initially were

---

[54] In 2000, as the start year of the technology convergence analysis.

technologically better off. The latter shall inevitably lead to the 'digital (technology) gaps' narrowing and the gradual eradication of different forms of exclusion from the access to and use of ICTs (Lechman 2012a, 2012b).[55] Additionally, we target potentially emerging technology convergence clubs to determine whether all of the analysed countries have been included in the technology convergence process, or conversely, some of them have been left out of the technology convergence club.

The approach to technology convergence analysis that we suggest is not very common, and the empirical evidence in the field remains relatively poor. Some evidence can be gleaned from the works of Comin and Hobijn (2004), Comin et al. (2006), Castellacci (2006, 2008), Castellacci and Archibugi (2008), Castellacci (2011) and Lechman (2012a, 2012b). Comin and Hobijn (2004) provide extensive analysis of technology convergence over the period 1788–2001. Their study covers 20 technologies in 23 different countries, and they test the convergence hypothesis applying beta- and sigma-convergence procedures. Comin et al. (2006) present similar exercises to Comin and Hobijn (2004). They test beta- and sigma-convergence using a CHAT (Cross-Country Historical Adoption of Technology) dataset, additionally separating the within-technologies and across-technologies effects. In addition to technology convergence testing, Castellacci (2006, 2008), Castellacci and Archibugi (2008) detect technology convergence clubs. Castellacci (2008) reports on technology convergence and technology convergence clubs for 149 world countries over the period 1990–2000. He additionally tests for 'technological capabilities' that may enhance or hinder the process of closing the cross-country technology gaps. Additional evidence on the process of closing technology gaps is also reported by Castellacci (2011). Castellacci and Archibugi (2008), using data from the ArCo database (Archibugi and Coco 2004a, 2004b, 2005), provide similar evidence over the analogous time period; however, they include analysis for 131 countries. The empirical analysis found in the works of Lechman (2012a, 2012b), reports on technology convergence exclusively for Information and Communication Technologies for 145 countries over the period 2000–2010. Technology convergence is tested by adopting the beta- sigma-, and quantile-convergence approaches.

4.6.1 Technology Convergence

To meet the major aims of Sect. 4.6.1, we propose the extension of our basic empirical sample (17 low-income and 29 lower-middle-income countries) and include an additional 25 upper-middle-income and 42 high-income economies.[56] Hence, our empirical sample encompasses 113 countries, for which the technology

---

[55] For the conceptual framework, see Chap. 3, Sect. 3.4.

[56] We base these on the World Bank 2013 country classifications (see: http://data.worldbank.org/news/new-country-classifications, accessed: May 2014).

convergence behaviour is explored over the period 2000–2012. For purposes of clarity in the further research provided in the following Sect. 4.6, we suggest labelling the group of low-income and lower-middle-income economies as 'developing countries' and the group of upper-middle-income and 42 high-income economies as 'developed countries'.

We examine the technology convergence process with respect to 4 different ICT indicators, namely: Mobile Cellular Subscriptions ($MCS_{i, y}$), Fixed Internet Subscriptions ($FIS_{i, y}$), Fixed Broadband Subscriptions ($FBS_{i, y}$) and Internet Users ($IU_{i, y}$). To check for the technology convergence with respect to $MCS_{i, y}$, the sample covers 111 countries[57] (Comoros and Eritrea were excluded)[58]; with regard to $FIS_{i, y}$ and $IU_{i, y}$, 113 countries; and with regard to $FBS_{i, y}$, 107 countries (Honduras, Congo, Swaziland, Eritrea, Togo, and Bangladesh were excluded)[59]; the analysis time period is set for 2005–2012.[60] Intentionally, we do not analyse the technology convergence process regarding change in wireless-broadband technologies adoption, as the time span for the analysis would have to be limited to the period 2009–2012, which we claim to be too short to provide reliable evidence on the convergence process. However, to meet the main targets of Sect. 4.6 and report on increasing (or decreasing) cohesion among world countries in terms of $WBS_{i, y}$, we perform an alternative analysis that allows the possibility to detect the emerging worldwide trends in this respect (see the final part of Sect. 4.6).

Table 4.15 displays the overall trends in changes in ICT development over 2000–2012, both in developing and developed countries (see also Appendix E). The basic conclusion from the statistics provided in Table 4.15 is that with respect to each considered technology, the average annual growth dynamics were extraordinarily high, which profoundly reshaped the world landscape with regard to ICT access and use. Thus, with some exceptions, the 'technologically poor' became 'technologically rich', which may be empirically supported by the elementary analysis of the core ICT indicator summary statistics. The most vital changes are reported in terms of mobile cellular telephony penetration rates. While in 2000, the average $MCS_{average,2000}$ is reported as 20.4 per 100 inhabitants, in 2012, it grew to 106.6 (*sic!*), which accounts for approximately 18.6 % of the average annual growth rate. Similar in average annual dynamics are the changes reported for the $IU_{i, y}$ indicators, which reveals a tremendous boom in the accessibility of Internet network. In 2000, the average $IU_{average,2000} = 9.8$ %, while in 2012, it increased to $IU_{average,2012} = 45.6$ %. The latter indicates that in 2012, nearly 50 % of the individuals in the countries within this scope had access to the Internet, regardless of the type of connection (fixed or wireless, narrowband or broadband). In addition, in the cases of both $MCS_{i, y}$ and $IU_{i, y}$, dramatic drops in cross-country inequalities are reported (see the changes in Gini coefficients in Table 4.15 and Figs. D.1 and

---

[57] All countries are listed in Appendix C.

[58] Significant lacks in data time series.

[59] Significant lacks in data time series.

[60] For Madagascar and Nepal, the data are available beginning with the year 2006.

Table 4.15 Core ICT indicators summary statistics

ICT indicator	# of observations	Mean	Std. Dev.	Min. val. (observed)	Max. val. (observed)	Gini coefficient[a]	Average annual growth rate (%)
MCS(2000)	111	20.4	24.6	0.018	76.4	0.61	18.6
MCS(2012)	111	106.6	35.3	10.3	187.4	0.18	
FIS(2000)	113	4.3	7.3	0.001	37.2	0.73	15.3[b]
FIS(2012)	113	12.2	12.2	0.015	40.2	0.54	
FBS(2005)	107	4.9	7.6	0.00008	27.9	0.72	23.3
FBS(2012)	107	12.4	12.1	0.007	39.8	0.53	
IU(2000)	113	9.8	14.4	0.00019	52	0.68	18.9
IU(2012)	113	45.6	28.7	0.8	96.2	0.36	
WBS(2009)	72	14.1	21.04	0.004	88.5	0.69	38.07[c]
WBS(2012)	111	30.1	30.5	0.003	126.06	0.54	

113 countries

Source: Author's calculations

[a]See Appendix D to see Gini coefficients time trends and Lorenz curves for respective ICT indicators

[b]Arithmetic average—negative growth rates reported in some cases

[c]Period 2010–2012; estimates for 102 countries

D.2 in Appendix D for the Lorenz curves), which might suggest growing cohesion among the analysed countries in this respect. The growth in both fixed-narrowband ($FIS_{i, y}$) and fixed-broadband ($FBS_{i, y}$) network penetration rates was comparable, and in 2012, the respective averages achieved $FIS_{average,2012} = 12.2$ and $FBS_{average,2012} = 12.4$. The reported average annual growth rates of $FIS_{i, y}$ and $FBS_{i, y}$ are 15.3 % and 23.3 %, respectively. However, the most disruptive changes in the access to the Internet were brought about by the introduction of wireless-broadband connections (since 2009[61]), which invaded the telecommunication markets both in developing and developed countries. The estimated average annual growth of $WBS_{i, y}$ is approximately 38.07 %,[62] which accounts for the best result compared to the remaining four ICT indicators, which might have affected the cross-country disparities in this respect.

Next, addressing the main scope of Sect. 4.6.1, the attention is brought to the empirical analysis of the technology beta(β)-convergence and technology sigma (σ)-convergence. We define the technology beta-convergence (β-convergence) as the process that examines whether initially backward countries reveal higher average annual growth rates in ICT deployment compared to advanced countries, and the technology sigma-convergence (σ-convergence) is defined as the process that reports on diminishing cross-country disparities in terms of the access to and use of ICTs. The technology absolute β-convergence formal specification follows:

$$g_z = a_z + b_z v_{z,y_0} + \varepsilon_z, \qquad (4.2)$$

where z denotes ICT indicators, y_0 is the initial year, v_{z,y_0} is the level of respective ICT indicators in y_0, and ε_z is random error term. The b coefficient[63] in Eq. (4.2) indicates the technology convergence coefficient that, indirectly, explains the speed of convergence. If $b_z < 0$, the technology absolute β-convergence is confirmed, and divergence occurs otherwise. If $b_z = 0$, either convergence or divergence is reported. The σ-convergence, reporting on changing cross-country disparities, may be measured by standard deviation or—if we aim to capture the influence of the mean of the examined population—by the coefficient of variation. Hence, we deploy these two classical measures of dispersion to verify the technology sigma-convergence hypothesis. We assume that the standard deviation (dispersion) for i-country in n-country set and y-year follows:

$$\sigma_{z,y} = \left[\frac{1}{n} \sum_{i=1}^{n} \left(log \left(\frac{a_z}{a^x} \right) \right)^2 \right]^{1/2}, \qquad (4.3)$$

---

[61] The year of introduction of wireless-broadband technologies varies in different countries. In some regions, this type of network was already available in 2007.

[62] Note: the estimate covers only a 3-year period (*sic!*).

[63] Also explaining partial correlation between variable growth rate and its initial level.

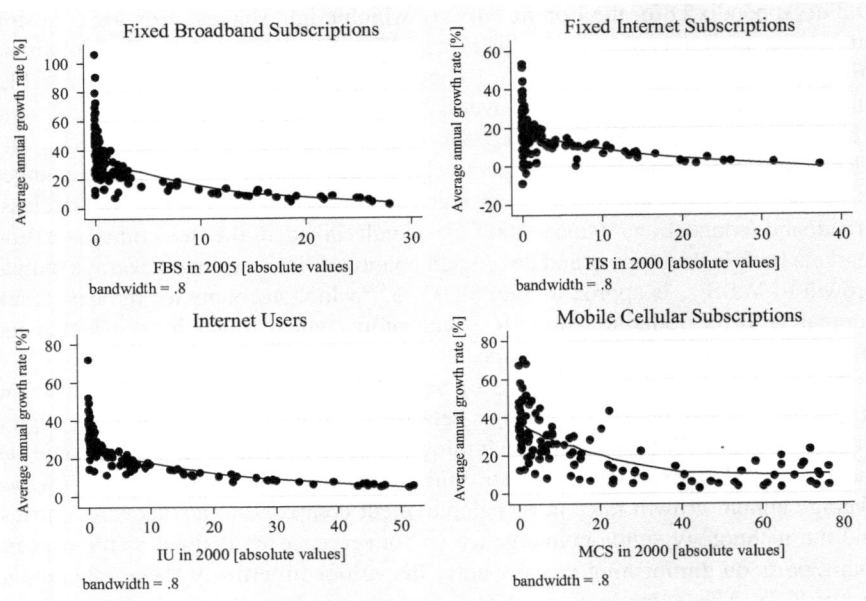

Fig. 4.18 Technology β-convergence. 113 countries. Note: for visualization—nonparametric approximation applied. For FBS—estimates since 2005. For MCS—111 countries (Comoros and Eritrea excluded—significant lacks in data time series). For FBS—107 countries (Bangladesh, Congo, Honduras, Eritrea, Swaziland and Togo excluded—significant lacks in data time series); for Madagascar and Nepal—initial values for 2006 (*Source:* Author's elaboration)

if $a^x \equiv \frac{1}{n} \sum_{z=1}^{n} log(z_i)$ and assuming that z stands for the respective ICT indicator, the technology σ-convergence hypothesis is thus verified positively if $\sigma_{z,t} \rightarrow 0$ is satisfied. If the technology σ-convergence is tested by use of the coefficient of variation, the coefficient of variation may be expressed as $cv_{z,y} = \frac{\sigma_{z,y}}{\theta_{z,y}}$, where $\theta_{z,y}$ is the mean of the tested variable over the whole sample.

Figure 4.18 displays the scatterplot for the average annual growth rates of respective ICTs against the initial level of ICT deployment (in 2000–y_0). To confirm the hypothesis on technology β-convergence, we expect to determine that countries with initially lower levels of ICT development, by taking advantage of their relative technological backwardness (Gerschenkron 1962; Abramovitz 1986), perform better in terms of average annual growth compared to the technologically advanced countries in y_0. The graphical inspection provided in Fig. 4.18 suggests that, in the analysed country sample, the presence of technology β-convergence is unambiguous with respect to all four ICT indicators. Although the tendencies for absolute technology convergence are easily detectable in Fig. 4.18, the relationships for the respective ICT indicators call for closer examination by means of a more formal test. To this end, we adopt Eq. 4.2 and estimate the regression coefficients, assuming that $b_z < 0$ indicates the existence of

Table 4.16 Technology β-convergence. 113 countries

	$MCS_{i, y}$	$FIS_{i, y}$	$FBS_{i, y}$	$IU_{i, y}$
OLS estimates				
b_z coefficient	−7.31	−2.65	−4.91	−4.93
Robust SE	0.21	0.373	0.492	0.239
Ramsey test (prob > F)	1.09 (0.358)	0.82 (0.482)	5.8 (0.0011)	2.53 (0.06)
R-squared	0.96	0.323	0.562	0.83
β_z parameter	17.6 %	10.7 %	25.3 %	14.8 %
HL_{ICT}(# of years)	3.92	6.42	2.73	4.57
# of time periods	12	12	7	12
Robust regression estimates				
b_z coefficient	−7.61	−2.52	−5.17	−5.4
SE	0.11	0.283	0.301	0.136
prob > F	0.000	0.000	0.000	0.000

Source: Author's estimates. Note: all estimates for significance at 95 %. For robust regression—biweight iteration = 7 (set as default). Constant included—not reported. For MCS—111 countries (Comoros and Eritrea excluded—significant lacks in data time series). For FBS—107 countries (Bangladesh, Congo, Honduras, Eritrea, Swaziland and Togo excluded—significant lacks in data time series; for Madagascar and Nepal—initial values of FBS for 2006); estimates for the period—2005–2012

unconditional technology β-convergence over the analysed sample. We run the regression analysis using the OLS estimation. However, bearing in mind that the OLS estimation might be sensitive to the influence of outliers, we additionally adopt the robust regression approach (Härdle 1984; Rousseeuw and Leroy 2005). Table 4.16 summarizes regression analysis outcomes, providing cross-country evidence on unconditional technology β-convergence. The results in Table 4.16 indicate that the hypothesis on technology β-convergence may be positively verified, as in the case of each ICT indicator, the regression coefficients hold a negative sign and are statistically significant at the 95 % level of significance. Estimates generated from the robust regressions are very close in value to those obtained from OLS estimates, which confirms the validity of the previous estimates. The highest b_z coefficient—$b_{MCS} = (−7.31)$ is reported for the technology convergence with respect to mobile cellular telephony (thereafter labelled as MCS-β-convergence). It indicates that the speed of MCS-β-convergence is the highest compared to the remaining three ICTs. Given the value of estimated b_{MCS}, we conclude that the MCS-β-convergence proceeds at the annual rate of 17.6 %, which indicated that the cross-country disparities in mobile telephony penetration rates may be halved within 3.92 years.[64] The MCS-β-convergence

---

[64] The regression coefficients generated from the robust regression are at a very comparable level to those obtained from the OLS estimates. Hence, we do not calculate the speed of convergence and the specific half-times considering the robust regression coefficients, as they would suggest the analogous qualitative conclusions as with those derived from the simple OLS analysis.

results strongly reflect the worldwide tendency for initially technologically backward countries to grow faster than advanced economies; hence, the catching-up process emerges (Baumol 1986), and the technology gaps between 'technology poor' and 'technology rich' are gradually eradicated (James 2009). Similar in terms of strength and statistical significance, technology β-convergence is reported in the case of ICT indicators explaining the use of Internet network—$IU_{i, y}$. The regression coefficient resulted with $b_{IU} = (-4.93)$ and is statistically significant at the 95 %. The speed of IU-β-convergence is approximately 14.8 %, indicating that the cross-country inequalities with respect to $IU_{i, y}$ shall be halved within 4.67 years. The observed countries' technology β-convergence behaviour with regard to the $FIS_{i, y}$ and $FBS_{i, y}$ indicators slightly differs from the what was reported in the case of $MCS_{i, y}$ and $IU_{i, y}$. The estimated regression coefficient for $FIS_{i, y}$ is $b_{FIS} = (-2.65)$, which still confirms the existence of convergence (in this case labeled as FIS-β-convergence); however, the R-squared of the model is only 0.32, which suggests that the model has little explanatory power, and the variability of the explanatory variable—the level of fixed-narrowband network penetration rate in 2000—in barely 32 % of cases explains the countries' average annual growth rates.

As the $\beta_{FIS} = 10.7\%$, the specific half-time accounts for 6.42 years, which yields the longest time period necessary for the cross-country disparities to be halved in this respect, compared to the MCS-β-convergence and IU-β-convergence. Finally, the estimates for the $FBS_{i, y}$ explaining whether cross-country disparities are diminishing with regard to fixed broadband Internet penetration rates, might suggest that the speed of FBS-β-convergence is substantially higher if compared to the reported speed of FIS-β-convergence. Important to note is that the number of time periods considered for the analysis of FBS-β-convergence is only 7 (the years 2005–2012), while in the remaining cases, it is 12 time periods (the years 2000–2012). The latter heavily affects the calculations of the rate of FBS-β-convergence and consequently the number of years required to halve the cross-country disparities in this regard.

The analysis of technology β-convergence provides a general and intuitively interpreted idea of the respective ICT indicator growth behaviour of the average of the distribution but is not very informative with respect to the changes in the distribution over the analysed time period. Hence, we claim that considering the convergence analysis a modified approach might be adequate, especially when the empirical distribution of examined ICT variables are highly skewed (thus, are not normally distributed—compare Appendix E), the OLS estimates may be biased and inefficient. To address this problem, Koenker and Bassett (1978) suggest the adoption of nonparametric quantile regression. The quantile regression approach is highly useful when original variable distribution is highly skewed (asymmetric). Standard β-convergence estimates allow for variable behaviour assessment but based on the conditional mean, while the quantile regression (q-regression) introduced estimates in noncentral locations (Koenker 2000, 2004, 2005; Hao and Naiman 2007; Laurini 2007). Using the quantile regression approach, it is possible to determine any number of quantiles for estimations, which allows the modelling of variable behaviours in any pre-defined location of variable distribution (including the tails of the distribution), and surpass the regression to the mean problem (Koenker 2000). Hence, with respect to the examined ICT variables, we formulate the following regression to report on the technology quantile (Q)-convergence (Castellacci 2006; Lechman 2012a, 2012b):

$$g_{z,j} = a_z + b_{z,j}v_{z,j,y_0} + \varepsilon_{z,j}, \tag{4.4}$$

where j denotes the j_{th} quantile of z-ICT indicator. Under the assumption that technology convergence is unconditional, we estimate the cross-country regressions reflecting the convergence process with respect to consecutive ICT indicators at different quantiles of the respective ICT variable distribution, which broadens the general view of the examined technology β-convergence. Arbitrarily, we estimate the $b_{z,j}$, which corresponds to consecutive quantiles—the 20th, 40th, 60th and 80th—of each ICT variable distribution. The results of the technology Q-convergence estimates are displayed in Table 4.17.

The brief analysis of regression coefficients $b_{z,j}$ reported for different quantiles of each ICT variable distribution suggests that the speed of convergence is greater at the upper quantiles and the opposite otherwise. In the case of MCS-Q-convergence, there are relatively low differences between the $b_{MCS,20th}$ and $b_{MCS,80th}$, only 0.64 in absolute terms, and the reported pseudo-R^2 are also similar in the lower and upper quantiles. However, despite the observed differences among the regression coefficients for each quantile, the $b_{MCS,20th} = (-7.02)$, which implicates the initially heavily backward countries with respect to mobile telephony adoption, dynamically catches up with the advanced economies. The coefficients that are estimated for the IU-Q-convergence may be similarly interpreted as in the case of $b_{MCS,j}$; however, in absolute terms, greater disparities between the 20th and 80th quantiles are revealed. Accordingly, the pseudo-R^2 is significantly lower for the 20th quantile than for the 80th quantile. The latter may imply that the technology convergence with regard to $IU_{i,y}$ indicator is much weaker among technologically poorer countries, while the convergence tendencies become stronger over the upper quantiles. However, leaving aside the differences in coefficients for the 20th and 80th, for example, there is still no doubt that even countries that in the year 2000 performed poorly in terms of Internet penetration rates rapidly improved in this regard over the period 2000–2012, reaching parity with technologically richer economies. The cross-quantile analysis of estimated $b_{FIS,20th}$, $b_{FIS,40th}$, $b_{FIS,60th}$ and $b_{FIS,80th}$ demonstrates that the weakest and the slowest Q-convergence occurs in the two lower quantiles (20th and 40th). This is unquestionably caused by the low dynamics of fixed narrowband network adoption in the poorer countries, which did not make significant progress in this regard, being stuck in the low-penetration trap. The situation dramatically changes in the 80th quantile, where the strong and dynamic technology convergence is reported. Similarly, as in the standard technology β-convergence, the estimated coefficients revealing the high rate of convergence for $FBS_{i,y}$ should be carefully interpreted, as the analysis covers exclusively 7 time periods, instead of 12 (see discussion on the technology FIB-β-convergence).

Table 4.17 Technology Q-convergence. 113 countries

	MCS$_{i, y}$	FIS$_{i, y}$	FBS$_{i, y}$	IU$_{i, y}$
20_quantile				
$b_{z,20th}$	−7.02	−1.72	−3.56	−4.19
	(0.22)	(0.37)	(0.51)	(0.32)
pseudo-R^2	0.78	0.125	0.355	0.55
40_quantile				
$b_{z,40th}$	−7.42	−2.28	−4.91	−5.27
	(0.19)	(0.32)	(0.39)	(0.17)
pseudo-R^2	0.82	0.22	0.44	0.63
60_quantile				
$b_{z,60th}$	−7.63	−3.25	−6.28	−5.6
	(0.22)	(0.44)	(0.71)	(0.29)
pseudo-R^2	0.84	0.28	0.46	0.66
80_quantile				
$b_{z,80th}$	−7.66	−4.29	−7.93	−6.39
	(0.15)	(0.75)	(1.03)	(0.32)
pseudo-R^2	0.85	0.32	0.45	0.71
# of time periods	12	12	7	12

Source: Author's estimations. Note: bootstrapped quantile regression run for 100 replications. All results for significance at 95 %. In parenthesis—SE reported. For MCS—111 countries (Comoros and Eritrea excluded—significant lacks in data time series). For FBS—107 countries (Bangladesh, Congo, Honduras, Eritrea, Swaziland and Togo excluded—significant lacks in data time series; for Madagascar and Nepal—initial values of FBS for 2006); estimates for the period—2005–2012

Presumably, if we had a longer data series, the conclusion on the rate of FBS-β-convergence would hardly differ from the results obtained for FIS-β-convergence. What warrants special attention is that the estimated Q-convergence regression parameters are negative regardless of the quantile, which may indicate that no technology convergence clubs are formed and the technology divergence does not occur (for the detailed technology-club empirical evidence).

Following Barro and Sala-i-Martin (1992), Quah (1993) and Friedman (1992), we argue that the presence of the β-type convergence is necessary but not a sufficient condition for the σ-convergence. The positive verification of the beta-convergence hypothesis does not mean that cross-country inequalities will automatically fall. As σ-convergence allows the ability to conclude directly on the changing distribution of variables (Young et al. 2008) over the sample, we decide to use this approach as complementary analysis to the β-convergence. Some argue (see, e.g., Quah 1993) that the use of the sigma-convergence approach is more conclusive compared to beta-convergence, as it directly reports on the decreasing (or increasing) cross-country disparities in a given dimension. Figure 4.19 provides preliminary evidence on technology σ-convergence, measured by standard

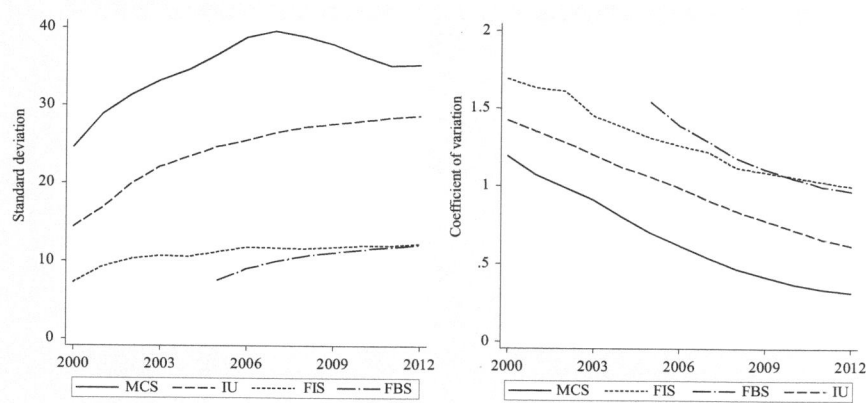

Fig. 4.19 Technology σ-convergence. 113 countries. Note: dispersion expressed as standard deviation (*left chart*) and coefficient of variation (*right chart*) (*Source*: Author's elaboration)

deviation ($\sigma_{z,y}$) and, alternatively, by the coefficient of variation ($cv_{z,y}$); and Table 4.17 displays detailed technology σ-convergence results. The time and country coverage are analogous to those applied in the technology β-convergence analysis.

Based on the charts included in Fig. 4.19, the conclusion may be drawn that over the analysed period, the results of technology σ-convergence are ambiguous. The previous conclusion is strongly supported by the numerical evidence summarized in Table 4.18. Over the examined period, the steadily increasing standard deviation (the upward trends) would suggest the rejection of the hypothesis on the technology σ-convergence (with respect to all four ICT indicators—$MCS_{i,\,y}$, $FIS_{i,\,y}$, $FBS_{i,\,y}$ and $IU_{i,\,y}$). Considering the changes in $\sigma_{MCS,y}$ since 2009, the standard deviation slightly decreases; however, in the terminal year (2012), it was still significantly higher than in 2000, which accounts for the growing cross-country absolute disparities with respect to the mobile cellular telephony penetration rates. The examination of the trends of standard variations of $FIS_{i,\,y}$, $FBS_{i,\,y}$ and $IU_{i,\,y}$ over the period 2000–2012 also yields the conclusion on the absence of technology σ-convergence.

Conversely, as pictured on the right chart in Fig. 4.19, there is a sharp decline in the coefficients of variation ($cv_{z,y}$) with respect to all four ICT indicators. The downward trends of ($cv_{MCS,y}$), ($cv_{FIS,y}$), ($cv_{FBS,y}$) and ($cv_{IU,y}$) would allow confirmation of the hypothesis on the presence of technology σ-convergence, as exhibited significant drops in ($cv_{z,y}$) coincide with the strong ups of ($\sigma_{z,\,y}$). The reasonable explanation of the divergent outcomes may be that we face two parallel processes generated by fast ICT diffusion worldwide. First, we observe growing cross-country heterogeneity with respect to the level of access to and use of basic ICTs; in absolute terms, the technology gaps have widened (to compare, see statistics in Table 4.15 and the changes in ICT variable distributions in Appendix E). Hence, the calculated standard deviations demonstrate the upward trend over the period

Table 4.18 Technology σ-convergence. 113 countries

Standard deviation	$MCS_{i,y}$	$FIS_{i,y}$	$FBS_{i,y}$	$IU_{i,y}$	Coefficient of variation	$MCS_{i,y}$	$FIS_{i,y}$	$FBS_{i,y}$	$IU_{i,y}$
σ_{2000}	24.6	7.3		14.4	cv_{2000}	1.19	1.69		1.42
σ_{2001}	28.8	9.3		16.8	cv_{2001}	1.07	1.63		1.35
σ_{2002}	31.3	10.3		19.9	cv_{2002}	0.99	1.61		1.28
σ_{2003}	33.2	10.7		22.1	cv_{2003}	0.91	1.45		1.2
σ_{2004}	34.6	10.6		23.4	cv_{2004}	0.8	1.38		1.12
σ_{2005}	36.6	11.2	7.58	24.7	cv_{2005}	0.7	1.31	1.54	1.06
σ_{2006}	38.8	11.8	9.03	25.5	cv_{2006}	0.62	1.26	1.39	0.99
σ_{2007}	39.6	11.7	10	26.5	cv_{2007}	0.54	1.22	1.29	0.91
σ_{2008}	38.9	11.6	10.7	27.2	cv_{2008}	0.47	1.12	1.18	0.84
σ_{2009}	37.9	11.8	11.1	27.6	cv_{2009}	0.42	1.09	1.11	0.78
σ_{2010}	36.4	12	11.5	28	cv_{2010}	0.37	1.06	1.05	0.72
σ_{2011}	35.2	12	11.8	28.4	cv_{2011}	0.34	1.03	1	0.66
σ_{2012}	35.3	12.3	12.1	28.7	cv_{2012}	0.32	1	0.97	0.62

Source: Author's estimates. Note: to measure dispersion—standard deviation (σ) and coefficient of variation (cv) applied; for FBS –estimates since 2005. For MCS—111 countries (Comoros and Eritrea excluded—significant lacks in data time series). For FBS—107 countries (Bangladesh, Congo, Honduras, Eritrea, Swaziland and Togo excluded—significant lacks in data time series); for Madagascar and Nepal—initial values for 2006

2000–2012. However, as recognized, the mean values of $MCS_{i, y}$, $FIS_{i, y}$, $FBS_{i, y}$, and $IU_{i, y}$ have dramatically increased (to compare, see statistics in Table 4.15); hence, the coefficients of variation fell. The in-depth examination of the technology σ-convergence process, however, brings another possible explanation of the mixed results. It is possible that the countries that were initially relatively poorer due to the rapid diffusion of ICTs finally, in 2012, performed better compared to advanced countries with respect to, e.g., mobile cellular telephony penetration rates. The later would suggest the emergence of the unique 'technological leapfrogging' effect that consists of bypassing stages on the development path (Steinmueller 2001; James 2012) and fosters rapid catching-up with the advanced countries. Hobday (1995) argues that the rapid catching-up process potentially leads to the 'leapfrog' of the developed countries by the developing countries, so that initially poor countries in the terminal year of analysis perform better compared to developed countries in the selected dimension. Such an 'effect' is actually reported in the case of mobile cellular telephony penetration rates. For example, countries such as Viet Nam, Ukraine, Cambodia, El Salvador, Egypt and Mauritius in 2012 reached $MCS_{i, y}$ penetration rates far exceeding many of those of the developed countries (to compare, see Appendix B).

4.6.2 Technology Convergence Clubs

In Sect. 4.6.1, the in-depth insight into technology *beta* and *sigma* convergence was provided. We have found strong evidence for technology *beta*-convergence (additionally supported by the Q-convergence specification) and technology *sigma*-convergence but only *if* the dispersion is expressed as the coefficient of variation. For a more detailed investigation of the technology convergence behaviour of the countries in focus, we aim to test the convergence clubs hypothesis (Baumol 1986). Reformulating the Baumol (1986) definition of the convergence club, we claim that the technology convergence club is the group of countries that demonstrate strong convergence tendencies regarding the access to and use of ICTs. To check the presence of the convergence clubs, we use two formal approaches. The first, proposed by Baumol and Wolff (1988), formulates a prerequisite for the existence of convergence clubs by using the augmented version of the traditional Sala-i-Martin (1995) convergence equation. Hence, to test the technology convergence club hypothesis, we estimate the following regression:

$$g_{ICT} = a + b_1 v_{ICT, y_0} + b_2 v_{ICT, y_0}^2 + \varepsilon_{ICT}. \tag{4.5}$$

If $b_1 > 0$ and $b_2 < 0$ are satisfied, the convergence clubs may emerge. Countries potentially belonging to the "club" are those that have 'passed' the specific *threshold*, accounting for the *critical value* of the initial level of technological development. The threshold value is defined as the maximum of the function expressed in

Eq. (4.5), which is $v_{ICT,max} = \frac{-b_1}{2b_2}$. This approach suggests that countries belonging to 'the clubs' exhibit technology convergence, while countries 'outside 'the club' rather tend to diverge from each other. Chatterji (1992), Chatterji and Dewhurst (1996) proposed an alternative approach to test the convergence hypothesis. Chatterji and Dewhurst (1996) presume that convergence clubs can emerge only *if*, initially, significant gaps among leading regions and the rest of countries are reported. Their approach to convergence club identification encompasses the analysis of changes in relative gaps between countries. However, a significant drawback of their concept lies in the rigid assumption that, both in initial and terminal year, the leading country does not change, which is not realistic. The Chatterji and Dewhurst (1996) formal specification of the convergence club hypothesis yields the following equation:

$$G_{i,y}^{ICT} = \Psi_1\left(G_{i,y_0}^{ICT}\right) + \Psi_2\left(G_{i,y_0}^{ICT}\right)^2 + \Psi_3\left(G_{i,y_0}^{ICT}\right)^3 + \varepsilon_i^{ICT}, \qquad (4.6)$$

where $G_{i,y}^{ICT}$ explains the technology gap between countries in terminal year (y) with respect to certain ICT, the G_{i,y_0}^{ICT}-technology gap[65] between two countries in initial year (y_0) with respect to the same technology, and Ψ_{1-3} stands for the regression coefficients.[66]

Therefore, to challenge the target of this section, we verify the technology convergence club hypothesis by adopting the Baumol and Wolff (1988) formal specification. First, we test for nonlinearities to check whether the 2° polynomials might belong to the technology β-convergence model. To this aim, we perform the Wald test for consecutive ICT variables $MCS_{i,y}$, $FIS_{i,y}$, $FBS_{i,y}$ and $IU_{i,y}$. The results are summarized in Table 4.19.

The null hypothesis that the regressions are linear is rejected at the 5 % significance level in the case of $MCS_{i,y}$, $FIS_{i,y}$ and $IU_{i,y}$. The Wald test results suggest that only in the FBS-β-convergence equation is nonlinearity possibly the case. Afterwards, we deploy the refined Baumol and Wolff (1988) specification and

Table 4.19 Test for nonlinearities in technology β-convergence model. Baumol and Wolff specification (1988)

	$MCS_{i,y}$	$FIS_{i,y}$	$FBS_{i,y}$	$IU_{i,y}$
Wald test	0.36	2.21	13.88	0.62
Prob > F	0.548	0.139	0.0003	0.43
Need for polynomial in the model	no	no	yes	no

Source: Author's estimations

---

[65] The technology gap is calculated as: $G_{i,y}^{ICT} = ln\left(\frac{ICTvalue_{[leader,y]}}{ICTvalue_{[i,y]}}\right)$.

[66] For more theoretical details, see Chap. 3.

Table 4.20 Technology club convergence. 113 countries. Baumol and Wolff's (1988) specification (2SLS estimates)

	$MCS_{i, y}$	$FIS_{i, y}$	$FBS_{i, y}$	$IU_{i, y}$
b_1	−4.8 (1.21)	−2.95 (0.87)	−6.9 (1.26)	−4.12 (0.51)
b_2	−0.83 (0.40)	−0.20 (0.95)	−0.82 (0.43)	−0.42 (0.39)
ICT threshold	−1.92	−0.29	−2.8	−0.86
Ramsey test (Prob > F)	1.44 (0.24)	0.12 (0.94)	1.32 (0.27)	2.35 (0.07)
R-squared	0.87	0.32	0.57	0.73
# of countries	111	113	107	113

Source: Author's estimations. Note: 2SLS estimates (instruments—1-year lagged v_{ICT, y_0} and lagged v_{ICT, y_0}^2); in parenthesis—robust SE. All estimates for significance at 95 %. For MCS—111 countries (Comoros and Eritrea excluded—significant lacks in data time series). For FBS—107 countries (Bangladesh, Congo, Honduras, Eritrea, Swaziland and Togo excluded—significant lacks in data time series; for Madagascar and Nepal—initial values of $FBS_{i, y}$ for 2006); estimates for the period—2005–2012

estimate the ICT-specific regressions defined in Eq. (4.5), to verify the hypothesis on the presence of the technology convergence clubs. To control for possible collinearity, which might have emerged due to the inclusion of the squared terms of the explanatory variable, we use the Two-Stages Least Squares (2SLS) approach, using lagged values of explanatory variables as instruments. Table 4.20 summarizes the regression outcomes, and Fig. 4.20 provides additional graphical support to the results of the econometric analysis.

The existence of technology convergence clubs requires that $b_1 > 0$ and $b_2 < 0$, and both coefficients are statistically significant. If the condition is satisfied, the maximum of the Eq. (4.5) may be calculated, which specifies the ICT threshold value. The ICT threshold indicates the critical value of initial ICT deployment that allows for the division of the whole sample into two groups: countries that demonstrate convergences tendencies and hence form the technology convergence club and countries that are excluded from the 'club' and instead diverge from the others. Countries that may be classified as technology convergence club members are those that, in the initial year (y_0), achieved the ICT threshold. The countries that did not manage to reach the ICT threshold in (y_0) and so converge with advanced countries are those where technology development was so low that the ability to take advantage of the relative backwardness and catch up was impeded. As can be concluded from Table 4.20, in each case, the regression coefficients—b_1 and b_2—are negative, which suggests that no technology convergence clubs emerge with respect to any of analysed ICT indicators. As both regression coefficients hold negative signs, the calculated ICT thresholds also resulted in negative values. Thus, the empirical evidence clearly demonstrates that, using the approach proposed by Baumol and Wolff (1988), no technology convergence clubs have been identified;

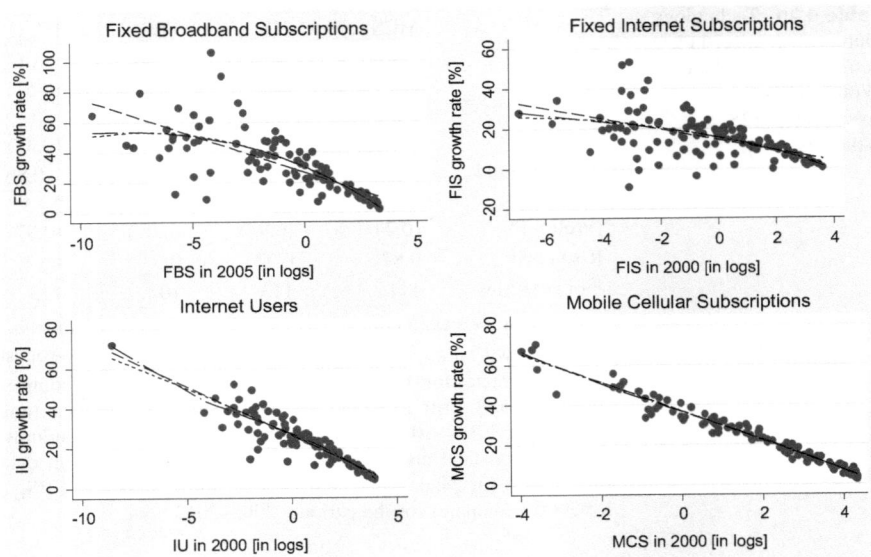

Fig 4.20 Technology convergence clubs. Notes: *Long-dash line*—linear prediction; *short-dash-line*—quadratic prediction; *long-dash-dot line*—cubic prediction; *dots*—empirical observations (*Source*: Author's elaboration)

hence, no divergence tendencies emerge and no countries are excluded from the convergence process. The evidence again supports the supposition that the fast and dynamic diffusion of basic ICTs in developing countries over the period 2000–2012 had a strong positive impact on diminishing cross-country disparities in this respect. The 'technologically poor' countries managed to catch up rapidly with 'technologically rich' countries, as overall inequalities fell (to compare, see Table 4.15), and no countries were left behind.

As demonstrated using the Baumol and Wolff (1988) approach, the hypothesis on the existence of technology convergence clubs was rejected. To re-examine the technology convergence club, we consequently adopt the alternative approach formalized by Chatterji (1992). As Chatterji defines convergence as the process of cross-country gaps narrowing in the selected dimension, our identification of technology convergence clubs suggests solving the Eq. (4.6) with respect to the selected ICT indicators—$MCS_{i, y}$, $FIS_{i, y}$, $FBS_{i, y}$ and $IU_{i, y}$. Adopting the Chatterji (1992) specification, we seek to answer the question whether, over the period 2000–2012, we observe cross-country technology gaps narrowing or widening and if these results are consistent with the previous ones. First, using the Wald test, we check whether the 3° polynomials should belong to Eq. (4.6). The results are displayed in Table 4.20 and suggest that with the exception of the $FIS_{i, y}$ variable, the cubic specification might be appropriate to solve Eq. (4.6). To control for the multicollinearity that may arise due to the inclusion of different powers of the regressor in the model, we use the Two-Stage Least Squares (2SLS) approach.

Table 4.21 Test for nonlinearities in technology convergence clubs model (Chatterji specification 1992)

	$MCS_{i,y}$	$FIS_{i,y}$	$FBS_{i,y}$	$IU_{i,y}$
Wald test	11.22	1.22	6.93	3.64
Prob > F	0.000	0.29	0.001	0.029
Need for polynomial in the model	yes	no	yes	yes

Source: Author's estimations

Table 4.22 Technology club convergence. 113 countries. Chatterji and Dewhurst specification (1996) (2SLS estimation)

	$MCS_{i,y}$	$FIS_{i,y}$	$FBS_{i,y}$	$IU_{i,y}$
Ψ_1	−2.59 (6.79)	−0.20 (2.1)	−1.05 (6.41)	0.15 (0.47)
Ψ_2	1.06 (2.79)	0.05 (0.05)	0.35 (1.61)	0.02 (0.12)
Ψ_3	−0.10 (0.27)	0.008 (0.04)	−0.01 (0.10)	0.002 (0.009)
G2	24.25	−81.25	misspecification returned	misspecification returned
G3	−13.6	18.75	misspecification returned	misspecification returned
Ramsey test (Prob > F)	0.35 (0.79)	0.07 (0.97)	1.49 (0.22)	0.41 (0.76)
R-squared	0.54	0.33	0.75	0.31
# of countries	111	113	107	113

Source: Author's estimations. Note: 2SLS estimates; instruments—lagged values of explanatory variables; in parenthesis—robust SE at 95 % significance. *All results are statistically insignificant.* Leaders in 2000: MCS—Iceland; FIS—Netherlands; FBS (for 2005)—Sweden; IU—Norway

The results of the empirical analysis are summarized in Table 4.21 and supported by graphical inspection in Fig. 4.22. To verify the technology convergence hypothesis, we estimate the ICT-specific cubic regression models, and afterwards, we calculate the equilibrium points[67] (G_{2,y_0}) and (G_{3,y_0}), which enable identification of the technology convergence club members.

To satisfy the condition of technology convergence club existence, the two equilibria points (G_{2,y_0}) and (G_{3,y_0}) would have to be indicated. However, as displayed in Table 4.22, all estimated regression coefficients are statistically insignificant at the 95 % level of significance, and the calculated equilibrium points, which potentially should indicate the technology club members, are mis-specified. The latter also indicates that no internal equilibria emerge, and none of the analysed

---

[67] See Chap. 3 for technical details.

countries was identified as locked in the low-equilibrium trap, unable to catch-up with the technologically advanced economies.

The visual inspection of the technology convergence clubs supports the statistically insignificant estimates of ICT-specific regression. The fitted curves do not actually cut the 45° theoretical line[68] (the 45° theoretical line is cut at the very low—close to zero—level of the initial technology gaps), which may suggest the existence of internal equilibria points. Based on the graphical evidence, it may be concluded that over the period 2000–2012, with respect to four analysed ICT indicators ($MCS_{i, y}$, $FIS_{i, y}$, $FBS_{i, y}$ and $IU_{i, y}$) significant drops in technology gaps between each i-country and the leading country are observed. The most prominent 'gaps closing' is revealed in the case of mobile cellular telephony and Internet users. The location of both fitted curves (the MCS-curve and IU-curve) suggests that all countries were rapidly converging toward the leading economy, regardless of the size of the gaps in y_0. The average[69] MCS-gap[70] in 2000 was at approximately $G^{MCS}_{average, 2000} = 2.6$, while in the terminal year, $G^{MCS}_{average, 2012} = .008$. The analogous averages[71] with respect to the IU were $G^{IU}_{average, 2000} = 3.2$ and $G^{IU}_{average, 2012} = 1.1$. What is worth noting is that with regard to both $MCS_{i, y}$ and $IU_{i, y}$, none of the analysed countries experienced growth in the respective technology gap (see Fig. 4.21 to compare) but rather significant increases in terms of the latter. Such a conclusion may be easily drawn based on visual inspection of Fig. 4.21. Countries located *below* the 45° theoretical line are those where the *decreases* in gaps with respect to the technology in question are reported and the reverse otherwise. In case of the $MCS_{i, y}$ and $IU_{i, y}$, no country is located above the 45° line; hence, the technology gaps decreased in general. With respect to the consecutive ICT indicators indicating the access to fixed narrowband ($FIS_{i, y}$) and fixed broadband ($FBS_{i, y}$) networks, drops in technology gaps are also reported; however, the empirical evidence has demonstrated that the changes in the size of the gaps are less pervasive if compared to the $MCS_{i, y}$ and $IU_{i, y}$. The average FIS gap[72] in y_{2000} was $G^{FIS}_{average, 2000} = 3.9$; in y_{2012}, $G^{FIS}_{average, 2012} = 2.2$. The respective values[73] for $FBS_{i, y}$ were $G^{FBS}_{average, 2005} = 4.1$ and $G^{FBS}_{average, 2012} = 2.2$. In the case of the FIS-gaps, there were four (out of 113) countries where slight growth in gaps was noted, namely, Eritrea, Swaziland, Madagascar and Brunei Darussalam.

---

[68] To be specific, the 45° line is cut by the fitted curves at the very low—close to zero—level of the initial technology gaps.

[69] Author's estimates.

[70] The gap is expressed as $G^{ICT}_{i, y} = ln\left(\frac{ICTvalue_{|leader, y|}}{ICTvalue_{|i, y|}}\right)$.

[71] Author's estimates.

[72] Author's estimates.

[73] Author's estimates.

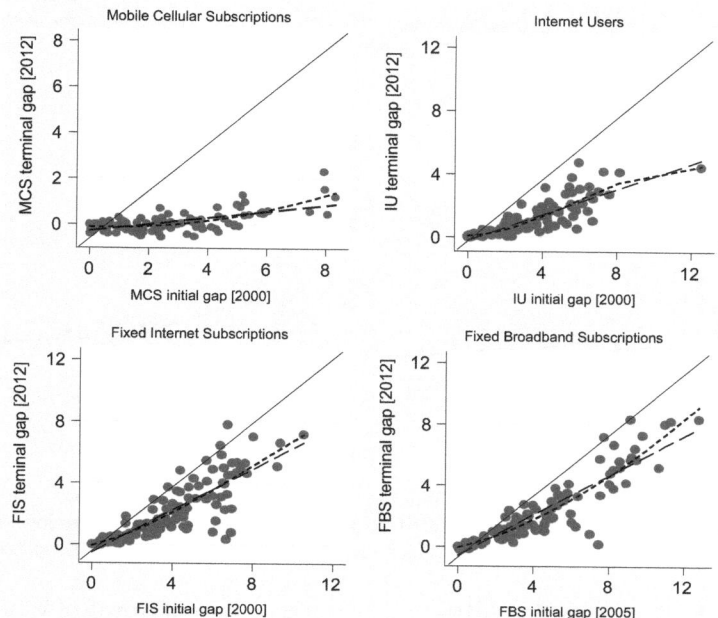

Fig 4.21 Initial technology gaps (in 2000 (For FBS–2005)) vs. terminal technology gaps (in 2012). 113 countries. Note: for FBS$_{i,\,y}$ initial technology gap for 2005. Linear vs. inverse relationship; growth rate versus initial level of ICTs. Note: *short-dash*—cubic prediction; *long-dash*—linear prediction (*Source*: Author's elaboration)

4.6.3 Brief Evidence on How Wireless Broadband Networks Expanded Worldwide Over the Period 2010–2012

The main purpose of Sect. 4.6.3 is to unveil the main tendencies of wireless-broadband solution adoption worldwide and to provide in-depth comparison of the developing and developed countries in this respect. As the wireless-broadband networks yield to a relatively new technological solution, the first data on wireless broadband subscriptions (WBS$_{i,\,y}$) are available beginning with the year 2007[74]— hence, the time series are claimed to be too short to return reliable and conclusive estimates of the conventional convergence models. The evidence below explores the major tendencies in wireless broadband network penetration rates over the period 2009–2012 in 72 countries and additionally over the period 2010–2012 in 102 countries (for detailed statistics, see Appendix B), reporting on the growing cross-country cohesion in this respect. The rapid growth of WBS$_{i,\,y}$ penetration rates is observed both in developing and developed countries, and over the period 2010–2012, the average annual growth rate was approximately 38.07 % (see

---

[74] According to ITU 2014 ICT Eye.

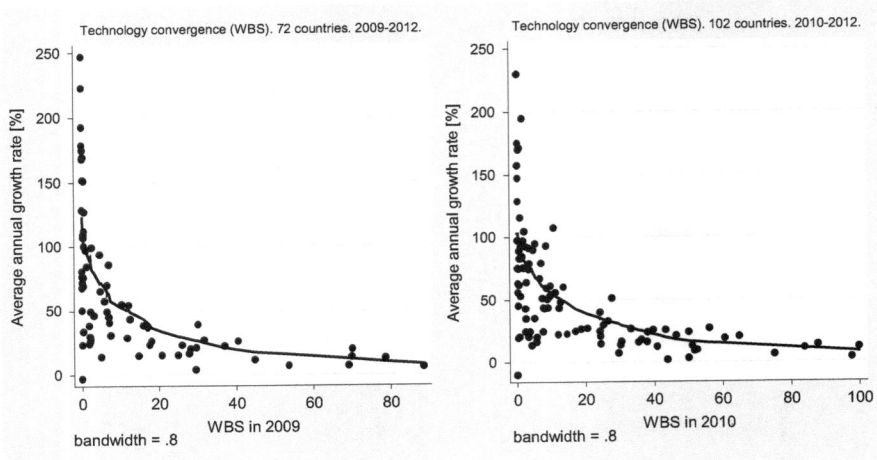

Fig. 4.22 Technology convergence—Wireless Broadband Subscriptions. Note: evidence for 2009–2012—72 countries; evidence for 2010–2012—102 countries. For visualization—nonparametric approximation applied. $WBS_{i, y}$ values—absolute values (*Source*: Author's elaboration)

Table 4.15). Over the period 2010–2012, the most prominent growth of $WBS_{i, y}$ was reported for Rwanda—230 %, Botswana—194 %, Ethiopia—175 %, Nigeria— 171 %, Swaziland—170 %, Lao P.D.R—157 %, Senegal—147 % and Kenya— 129 %. However, if we observe the bottom of the ranking, sharp differences in the $WBS_{i, y}$ annual growth rate are noted. Among the countries that revealed the slowest progress with respect to growth in $WBS_{i, y}$ penetration rates are countries such as Israel, Korea, Norway, Singapore and Japan. As expected, it indicates the general tendency of relatively more backward countries to grow faster than the advanced economies; hence, the convergence process occurs, and the developing countries shall rapidly catch-up with the developed countries with respect to wireless broadband penetration rates. Figure 4.22 plots the $WBS_{i, y}$ average growth rates against the $WBS_{i, y}$ penetration rates in the initial year (see upper charts) and the $WBS_{i, y}$ penetration rates in 2012 against the $WBS_{i, y}$ penetration rates in the initial year (see lower charts). The general picture arising from the charts in Fig. 4.22 suggests strong technology convergence with respect to wireless broadband penetration rates, both for the period 2009–2012 and 2010–2012. The elementary econometric evidence on the WBS-convergence over the period 2009– 2012 indicates that the speed of convergence is approximately 92 % so that the cross-country disparities shall be halved in 0.75 years (approximately nine months—*sic!*),[75] and over the period 2010–2012, the values yield 135 % and

---

[75] The regression (OLS with robust SE) coefficient was (−14.96) and statistically significant at the 95 % level of significance. The R-squared of the model was 0.47.

0.51 years, respectively (approximately 6 months—*sic!*).[76] However, it is important to note that these results report on observed technology convergence behaviour of countries over an extremely short time period, so they probably lack robustness, and the results generated from the convergence model coefficients might be heavily overestimated.

Additional evidence may be reported by observing the cross-country averages and inequalities in wireless broadband network adoption measured by the Gini coefficient. In 2009, the $WBS_{average,2009} = 14.1$ per 100 inhab., while in 2012, the $WBS_{average,2009} = 30.1$ per 100 inhabitants; thus, the number has doubled over the 4-year period. However, despite the reported prominent growth in $WBS_{i, y}$ penetration rates, the inequalities in access to wireless broadband are slowly diminishing. In 2009, the Gini coefficient was 0.69 (the estimate for 72 countries); in 2012, 0.54 (the estimate for 102 countries). Hence, although the cross-country inequalities are gradually decreasing, significant gaps persist (see Appendix D), and quite a number of developing countries significantly lag behind in wireless broadband diffusion. Despite strong convergence tendencies and rapid expansion of wireless broadband networks that may be identified due to the extraordinary pace of growth, the average $WBS_{i, y}$ penetration rates both in low-income and lower-middle-income economies are still relatively low. In 2012, the average[77] $WBS_{i, y}$ penetration rate was approximately 4.00 per 100 inhabitants in low-income economies; in lower-middle-income economies, 9.6. Compared to the values of 20.6 and 59.4 in upper-middle-income and high-income countries, respectively, these are relatively poor results and clearly demonstrate the differences in access to wireless broadband networks among various income groups.

4.6.3.1 Final Remarks

The primary goal of Sect. 4.6 was to verify the hypothesis on technology convergence (beta and sigma) and to test whether countries in the scope of the study form specific technology convergence clubs. Regardless of the method applied, our empirical evidence supports the supposition on strong technology convergence tendencies among world countries over the period 2000–2012. Furthermore, detailed analysis of country technology convergence behaviour did not confirm the technology convergence club hypothesis. In other words, we have demonstrated that with respect to basic ICT deployment, countries dynamically converge, and as no technology convergence clubs were identified, no technology divergence tendencies were recognized. The success of economically backward economies in rapidly catching up with the developed world in terms of ICT adoption was predominantly driven by long-term, continuous growth patterns undisrupted by sudden ups and downs with respect to increasing access to and use of basic ICTs. The latter is of unique importance, as it demonstrates that, with regard to ICTs, the

---

[76] The regression (OLS with robust SE) coefficient was (-14.09) and statistically significant at the 95 % level of significance. The R-squared of the model was 0.39.

[77] Author's calculations.

'Great Escape' (Deaton 2013) is possible, that even the most technologically backward countries were able to shift rapidly from 'technologically poor' to 'technologically rich', and that hardly any of the analysed countries suffered from the 'low technology trap'. The convergence process additionally resulted in significant drops in cross-country disparities in terms of access to and use of basic ICTs; hence, the technology gaps diminished worldwide. What calls for special attention is that the technology convergence was discovered although the countries in focus are highly heterogeneous, perform differently in terms of socioeconomic achievements and have various noneconomic characteristics.

4.7   Summary

The chapter was designed to demonstrate a detailed analysis of country-specific ICT diffusion patterns, detect technological substitution, and examine the process of technology convergence and technology convergence clubs formation. The sample covered 17 low-income and 29 lower-middle-income economies over the period 2000–2012. Throughout our study we have unveiled that during analyzed period, a vast majority of economically backward countries had made enormous progress in terms of average mobile cellular telephony adoption. Regarding low-income economies, in 2000 the average $MCS_{i, y}$ was at around 0.45 per 100 inhab., while until 2012 it increased reaching almost $MCS_{i, y} = 52$ per 100 inhab. In lower-middle-income countries the boost in mobile cellular telephony was even more spectacular, as in 2012 the average accessibility of this form of communication was at about $MCS_{i, y} = 95$ per 100 inhab. The latter was predominantly enhanced by broad development of pre-paid systems, telecommunication market liberalization, which allowed for gradual drops in prices. Among 2000–2012, the state of development of fixed narrowband, fixed broadband and wireless broadband architecture, in both income groups, reflected the significant efforts toward achieving high penetration rates. Nevertheless, despite the reported spectacular achievements, especially in wireless broadband deployment, in 2012, many low-, and lower-middle-income countries were still heavily deprived of universal access to ICT devices that enable Internet connections. This was extremely limited due to high prices for leasing lines; and additionally, country-specific barriers such as unfavourable location, poorly developed infrastructure, and permanent power supply problems significantly impede the broader introduction of fixed and wireless broadband technologies. As a consequence the unbound access to Internet network is still a 'luxury good' in economically backward countries. Our empirical evidence has also demonstrated that the phenomenal process of rapid growth mobile telephony and wireless network penetration rates have resulted in strong fixed-to-mobile telephony and fixed-to-wireless Internet connection substitution effects. However, importantly to note that although the results of numerical analysis clearly demonstrate the technological substitution effects, the latter shall be interpreted carefully. It might be arguable to claim that with respect to low- and lower-middle-income countries, the *typical* technological substitution process is actually *not* observed. Notably once the mobile cellular telephony and wireless broadband networks emerged across analyzed economies, people moved *straight* towards new technologies, having never adopted any form of 'old' ICT. The latter shows the technological leapfrogging process consisting of the rapid jump over several development stages, moving directly from the 'old' to the 'new' technologies. Finally, we have examined the technology convergence and intended to identify technology convergence clubs. Notably, regardless of the method applied, our evidence supports the supposition on fast technology convergence tendencies among world countries over the period 2000–2012. Furthermore, the examination of country technology convergence behaviour did not confirm the technology convergence club hypothesis, which shows that with respect to basic ICT deployment no technology divergence tendencies were recognized.

References

Abramovitz, M. (1986). Catching up, forging ahead, and falling behind. *The Journal of Economic History, 46*(02), 385–406.

Albon, R. (2006). Fixed-to-mobile substitution, complementarity and convergence. *Agenda, 13*(4), 309–322.

Antonelli, C. (1991). *The diffusion of advanced telecommunications in developing countries.* Paris: OECD.

Atkinson, A. B. (1970). On the measurement of inequality. *Journal of Economic Theory, 2*(3), 244–263.

Ausubel, J. H. (1991). Rate-race dynamics and crazy companies: The diffusion of technologies and social behavior. *Technological Forecasting and Social Change, 39*(1-2), 11–22.

Banerjee, A., & Ros, A. J. (2004). Patterns in global fixed and mobile telecommunications development: A cluster analysis. *Telecommunications Policy, 28*(2), 107–132.

Barro, R., & Sala-i-Martin, X. X. (1992). Convergence. *Journal of Political Economy, 100*, 223–251.

Baumol, W. J. (1986). Productivity growth, convergence, and welfare: What the long-run data show. *The American Economic Review, 76*, 1072–1085.

Baumol, W. J., & Wolff, E. N. (1988). Productivity growth, convergence, and welfare: Reply. *The American Economic Review, 78*, 1155–1159.

Briglauer, W., Schwarz, A., & Zulehner, C. (2011). Is fixed-mobile substitution strong enough to de-regulate fixed voice telephony? Evidence from the Austrian markets. *Journal of Regulatory Economics, 39*(1), 50–67.

Castellacci, F. (2006). Convergence and divergence among technology clubs. In DRUID Conference, Copenhagen (Vol. 30, No. 07). Accessed June from http://www.druid.dk/wp/pdf_files.org/06-21.pdf

Castellacci, F. (2008). Technology clubs, technology gaps and growth trajectories. *Structural Change and Economic Dynamics, 19*(4), 301–314.

Castellacci, F., & Archibugi, D. (2008). The technology clubs: The distribution of knowledge across nations. *Research Policy, 37*(10), 1659–1673.

Castellacci, F. (2011). Closing the technology gap? *Review of Development Economics, 15*(1), 180–197.

Caves, K. W. (2011). Quantifying price-driven wireless substitution in telephony. *Telecommunications Policy, 35*(11), 984–998.

Chatterji, M. (1992). Convergence clubs and endogenous growth. *Oxford Review of Economic Policy, 8*(4), 57–69.

Chatterji, M., & Dewhurst, J. L. (1996). Convergence clubs and relative economic performance in Great Britain: 1977–1991. *Regional Studies, 30*(1), 31–39.

Comin, D., & Hobijn, B. (2004). Cross-country technology adoption: Making the theories face the facts. *Journal of Monetary Economics, 51*(1), 39–83.

Comin, D., Hobijn, B., & Rovito, E. (2006). *Five facts you need to know about technology diffusion* (No. w11928). National Bureau of Economic Research.

Communications Commission of Kenya. (2013). *Quarterly sector statistics report.* Second quarter of the financial YEAR 2013/14 (OCT-DEC 2013).

Deaton, A. (2013). *The great escape: Health, wealth, and the origins of inequality.* New Jersey: Princeton University Press.

Friedman, M. (1992). Do old fallacies ever die? *Journal of Economics Literature, 30,* 2129–2132.

Gagnaire, M. (1997). An overview of broad-band access technologies. *Proceedings of the IEEE, 85*(12), 1958–1972.

Galperin, H. (2005). Wireless networks and rural development: Opportunities for Latin America. *Information Technologies and International Development, 2*(3), 47–56.

Garbacz, C., & Thompson, H. G., Jr. (2005). Universal telecommunication service: A world perspective. *Information Economics and Policy, 17*(4), 495–512.

Garbacz, C., & Thompson, H. G., Jr. (2007). Demand for telecommunication services in developing countries. *Telecommunications Policy, 31*(5), 276–289.

Geroski, P. A. (2000). Models of technology diffusion. *Research Policy, 29*(4), 603–625.

Gerschenkron, A. (1962). Economic backwardness in historical perspective. In: *Economic backwardness in historical perspective: A book of essays.* Cambridge: The Belknap Press.

Gini, C. (1912). *Variabilità e mutabilità.* [Reprinted in Memorie di metodologica statistica, E. Pizetti, & T. Salvemini (Ed.)]. Rome: Libreria Eredi Virgilio Veschi, 1.

Gruber, H. (2001). Competition and innovation: The diffusion of mobile telecommunications in Central and Eastern Europe. *Information Economics and Policy, 13*(1), 19–34.

Grzybowski, L. (2011). *Fixed-to-mobile substitution in the European union.* Working Paper, University of Cape Town, South Africa.

Gunasekaran, V., & Harmantzis, F. C. (2007). Emerging wireless technologies for developing countries. *Technology in Society, 29*(1), 23–42.

Hamilton, J. (2003). Are main lines and mobile phones substitutes or complements? Evidence from Africa. *Telecommunications Policy, 27*(1), 109–133.

Hao, L., & Naiman, D. Q. (2007). *Quantile regression quantitative applications in the social sciences* (Vol. 149). Thousand Oaks, CA: Sage.

Härdle, W. (1984). Robust regression function estimation. *Journal of Multivariate Analysis, 14*(2), 169–180.

Hobday, M. (1995). *Innovation in East Asia: The challenge to Japan* (p. 33). Aldershot: Edward Elgar.

ITU. (2006). *ICT and telecommunications in the least developed countries.* Mid-Term Review for the Decade 2001–2010. Geneva, Switzerland.

ITU. (2010). *Core ICT indicators.* Partnership on Measuring ICT for Development. International Telecommunication Union, Geneva.

ITU. (2011a). *Measuring information society.* International Telecommunication Union, Geneva.

ITU. (2011b). *The role of ICT in advancing growth in least developed countries.* Trends, challenges and opportunities. Geneva, Switzerland.

ITU. (2011c). *ICT and telecommunications in least developed countries.* Review of progress made during the decade 2000-2010. Geneva, Switzerland.

ITU, UNESCO. (2011). *Broadband: A platform for progress.* A report by the broadband commission for digital development. Geneva; Paris: ITU/UNESCO.

ITU. (2012). *Wireless broadband Masterplan until 2020 for the socialist republic of Vietnam.* Regional initiatives – Asia- Pacific. Geneva, Switzerland.

ITU. (2013). *World telecommunication/ICT indicators database 2013* (17th edition).

ITU. (2014a). *ITU ICT eye – ICT statistics, international telecommunication union 2014.* Accessed May 20, 2014.

ITU. (2014b). *Manual for measuring ICT access and use by households and individuals.* International Telecommunication Union, Geneva.

James, J. (2009). Leapfrogging in mobile telephony: A measure for comparing country performance. *Technological Forecasting and Social Change, 76*(7), 991–998.

James, J. (2012). The distributional effects of leapfrogging in mobile phones. *Informatics, 29*(3), 294–301.

James, J. (2014). Relative and absolute components of leapfrogging in mobile phones by developing countries. *Telematics and Informatics, 31*(1), 52–61.

Kauffman, R. J., & Techatassanasoontorn, A. A. (2005). Is there a global digital divide for digital wireless phone technologies? *Journal of the Association for Information Systems, 6*(12).

Koenker, R. (2000). Galton, Edgeworth, Frisch, and prospects for quantile regression in econometrics. *Journal of Econometrics, 95*(2), 347–374.

Koenker, R. (2004). Quantile regression for longitudinal data. *Journal of Multivariate Analysis, 91*(1), 74–89.

Koenker, R. (2005). *Quantile regression* (Vol. 38). Cambridge: Cambridge university press.

Koenker, R., & Bassett, G., Jr. (1978). Regression quantiles. *Econometrica: Journal of the Econometric Society, 84*, 33–50.

Laurini, M. (2007). A note on the use of quantile regression in beta convergence analysis. *Economics Bulletin, 3*(52), 1–8.

Lechman, E. (2012a). Technology convergence and digital divides. A country-level evidence for the period 2000-2010. *Ekonomia, Rynek, Gospodarka, Społeczeństwo, 31.*

Lechman, E. (2012b). *Cross national technology convergence. An empirical study for the period 2000-2010.* University Library of Munich, Germany.

Lee, K., & Lim, C. (2001). Technological regimes, catching-up and leapfrogging: Findings from the Korean industries. *Research Policy, 30*(3), 459–483.

Narayana, M. R. (2010). Substitutability between mobile and fixed telephones: Evidence and implications for India. *Review of Urban & Regional Development Studies, 22*(1), 1–21

Pentland, A., Fletcher, R., & Hasson, A. (2004). Daknet: Rethinking connectivity in developing nations. *Computer, 37*(1), 78–83.

Proenza, F. J. (2006). The road to broadband development in developing countries is through competition driven by wireless and internet telephony. *Information Technologies & International Development, 3*(2), 21.

Quah, D. T. (1993). Galton's fallacy and tests of the convergence hypothesis. *The Scandinavian Journal of Economics, 95*, 427–443.

Rodini, M., Ward, M. R., & Woroch, G. A. (2003). Going mobile: Substitutability between fixed and mobile access. *Telecommunications Policy, 27*(5), 457–476.

Rousseeuw, P. J., & Leroy, A. M. (2005). *Robust regression and outlier detection* (Vol. 589). New York: Wiley.

Sala-i-Martin. (1995). *Regional cohesion: Evidence and theories of regional growth and convergence.* Discussion paper No. 1075, Centre for Economic Policy Research, November.

Sharif, M. N. (1989). Technological leapfrogging: Implications for developing countries. *Technological Forecasting and Social Change, 36*(1), 201–208.

Srinuan, P., Srinuan, C., & Bohlin, E. (2012). Fixed and mobile broadband substitution in Sweden. *Telecommunications Policy, 36*(3), 237–251.

Steinmueller, W. E. (2001). ICTs and the possibilities for leapfrogging by developing countries. *International Labour Review, 140*(2), 193–210.

Sung, N., & Lee, Y. H. (2002). Substitution between mobile and fixed telephones in Korea. *Review of Industrial Organization, 20*(4), 367–374.

United Nations Conference on Trade and Development. (2007). *Information economy report 2007-2008. Science and technology for development: The new paradigm of ICT.* Geneva, Switzerland.

Vogelsang, I. (2010). The relationship between mobile and fixed-line communications: A survey. *Information Economics and Policy, 22*(1), 4–17.

Ward, M. R., & Zheng, S. (2012). Mobile and fixed substitution for telephone service in China. *Telecommunications Policy, 36*(4), 301–310.

Warren, M. (2007). The digital vicious cycle: Links between social disadvantage and digital exclusion in rural areas. *Telecommunications Policy, 31*(6), 374–388.

Waverman, L., Meschi, M., & Fuss, M. (2005). The impact of telecoms on economic growth in developing countries. *The Vodafone Policy Paper Series, 2*(03), 10–24.

World Bank. (2006). *Information and communication for development: Global trends and policies.* Washington, DC: World Bank.

Wulf, J., Zelt, S., & Brenner, W. (2013, January). *Fixed and mobile broadband substitution in the OECD countries – A quantitative analysis of competitive effects.* In System Sciences (HICSS), 2013 46th Hawaii International Conference on (pp. 1454–1463). IEEE.

Young, A. T., Higgins, M. J., & Levy, D. (2008). Sigma convergence versus beta convergence: Evidence from US county-level data. *Journal of Money, Credit and Banking, 40*(5), 1083–1093.

What Matters for ICT Diffusion?

5

> *Diffusion process will be impeded if the innovation requires new kinds of knowledge on the part of the user, new types of behavior, and the coordinated efforts of a number of organizations. If an invention requires few changes in socio-cultural values and behavior patterns, it is likely to spread more rapidly*
>
> Edwin Mansfield (1986)

Abstract

The major targets of the following chapter are twofold. First, adopting a newly developed approach, it traces 'critical mass' effects with regard to ICT diffusion (Mobile Cellular Telephony and Internet) in 17 low-income countries and 29 lower-middle-income countries over the period 2000–2012. To this end, it identifies respective critical penetration rates and a 'technological take-off' interval, which is defined as the period during which ICT diffusion enters an exponential growth phase along an S-shaped trajectory. Along these lines, we demonstrate country-specific socioeconomic and institutional conditions during the 'technological take-off' interval. Second, the chapter provides additional evidence on ICT diffusion determinants in low-income and lower-middle-income countries during the analogous period. It empirically traces the potential effect of selected factors on ICT spread. The analysis covers ten indicators, which are used to explain the level of mobile cellular telephony penetration rates, and nine indicators used to explain the level of Internet usage by individuals. Moreover, we have selected another eight indicators to demonstrate general socioeconomic and infrastructural features of examined countries.

Keywords

Critical mass • Technological take-off • ICT diffusion • ICT determinants

© Springer International Publishing Switzerland 2015   167
E. Lechman, *ICT Diffusion in Developing Countries*,
DOI 10.1007/978-3-319-18254-4_5

5.1   Introduction

During the last decade of the twentieth century and the first decade of the twenty-first century, the world has witnessed the unprecedentedly dynamic diffusion of new ICTs across even the most undeveloped countries. The empirical evidence reported in Chap. 4 revealed that many of the analysed countries experienced rapid and dynamic diffusion of ICTs, which resulted in extremely high penetration rates, especially with regard to access to and usage of mobile cellular telephony; other countries failed in this regard and remained stuck in a 'low-level trap' of not being able to actuate the diffusion process. The reasons for this might be traced by running country-specific analyses, which would provide extensive knowledge regarding why some countries succeeded, and others failed, in the complex process of broad ICT deployment. The following Chap. 5 is designed to understand, at least partially, why certain countries succeeded while others failed at ICT adoption and this challenging task requires context-specific thinking and a country-wise approach.

The Chap. 5 is made up of two major parts—Sects. 5.2 and 5.3, that present the results of the empirical analysis. Section 5.2 is aimed to trace the '*technological take-off*' and the '*critical mass*' effects, which allow for concluding on the critical penetration rates that fostered entering the exponential growth phase along the ICT diffusion path; and—explore country-specific social, economic and institutional conditions during the '*technological take-off*' interval. The latter analysis is complemented and enriched by the evidence demonstrated in Sect. 5.3 encompassing panel regression analysis that aims to identify which factors have positively affected—or conversely, impeded the ICT diffusion across analyzed countries. Finally, short Sect. 5.4 contains major conclusions.

5.2   Tracing the '*Technological Take-Off*' and the '*Critical Mass*' Effects

As highlighted in Chap. 4, over the period 2000–2012, most developing countries experienced significant shifts in access to mobile cellular telephony and use of Internet connections. In contrast, except for a few countries,[1] progress in the deployment of fixed-narrowband, fixed-broadband or wireless-broadband networks remained negligible over the analogous period. Thus, our continuing efforts are directed toward evaluating the '*technological take-off*' intervals and the '*critical mass*' effects regarding increases in access to and use of mobile cellular telephony ($MCS_{i,y}$) and Internet networks ($IU_{i,y}$).

---

[1] For details, see Chap. 4.

5.2.1 The Data

To meet the main targets of Sect. 5.2, we have arbitrary selected a bundle of factors that may help to explain the process of ICT diffusion in developing economies. The defined dataset covers ten indicators, which are used to explain level of mobile cellular telephony penetration rates, and nine to explain individual Internet usage levels. Moreover, we have selected another eight indicators[2] to demonstrate the general socio-economic and infrastructural features of the examined countries. The data were derived from various sources; however, most of the statistics were extracted from the World Telecommunication/ICT Indicators database 2013 (17th Edition) (International Telecommunication Union), World Development Indicators 2013 (World Bank 2014), Human Development Reports 2005–2013 (United Nation Development Program) and Measuring the Information Society reports 2009–2013 (International Telecommunication Union). Additional data were derived from the CIA World Factbook 2014, Freedom House 2013–2014, The Heritage Foundation 2014 and national telecommunication agencies.[3] All indicators used in the analysis are listed and explained in Table 5.1. The forthcoming Sect. 5.2.2 demonstrates the analysis outcomes, where the variables discussed in Table 5.1 are used.

5.2.2 Ready for the 'Technological Take-Off'?

Section 5.2.2 aims to challenge the identification of the 'critical mass' and the 'technological take-off' interval that emerged during the process of gradual ICT diffusion in low-income and lower-middle-income countries over the period 2000–2012.[4] Henceforth, it identifies the 'critical year', 'critical penetration rate', the 'technological take-off' interval that follows right after, along with the bundle of country-specific conditions during the first year of 'technological take-off' interval. To meet the main aims of this analysis, first, we designate ICT marginal growths ($\Omega_{MCS,i,y}$ and $\Omega_{IU,i,y}$), and the ICT replication coefficients ($\Phi_{MCS,i,y}$ and $\Phi_{IU,i,y}$) for each country separately. Figures 5.1, 5.2, 5.3, and 5.4 outline country-specific patterns of $\Omega_{MCS,i,y}$, $\Omega_{IU,i,y}$, $\Phi_{MCS,i,y}$ and $\Phi_{IU,i,y}$ (for detailed estimations, see

---

[2] We intentionally chose not to use any multidimensional ICT indicators, such as the Network Readiness Index (developed by the World Economic Forum) or the ICT Development Index (developed by the International Telecommunication Union). These measures, despite their simplicity and ability to show a country's overall performance in terms of ICT adoption and readiness to adopt and use the technologies, are not very informative for achieving the main goals of our analysis. The methodologies used to calculate the multidimensional indices are often modified, and hence, their values are lack comparability across time and conclusions drawn on that basis are limited and simplified.

[3] In some countries, the gaps in data coverage are significant, and the available statistics are poor with regard to completeness and time series. Henceforth, in the case of missing data, we provide the statistics for the most recent year for which reliable information was available.

[4] As explained in Chap. 4—if possible the period of analysis is extended for selected countries.

Table 5.1 Determinants of mobile cellular telephony and Internet users penetration rate

Determinants of mobile cellular telephony penetration rates	Determinants of Internet users penetration rates
• Mobile-cellular postpaid connection charge (in USD)—initial, one-time charge for a new postpaid subscription (source: ITU 2013) • Mobile-cellular prepaid connection charge (in USD)—initial, one-time charge for a new postpaid subscription (source: ITU 2013) • Number of mobile cellular prepaid connections charge per monthly GNI per capita (source: author's calculations) • Mobile-cellular prepaid—price of a 1-min local call (peak, on-net) (in USD)—the price per minute of call from a mobile cellular telephony to another of the same network[a] (source: ITU 2013) • Number of 1-min local calls (peak, on-net) per monthly GNI per capita (source: author's calculations) • Mobile-cellular prepaid—price of SMS (on-net) (in USD)—the price of sending one Short Message Service (SMS) message from mobile handset (source: ITU 2013) • Number of SMS (on-net) per monthly GNI per capita (source: author's calculations) • Mobile Cellular Sub-Basket—price of a standard basket of mobile usage per month, including 30 outgoing calls and 100 SMS in arbitrary determined ratios, expressed as percentage on monthly GNI per capita[b] (source: Measuring Information Society 2009, 2010, 2011, 2012, 2013) • Fixed telephony penetration rate—fixed telephony subscriptions (per 100 inhab.) (source: ITU 2013) • Type of competition on mobile telecommunication market—monopoly, partial competition or competition (source: ITU 2013)	• Fixed-narrowband subscriptions[c] (per 100 inhab.) (source: ITU 2013) • Fixed-broadband subscriptions[d] (per 100 inhab.) (source: ITU 2013) • Wireless-broadband subscriptions[e] (per 100 inhab.) (source: ITU 2013) • Fixed (wired)-broadband connection charge (in USD)—initial, one-time charge for new fixed-broadband Internet connection[f] (source: ITU 2013) • Fixed (wired)-broadband monthly subscription charge (in USD)—monthly subscription charge for fixed-broadband Internet service (source: ITU 2013) • Number of fixed-broadband monthly subscription charges per monthly GNI per capita (source: author's calculations) • Fixed-Broadband Sub-Basket—price of monthly subscription to an entry-level fixed-broadband plan expressed as percentage of monthly GNI per capita[g] (source: Measuring Information Society 2009, 2010, 2011, 2012, 2013) • Type of competition on Internet telecommunication market—monopoly, partial competition or competition (source: ITU 2013) • Internet Freedom—status (free, partly free or not free) of freedom of Internet and digital media; 0–100 points; encompasses three sub-indices: Obstacles to Access (infrastructural and economic barriers to access, legal and ownership control over internet service providers, and independence of regulatory bodies)—0–25 points; Limits on Content (legal regulations on content, technical filtering and blocking of websites, self-censorship, the diversity of online news media, and the use of ICTs for civic mobilization)—0–35 points; Violations of Users Right (surveillance, privacy, and repercussions for online activity)—0–40 points (source: Freedom House 2011, 2012, 2013)

Determinants of both mobile cellular telephony and Internet users penetration rates

• Liberalization of Telecommunication market—type of competition on the telecommunication market (full competition/partial competition/monopoly) (various sources)
• Gross Domestic Product per capita in PPP—in constant 2011 international US dollars (source: WDI 2013)

(continued)

Table 5.1 (continued)

Determinants of mobile cellular telephony penetration rates	Determinants of Internet users penetration rates

- Economic Freedom Index—status of economic freedom measured in 4 major areas[h]; scores— 0–100 (if 100—the country is fully free) (source: Heritage Foundation 2013)
- Democracy (Political Freedom)—status of the political regime: democracy[i] (score: 2); democracy with no alternation (score: 1); non-democracy (score: 0) (source: HDR 2010)
- Country Freedom—status (free, partly free or not free) of country freedom with regard to political rights[j] and civil liberties[k] (Freedom House 2014)
- School Enrollment, primary—gross enrollment in primary education regardless of age (%) (source: WDI 2013)
- Rural/urban population—proportion of country's total population living in rural/urban areas (source: WDI 2013)
- Population density—people per square kilometer of land area (source: WDI 2013)

Source: Author's compilation
[a]Refers to the prepaid tariffs
[b]For detailed description of the methodology used to calculate Mobile Cellular Sub-Basket—see Annex 2 in Measuring Information Society 2011 (ITU 2011)
[c]For details—see Chap. 4
[d]For details—see Chap. 4
[e]For details—see Chap. 4
[f]Refers to the cheapest available tariff
[g]For detailed description of the methodology used to calculate Fixed-Broadband Sub-Basket—see Annex 2 in Measuring Information Society 2011 (ITU 2011)
[h]Rule of Law, Government Size, Regulatory Efficiency and Market Openness
[i]The regime may be considered 'democracy' under the following major conditions: the chief executive must be chosen in free popular elections, the legislature shall be popularly elected, and—in free elections more than one political party shall compete and (Cheibub et al. 2010)
[j]Refers to electoral process, political pluralism and participation, functioning of government
[k]Refers to freedom of expression and belief, associational and organizational rights, rule of law, personal autonomy and individual rights

Appendices F and G), which allows for identifying those countries where the *'technological take-off'* was observed. The first thing to note is that the calculated values of $\Omega_{MCS,i,y}$, $\Omega_{IU,i,y}$, $\Phi_{MCS,i,y}$ and $\Phi_{IU,i,y}$ substantially differ across countries. However, despite essential differences, the majority of the economies included in the empirical sample meet the criteria defined in Eq. (3.39).[5] Thus, both $Y_{Crit,MCS}$ and *'technological take-off'* are observed with respect to $MCS_{i,y}$ and $IU_{i,y}$. Taking a closer look at the empirical evidence displayed in Fig. 5.1, we conclude that regarding mobile cellular telephony diffusion, the critical years ($Y_{Crit,MCS}$) that were followed by the characteristic *'technological take-off'* are reported for 16 (cut of 17 analysed) low-income countries. The only exception, where neither $Y_{Crit,MCS}$ nor *'technological take-off'* was found was Eritrea; the paths that demonstrated $\Omega_{MCS,ERI,y}$ and $\Phi_{MCS,ERI,y}$ in 2012 (the terminal year of the analysis) were still converging toward the intersection point. The in-depth analysis reveals that the

---

[5] In Chap. 3.

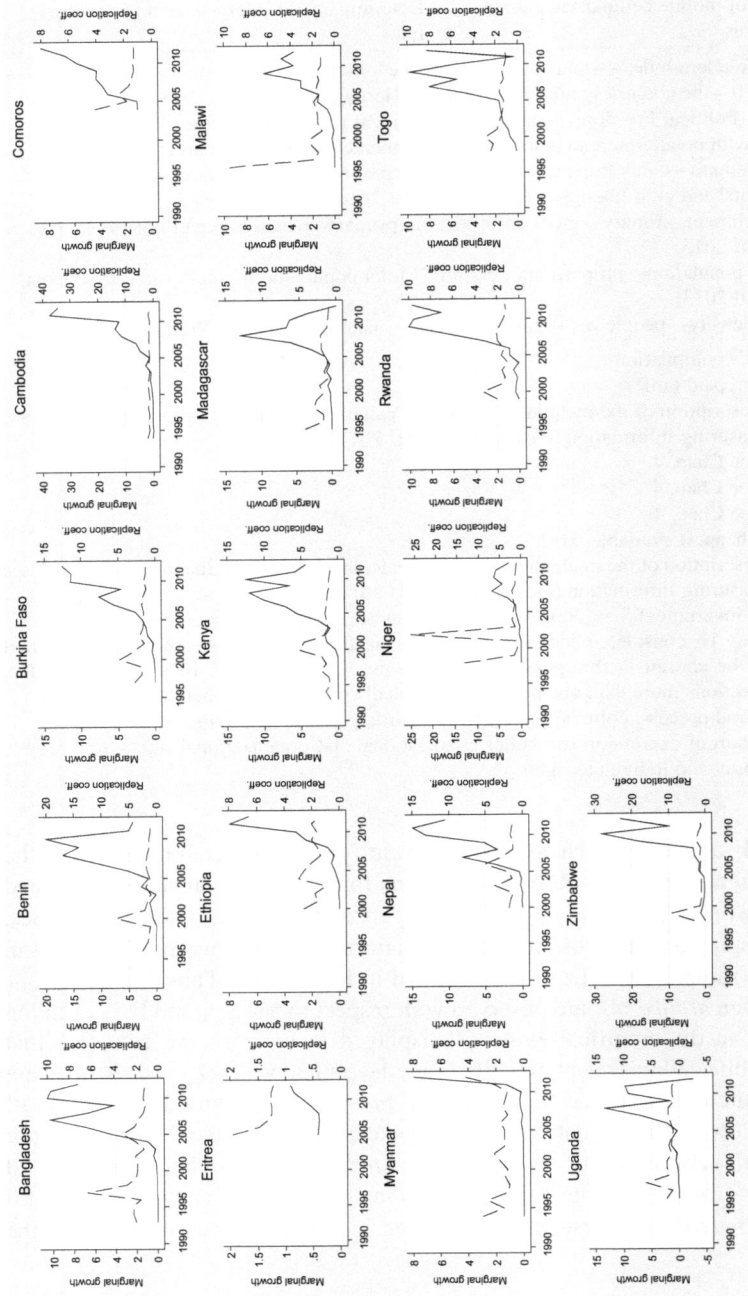

Fig. 5.1 $\Omega_{MCS,i,y}$ and $\Phi_{MCS,i,y}$—Mobile cellular telephony penetration rates. 17 low-income countries. *Source*: Author's elaboration

levels of both $\Omega_{MCS,i,y}$ and $\Phi_{MCS,i,y}$ vary significantly across countries. Over the period 2005–2012,[6] the highest *average* $\Omega_{MCS,i,2004-2012}$ are identified in Cambodia (15.2 per 100 inhab.), whereas the lowest are found in Eritrea (0.5 per 100 inhab.). Consequently, the countries that performed the best in terms of average $\Omega_{MCS,i,y}$ enjoyed the highest dynamics of $MCS_{i,y}$ diffusion, which resulted in their achieving highest $MCS_{i,y}$ penetration rates in 2012. In contrast, the countries that performed the worst in terms of $\Omega_{MCS,i,y}$, in 2012 were still considerably lagging behind with respect $MCS_{i,y}$ penetration rates. Finally, we see that during the 4-year period 2004–2007 that the vast majority of the analysed low-income countries (except Ethiopia and Myanmar) experienced the $Y_{Crit,MCS}$, which, shortly after, was followed by the *'technological take-off.'*[7] The comprehensive study of $\Omega_{IU,i,y}$ and $\Phi_{IU,i,y}$ (see Fig. 5.2), still in the group of low-income economies, documents that the results with respect to $IU_{i,y}$ diffusion are far less satisfactory compared with the evidence for $MCS_{i,y}$. The $Y_{Crit,IU}$ is registered exclusively in seven countries; in the remaining ten economies,[8] meanwhile, the critical year did not occur,[9] and thus, no *'technological take-off'* was observed. Although in Bangladesh, Cambodia, Kenya, Nepal, Rwanda and Uganda, the *'IU-technological take-off'* potentially emerges, the paths that display the changes in $\Omega_{IU,i,y}$ and $\Phi_{IU,i,y}$ are unstable (for Nepal, Rwanda and Uganda), and only in two countries is the initiation of *'IU-technological take-off'* signalled in 2012 (Bangladesh and Cambodia). Figures 5.3 and 5.4 display the evidence on $Y_{Crit,MCS}$, $Y_{Crit,IU}$, $\Omega_{MCS,i,y}$, $\Phi_{MCS,i,y}$, $\Omega_{IU,i,y}$ and $\Phi_{IU,i,y}$ in lower-middle-income countries during the period 2000–2012. Analysing the empirical results with respect to $MCS_{i,y}$, it is evident that irrespective of the strong variations in the $\Omega_{MCS,i,y}$ and $\Phi_{MCS,i,y}$ paths, each of the analysed countries experienced *'MCS-technological take-off'* that was preceded by the country-specific $Y_{Crit,MCS}$ (see Fig. 5.3). More detailed analysis reveals that Bolivia is the country where the $Y_{Crit,MCS}$ registered the earliest, in 1999. During the consecutive period 2000–2005, $Y_{Crit,MCS}$ was identified in the remaining 28 economies.[10] The results of $Y_{Crit,IU}$, $\Omega_{IU,i,y}$ and $\Phi_{IU,i,y}$ in lower-middle-income countries are plotted in Fig. 5.4. The evidence shows that across 26 countries (out of 29 in the scope), the $Y_{Crit,IU}$ occurred and was followed by immediate *'IU-technological take-off'* on the $IU_{i,y}$ diffusion pattern. Unfortunately, in Congo, Mauritania and Pakistan, the process of entering the exponential growth phase

---

[6] The year 2005 is identified as the first year for the *'technological take-off'* in analyzed countries; see later in this section.

[7] When the paths that explain the relationship between $\Omega_{MCS,i,y}$ and $\Phi_{MCS,i,y}$ are not stable, the *'technological take-off'* period may be different from that in the two consecutive years after the $Y_{Crit,MCS}$. In the low-income countries, this is the case for Benin, Cambodia, Myanmar and Nepal.

[8] In Comoros, in 2010, the value of $\Omega_{IU,Com,2010}$ exceeds $\Phi_{IU,COM,2010}$; however in 2 consecutive years, $\Omega_{IU,Com,y} < \Phi_{IU,Com,y}$ again; thus, we argue that the *'take-off'* is not reported.

[9] No intersection points between 'lines' displaying changes in $\Omega_{IU,i,y}$ and $\Phi_{IU,i,y}$ are identified.

[10] In 2000, two countries; in 2001 and 2003, six countries in each year; in 2002, three countries; in 2004, eight countries; and in 2005, three countries).

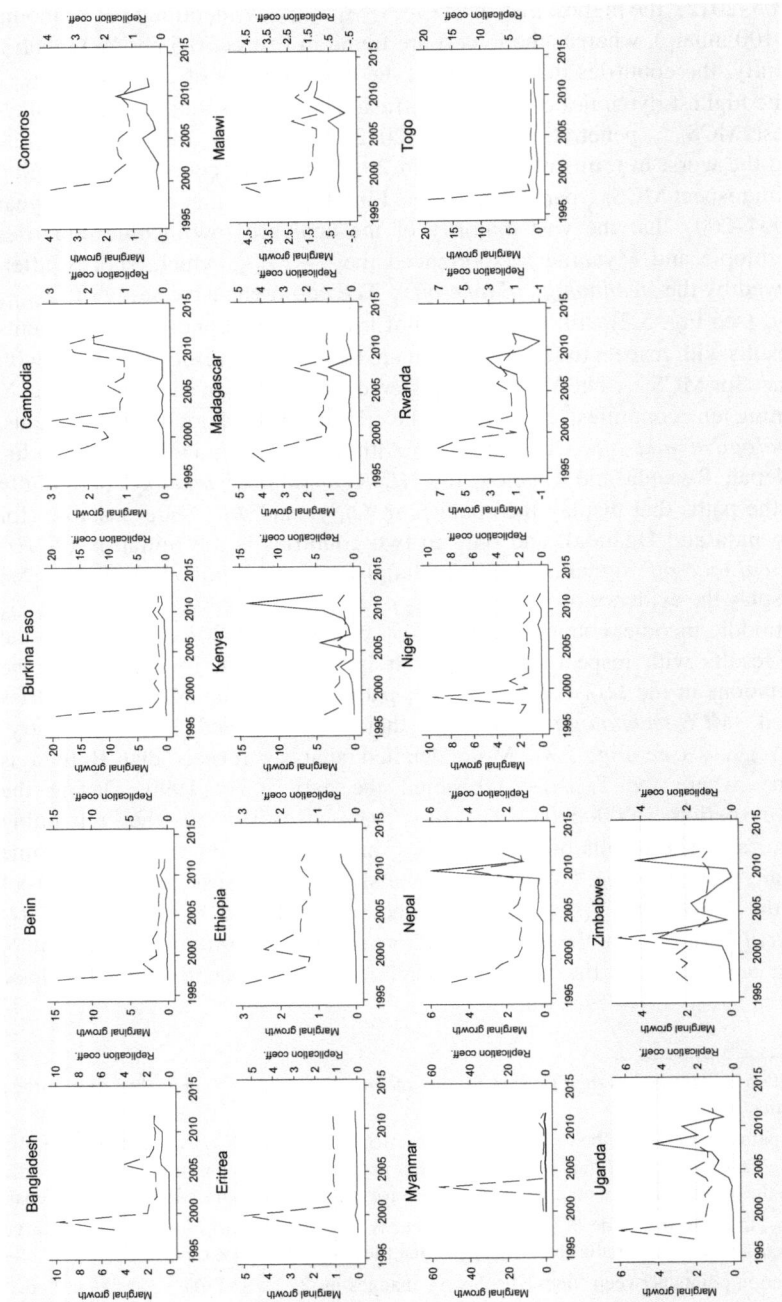

Fig. 5.2 $\Omega_{IV,i,y}$ and $\Phi_{IV,i,y}$—Internet users penetration rates. 17 low-income countries. *Source*: Author's elaboration

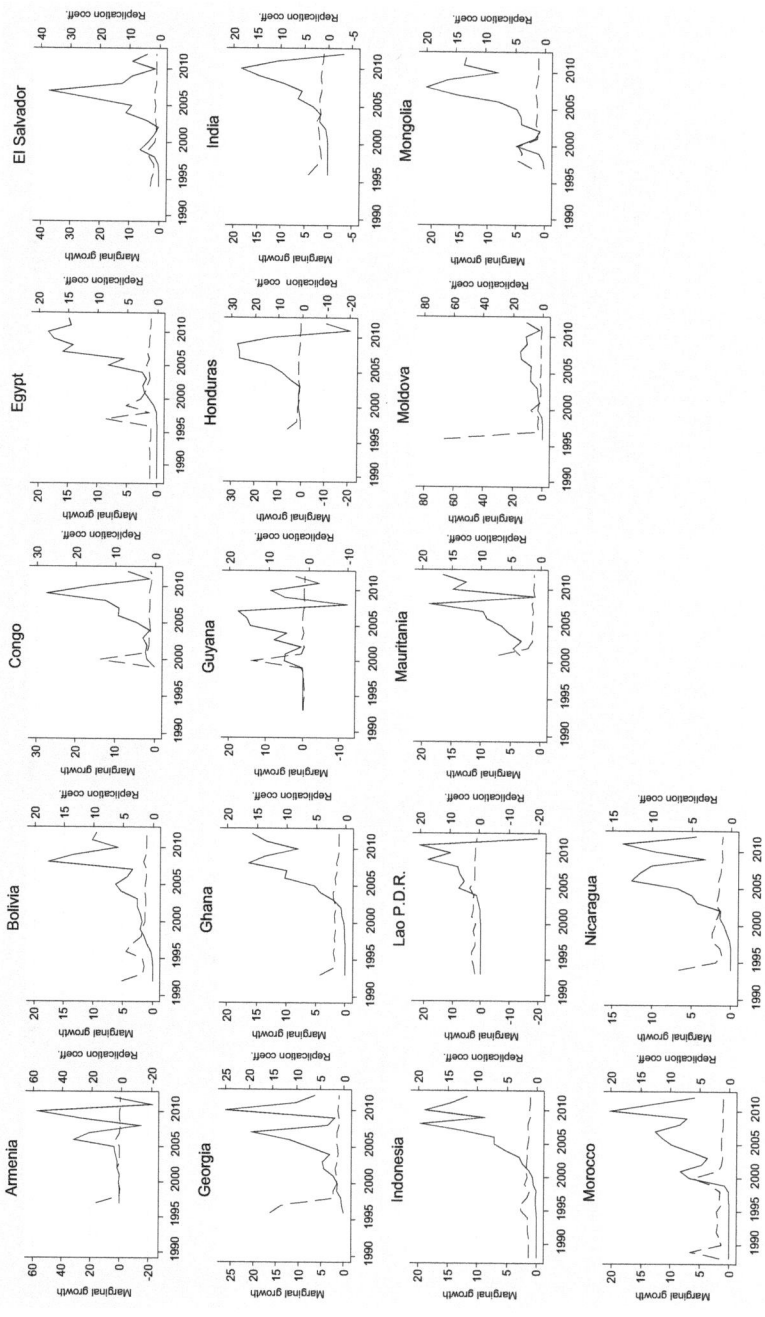

Fig. 5.3 (continued)

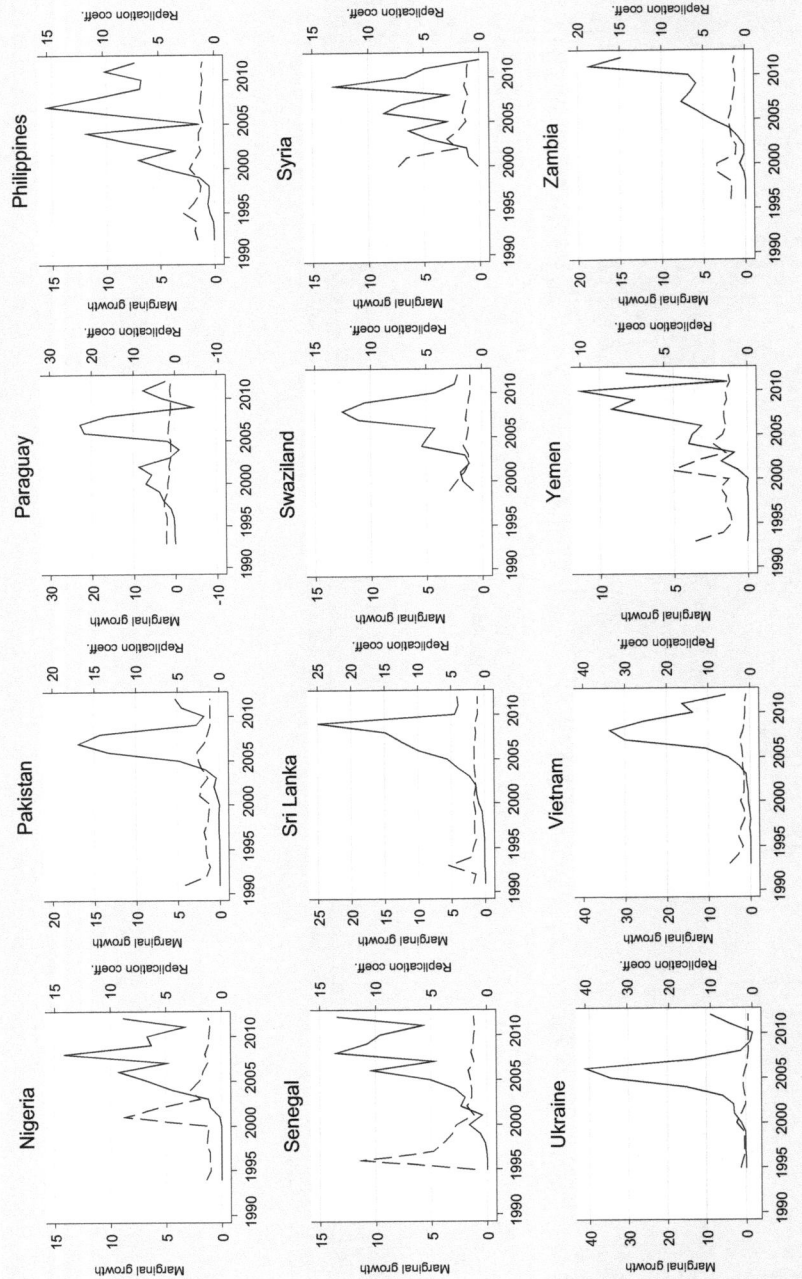

Fig. 5.3 $\Omega_{MCS,i,y}$ and $\Phi_{MCS,i,y}$—Mobile cellular telephony penetration rates. 29 lower-middle-income countries. *Source*: Author's elaboration

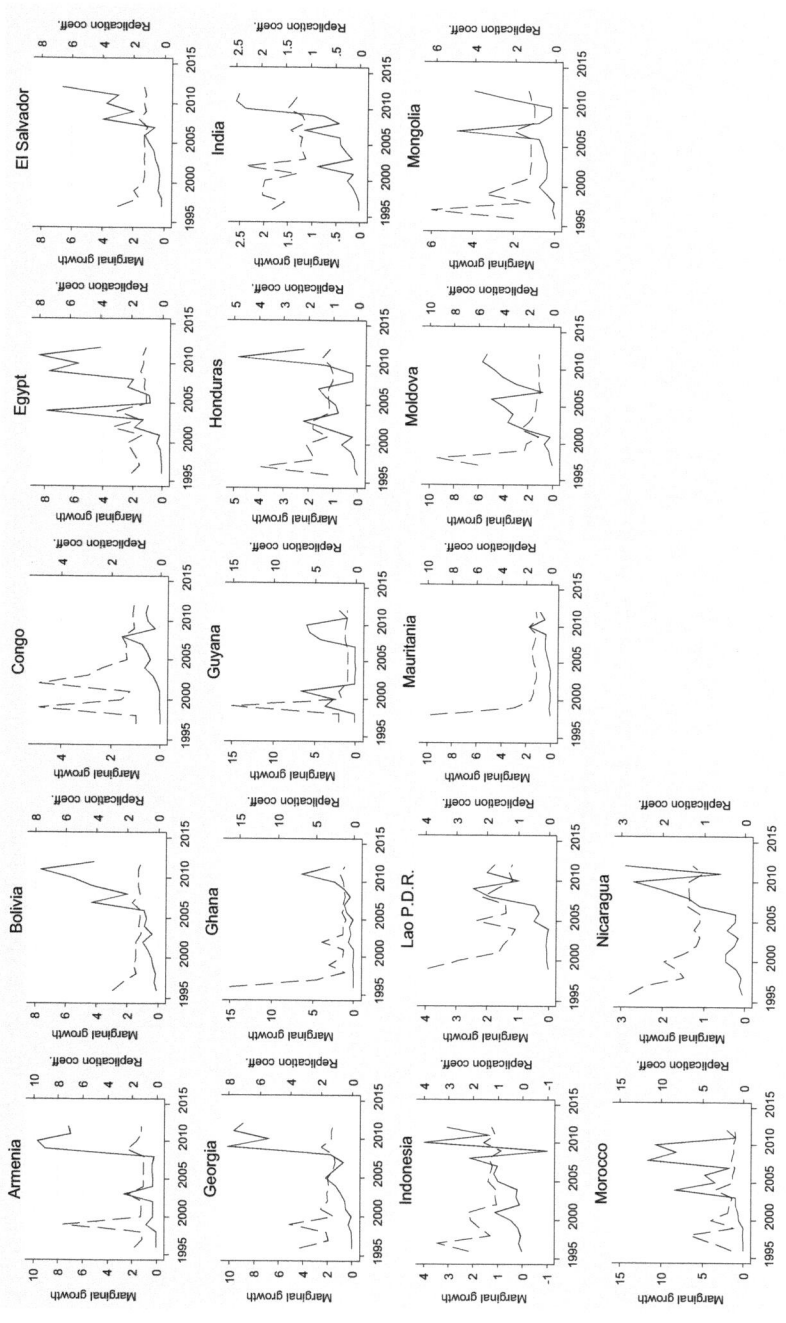

Fig. 5.4 (continued)

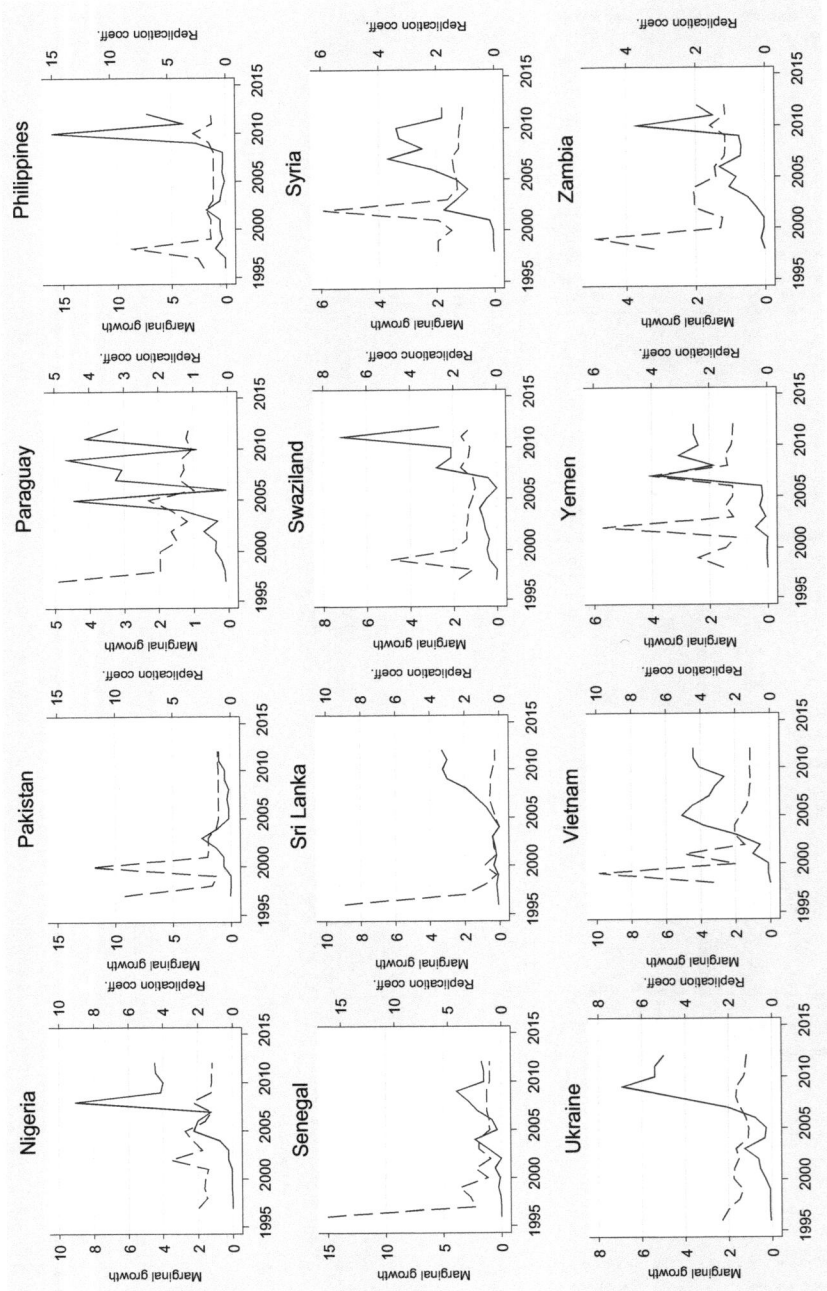

Fig. 5.4 $\Omega_{IU,i,y}$ and $\Phi_{IU,i,y}$—Internet users penetration rates. 29 lower-middle-income countries. *Source:* Author's elaboration

along the $IU_{i,\,y}$ diffusion trajectory was delayed. As a result, in 2012, those countries were still virtually locked in the 'low-level' trap, unable to speed up the ICT diffusion process. From the empirical evidence presented above, a few seminal findings emerge. The analysis of country-wise $\Omega_{ICT,i,y}$ and $\Phi_{ICT,i,y}$ demonstrates that in the early diffusion phase, the ICT replication coefficients are significantly higher compared with ICT marginal growth ($\Omega_{ICT,i,y} < \Phi_{ICT,i,y}$). As diffusion continues, the paths that display the changes in $\Omega_{ICT,i,y}$ and $\Phi_{ICT,i,y}$ gradually converge, so that eventually $\Omega_{ICT,i,y} > \Phi_{ICT,i,y}$. If $\Omega_{ICT,i,y} = \Phi_{ICT,i,y}$ is satisfied, both $Y_{Crit,ICT}$ and the *'technological-take-off'* are reported, which suggests that *'resistance to steady growth'* was overcome (Rostow 1990) and that it fostered exponential growth along the S-shaped diffusion pattern.

As countries experience the *'technological-take-off'*, the diffusion process speeds up, and ICT marginal growths are higher than ICT replication coefficients ($\Omega_{ICT,i,y} > \Phi_{ICT,i,y}$). Conversely, if during the initial phases of diffusion, the paths that demonstrate the changes in $\Omega_{ICT,i,y}$ and $\Phi_{ICT,i,y}$ tend to diverge rather than converge, and the condition $\Omega_{ICT,i,y} = \Phi_{ICT,i,y}$ is not satisfied; thus, $Y_{Crit,ICT}$ does not occur. Countries where $Y_{Crit,ICT}$ was not identified are those where the process of entering the exponential growth phase was restrained; these economies are locked in a 'low-level' trap are latecomers. The previous is reflected by the distinctly lower ICT penetration rates compared with those observed in the countries that forged ahead in the same area.

The remainder of this section is an attempt to answer the question: Under what conditions do countries break out of technological stagnation into exponential ICT growth?. To stay consistent with this target, we summarized the data on selected social, institutional and economic factors that could potentially have shaped the country's ability to accelerate ICT deployment. The data are collected for the first year of the *'technological take-off'* interval.[11] Tables 5.2 and 5.3 coherently summarize our findings on countries' individual characteristics that potentially may play a role in fostering the *'technological take-off'*. Respective tables also report the identified $Y_{crit,ICT}$ and the *'technological take-off'* intervals in examined countries. Following the conceptual specification provided in Chap. 3, we presume that the *'technological take-off'* interval is specified as the 2-year period that immediately follows $Y_{crit,ICT}$. The prime and striking conclusion that arises from the information included in Tables 5.2 and 5.3 is that the examined countries differ greatly on various dimensions. They vary not only in terms of observed $Y_{crit,ICT}$ and the *'technological take-off'* intervals but predominantly with respect to their socio-economic, institutional and political performances. The data displayed in the second column of Table 5.2 shows cross-country critical years ($Y_{crit,MCS}$), which is a starting point for our further analysis. This demonstrates *'how much was enough'* to enhance a specific chain reaction and boost additional $MCS_{i,\,y}$ deployment. In the low-income countries, observed $critMCS_{i,\,y}$ vary between 4.72 in

---

[11] If necessary data are not available for the first year of the *'technological take-off'*, we use the data from the nearest available year.

Table 5.2 '*MCS-technological take-off*'—country-specific conditions. Period 2000–2012[a]

Country	$Y_{Crit,MCS}$	'MCS-techno-logical-take-off' interval	$critMCS_{i,y}$	Mobile-cellular prepaid connection charge	Mobile-cellular prepaid—price of a 1-min local call	Mobile-cellular prepaid—price of SMS	MCS IPB (% of GNI per capita per month)	Fixed telephony penetration	Tele-communication market competition/ mobile	GDP PPP per capita	Economic freedom index	Investment freedom index[b]	Democracy (political freedom)	Country freedom status
Low-income countries														
Bangladesh	2005	2006–2007	6.29	13.0	0.03	0.01	3.38	0.78	C (2012)	1,757	52.9	30	0	Partly free
Benin	2004	2006–2007 (path unstable)	5.80	9.6	0.31	0.04	32.7	0.91	C	1,590	54.0	30	2	Free
Burkina Faso	2005	2006–2007	4.72	114.7	0.34	0.05	47	0.68	C	1,308	55.8	50	1	Partly free
Cambodia	2004	2006–2007 (path unstable)	6.55	n.a.	0.06 (2005)	0.02	11.10	0.18	C	2,136	56.7	50	1	Not free
Comoros	2006	2007–2008	5.98	38.2 (2006)	0.23 (2006)	0.12 (2006)	25.7	3.09	M	1,533	n.a.	n.a.	2	Partly free
Eritrea	Not reported	Not reported	–	100.1 (2009)	0.11 (2006)	0.06 (2009)	33 (2011)	0.94 (2010)	PC (de facto—M)	1,083 (2010)	36.2 (2012)	0 (2012)	1	Not free
Ethiopia	2009	2010–2011	4.78	14.3 (2009)	0.05	0.02	12.60	1.04	M	1,061	51.2	25	1	Partly free
Kenya	2004	2005–2006	7.31	33.0	0.37	0.06	23.60	1.2 (2007)	C	1,894	57.9	50	2	Partly free
Madagascar	2006	2007–2008	5.56	4.6 (2006)	0.14 (2006)	0.05 (2009)	46.60	0.69	C	1,477	61.1	70	2	Partly free
Malawi	2007	2008–2009	7.66	n.a.	0.19 (2008)	0.07	57.40	0.78	C	794	52.7	50	2	Partly free
Myanmar	2012	Not reported	10.3	n.a.	0.30 (2009)	0.05 (2009)	69.6 (2009)	0.86 (2009)	M	n.a.	38.7 (2012)	0 (2012)	0	Not free
Nepal	2007	2010–2011	12.6	15.5	0.02	0.01	7.80	3.1	PC	1,999	52.7	15	2	Partly free

Niger	2007	2008–2009	6.3	n.a.	0.37	0.05	59	3.06	C (since 2003)	843	52.9	50	2	Partly free
Rwanda	2007	2008–2009	6.4	n.a.	0.18	0.09	37.60	0.16	C	1,189	54.2	40	1	Not free
Togo	2004	2005–2006	6.2	23.7	0.26	0.07	60	1.13 (2006)	C	1,211	47.2	30	1	Not free
Uganda	2006	2007–2008	6.8	5.4 (2006)	0.23 (2006)	0.06 (2006)	36.70	0.53	C	1,160	63.1	50	1	Partly free
Zimbabwe	2005	2006–2007	12.9	3.03	0.20	0.07	53.4 (2009)	2.7 (2007)	C	1,568	33.5	10	1	Not free
Lower-middle-income countries													
Armenia	2004	2005–2006	6.7	17.3	0.08	0.05	3.8	19.7	C	5,297	n.a.	n.a.	2	Partly free
Bolivia	1999	2001–2002	5.1	n.a.	0.11	0.05 (2003)	5.6	6.04	C	4,315	68.0	90	2	Free
Congo (Rep.)	2001	2005–2006 (path unstable)	4.7	2.3 (2006)	n.a.	0.06	n.a.	0.44	C	4,978	n.a.	n.a.	1	Not free
Egypt	2001	2004–2005 (path unstable)	4.2	16.1	0.05 (2003)	0.04	3.4	13.4	C	8,332	55.5	55	1	Not free
El Salvador	2003	2004–2005	19.1	n.a.	0.08	0.09	4.4	14.6	C (2012)	6,693	71.2	70	2	Free
Georgia	2001	2002–2003	6.4	11.4 (2003)	0.13	0.27 (2007)	4.8	13.8	C	3,673	n.a.	n.a.	2	Partly free
Ghana	2004	2005–2006	8.1	8.3	0.19	0.06	12	1.5	C (since 1996)	2,521	56.5	50	2	Free
Guyana	2001	2003–2004 (path unstable)	10.1	23.2	0.19	0.03 (2006)	6.8	12.2	PC (2007)	5,124	50.3	50	1	Free
Honduras	2004	2005–2006	10.5	4.9	0.26	0.05	n.a.	7.1	C (2010)	3,952	55.3	30	2	Partly free
India	2004	2005–2006	4.7	2.2	0.02	0.03	2	4.4	C	3,328	54.2	50	2	Free

(continued)

Table 5.2 (continued)

Country	$Y_{CritMCS}$	'MCS-techno-logical-take-off' interval	critMCS$_{i,y}$	Mobile-cellular prepaid connection charge	Mobile-cellular prepaid—price of a 1-min local call	Mobile-cellular prepaid—price of SMS	MCS IPB (% of GNI per capita per month)	Fixed telephony penetration	Tele-communication market competition/mobile	GDP PPP per capita	Economic freedom index	Investment freedom index[b]	Democracy (political freedom)	Country freedom status
Indonesia	2002	2003–2004	5.4	10.3 (2002)	0.03	0.01 (2004)	3.8	3.7	C (2004)	6,035	55.8	50	2	Partly free
Lao P.D.R.	2005	2007–2008 (path unstable)	11.4	5.1 (2006)	0.08 (2006)	0.05 (2006)	6.10	1.6	PC (2005)	3,298	50.3	30	0	Not free
Mauritania	2002	2004–2005 (path unstable)	8.6	7.6 (2003)	0.17	0.03	14.1	1.3	C	2,250	61.8	70	0	Not free
Moldova	2001	2002–2003	5.6	12.5	0.20 (2003)	0.06	8.50	18.1	C	2,666	n.a.	n.a.	2	Partly free
Mongolia	2000–2003 (path unstable)	2003–2006 (path unstable)	6.44 (in 2000)	7.8	0.25	0.02	2.2 (2009)	5.6	PC (since 1999)	4,382	57.7	50	2	Free
Morocco	2000	2001–2006 (path unstable)	8.2	22.1	0.21	0.08 (2002)	11.8	4.1	C	4,710	63.9	70	0	Partly free
Nicaragua	2003	2004–2005	8.8	n.a.	0.48	0.06	16.8	3.9	C	3,644	61.4	50	2	Partly free
Nigeria	2004	2005–2006	6.7	3.8	0.29	0.11	15.6	0.87	C	4,157	48.4	30	2	Partly free
Pakistan	2005	2006–2007	8.1	2.5 (2005)	0.04	0.01	2.60	3.2	C	4,076	57.9	30	2	Not free
Paraguay	2005	2006–2007	32.0	9.8 (2005)	0.13	0.01	4.10	5.5	C	6,180	55.6	50	2	Partly free
Philippines	2000	2001–2004	8.3	3.3	0.15	0.02 (2002)	4.20	4.2	C (2002)	4,275	n.a.	n.a.	2	Free

Senegal	2002	(path unstable)	5.3	5.7	0.28	0.03	12.30	2.2	PC (since 1998)	2,049	58.9	50	2	Free
Sri Lanka	2003	2004–2005 (path unstable)	7.1	14.8	0.10	0.02	1.80	5.0	PC (2009)	5,559	62.5	50	2	Partly free
Swaziland	2003	2004–2005	7.8	6.6 (2003)	0.40	0.12	5.60	4.0	M	6,135	61.6	50	0	Not free
Syria	2003	2004–2006 (path unstable)	6.9	13.4 (2004); 31.1 (2005)	0.07	0.14	6.20	15.0	PC (since 2000)	n.a.	40.6	30	0	Not free
Ukraine	2001	2002–2003	4.6	20.0 (2003)	0.40 (2003)	0.06 (2003)	3.80	22.5	C	5,646	n.a.	n.a.	2	Partly free
Viet Nam	2004	2005–2006	5.9	3.1	0.14	0.02	6.40	9.9	C	3,485	48.1	30	0	Not free
Yemen	2004	2005–2008 (path unstable)	7.5	5.07 (2006)	0.03 (2004)	0.04 (2006)	6.70	4.5	C	4,295	53.8	50	1	Partly free
Zambia	2004	2005–2006	4.2	3.4	0.20 (2004)	0.05 (2004)	18.50	0.82	C	2,346	55	50	1	Partly free

Source: Author's elaboration. Note: Data for mobile cellular sub-basket are for 2008 (earlier not available)—reported if otherwise. Data for 'Democracy' are for 2010

[a]If necessary the period of analysis is extended

[b]Investment Freedom Index is a subindex of Economic Freedom Index

Table 5.3 '*IU-technological take-off*'—country-specific conditions. Period 2000–2012[a]

Country	$Y_{Crit,IU}$	'IU-technological-take-off' interval	$critIU_{i,y}$	Internet freedom	Obstacles to access	Limits on content	Violations of users right	$FIS_{i,y}$ in Y_{Crit}	$FBS_{i,y}$ in Y_{Crit}	$WBS_{i,y}$ in Y_{Crit}
Low-income countries										
Bangladesh	2012	2013–2014 (projected)	6.3	Partly free	13 (2013)	12 (2013)	24 (2013)	1.5	0.38	0.47
Benin	Not reported	Not reported	–	n.a.	n.a.	n.a.	n.a.	–	–	–
Burkina Faso	Not reported	Not reported	–	n.a.	n.a.	n.a.	n.a.	–	–	–
Cambodia	2012	2013–2014 (projected)	4.9	Partly free (2013)	14 (2013)	15 (2013)	18 (2013)	0.41	0.2	6.7
Comoros	Not reported	Not reported	–	n.a.	n.a.	n.a.	n.a.	–	–	–
Eritrea	Not reported	Not reported	–	n.a.	n.a.	n.a.	n.a.	–	–	–
Ethiopia	Not reported	Not reported	–	Not free	22	27	26	–	–	–
Kenya	2009	2010–2011	10.04	Partly free	10	7	12	0.02	0.02	0.01
Madagascar	Not reported	Not reported	–	n.a.	n.a.	n.a.	n.a.	–	–	–
Malawi	Not reported	Not reported	–	Partly free	16 (2013)	11 (2013)	15 (2013)	–	–	–
Myanmar	Not reported	Not reported	–	Not free	22	23	30	–	–	–
Nepal	2010 and 2012	2010–2012 (path unstable)	11.2 (2010)	n.a.	n.a.	n.a.	n.a.	0.47	0.2	n.a.
Niger	Not reported	Not reported	–	n.a.	n.a.	n.a.	n.a.	–	–	–
Rwanda	2008	2008–2012 (path unstable)	4.5	n.a.	n.a.	n.a.	n.a.	0.08	0.1	n.a.

Togo	Not reported	Not reported	–	n.a.	n.a.	n.a.	n.a.	–	–	–
Uganda	2008	2008–2012 (path unstable)	7.9	Partly free (2012)	11	8	15	0.06	0.1	n.a.
Zimbabwe	2003, 2005, 2011	2003–2012 (path unstable)	6.4 (2003)	Partly free (2012)	17	14	23	0.65	0.7	n.a.
Lower-middle-income countries										
Armenia	2009	2009–2012 (path unstable)	15.3	Free (2013)	8 (2013)	9 (2013)	12 (2013)	2.6	1.05	6.5
Bolivia	2007	2009–2010 (path unstable)	10.5	n.a.	n.a.	n.a.	n.a.	2.05	0.35	n.a.
Congo (Rep.)	Not reported	Not reported	–	n.a.	n.a.	n.a.	n.a.	–	–	–
Egypt	2007	2009–2010 (path unstable)	16.03	Partly free (2013)	14	12	33	3.6	0.64	n.a.
El Salvador	2008	2008–2012 (path unstable)	10.08	n.a.	n.a.	n.a.	n.a.	2.04	2	n.a.
Georgia	2008	2009–2010	10.01	Free (2013)	9	10	11	3.04	2.5	n.a.
Ghana	2010	2011–2012	7.8	n.a.	n.a.	n.a.	n.a.	0.21	0.2	6.9
Guyana	2008	2009–2010	18.2	n.a.	n.a.	n.a.	n.a.	2.8	0.65	n.a.
Honduras	2010	2011–2012	11.09	n.a.	n.a.	n.a.	n.a.	0.9	0.01	n.a.
India	2010	2011–2012	7.5	Partly free (2013)	13	9	17	1.5	0.91	0.94
Indonesia	2010	2010–2012 (path unstable)	10.9	Partly free (2013)	11	11	20	0.79	0.94	18.6
Lao P.D.R.	2009 and 2011	2009–2012 (path unstable)	6	n.a.	n.a.	n.a.	n.a.	0.31	0.06	n.a.
Mauritania	Not reported	Not reported	–	n.a.	n.a.	n.a.	n.a.	–	–	–
Moldova	2008	2009–2010	23.4	n.a.	n.a.	n.a.	n.a.	5.6	3.2	0.96

(continued)

Table 5.3 (continued)

Country	$Y_{Crit,IU}$	'IU-technological-take-off' interval	$critIU_{i,y}$	Internet freedom	Obstacles to access	Limits on content	Violations of users right	$FIS_{i,y}$ in Y_{Crit}	$FBS_{i,y}$ in Y_{Crit}	$WBS_{i,y}$ in Y_{Crit}
Mongolia	2007 and 2011	2007–2012 (path unstable)	9	n.a.	n.a.	n.a.	n.a.	0.66	0.28	n.a.
Morocco	2004	2004–2008 (path unstable)	11.6	Partly free (2013)	11 (2013)	7 (2013)	24 (2013)	0.37	0.21	n.a.
Nicaragua	2008	2009–2010	5.3	n.a.	n.a.	n.a.	n.a.	0.75	0.76	n.a.
Nigeria	2006	2008–2009 (path unstable)	5.5	Partly free (2013)	12	9	12	0.08	0.01	n.a.
Pakistan	Not reported	Not reported	–	Not free (2013)	19	18	26	–	–	–
Paraguay	2005 and 2007	2005–2009 (path unstable)	7.9	n.a.	n.a.	n.a.	n.a.	1.01	0.09	n.a.
Philippines	2009	2010–2011	9	Free (2013)	10	5	8	3.9	1.8	1.8
Senegal	2007	2008–2009	7.7	n.a.	n.a.	n.a.	n.a.	0.32	0.38	n.a.
Sri Lanka	2003 and 2005	2003–2009 (path unstable)	1.4	Partly free (2013)	16	18	21	0.43	0.01	n.a.
Swaziland	2008	2008–2011 (path unstable)	6.8	n.a.	n.a.	n.a.	n.a.	1.7	0.06	n.a.
Syria	2005	2006–2007	5.6	Not free (2013)	23	25	35	1.2	0.01	n.a.
Ukraine	2007	2008–2009	6.5	Free (2013)	7	8	12	2.9	1.7	n.a.
Viet Nam	2004	2005–2006	7.6	Not free (2013)	16	26	31	1.9	0.06	n.a.
Yemen	2008	2009–2010	6.9	n.a.	n.a.	n.a.	n.a.	1.4	0.11	n.a.
Zambia	2010	2010–2012 (path unstable)	10	n.a.	n.a.	n.a.	n.a.	0.13	0.07	0.26

Country	Fixed (wired)-broadband connection charge	Fixed (wired)-broadband monthly subscription charge	FBS IPB (% of GNI per capita per month)	Telecommunication market competition (fixed broadband connections/internet services)	GDP PPPper capita	Economic freedom index	Investment freedom index[b]	Democracy (political freedom)	Country freedom status
Low-income countries									
Bangladesh	4.2	4.2	7.3	C/C	2,363.8	53.2	55	0	Partly free
Benin	39.1	48.9	81.5	M/PC	1,686.5	55.7	60	2	Free
Burkina Faso	29.3	43.09	98.2	C/C	1,527.9	60.6	55	1	Partly free
Cambodia	4.9	23.7	34	C/C	2,790.4	57.6	60	1	Not free
Comoros	26.1	48.3	620 (2010)	n.a./M	1,493.2	45.7	15	2	Partly free
Eritrea	68.3	1,596	4,455.4	M/PC	1,179.8	36.2	0	1	Not free
Ethiopia	56.4	22.5	71	M/M	1,218.3	52	25	1	Partly free
Kenya	50.4 (2010)	38.7	57.6	C/C	2,079.9	57.5	45	2	Partly free
Madagascar	45.01	58.7	177.8	M/C (2008)	1,390.8	62.4	55	2	Partly free
Malawi	702.7 (2010)	30.2	169.7	C (2012)/PC (1998)	737.3	56.4	50	2	Partly free
Myanmar	1315.7 (2009)	28.2 (2009)	155 (2009)	PC/M	n.a.	38.7	0	0	Not free
Nepal	7.7	23.2	63.4	C/C	1,999.1	52.7	15	2	Partly free
Niger	39.1	58.2	193 (2011)	n.a./M (2003)	884.0	54.3	55	2	Partly free
Rwanda	175.9 (2009)	91.4	344.3	C/C (2010)	1,189.2	54.2	40	1	Not free
Togo	125.1 (2011)	43.6	101.2	M/C (2010)	1,313.9	48.3	25	1	Not free

(continued)

Table 5.3 (continued)

Country	Fixed (wired)-broadband connection charge	Fixed (wired)-broadband monthly subscription charge	FBS IPB (% of GNI per capita per month)	Telecommunication market competition (fixed broadband connections/internet services)	GDP PPPper capita	Economic freedom index	Investment freedom index[b]	Democracy (political freedom)	Country freedom status
Uganda	145.2 (2009)	131.2	600	C/C (2009)	1,218.9	63.8	50	1	Partly free
Zimbabwe	267.3 (2006)	2,672.9 (2006)	1,059 (2010); 56.3 (2011)	C/C	1,907.7	36.7	10	1	Not free
Lower-middle-income countries									
Armenia	n.a.	33.03	11.3	C/C	6,357.8	n.a.	n.a.	2	Partly free
Bolivia	42.4	34.9	28.9	n.a./C (2001)	5,152.5	53.6	20	2	Free
Congo (Rep.)	250.06 (2012)	400.1 (2012)	n.a.	n.a./C	5,631.3	n.a.	n.a.	1	Not free
Egypt	n.a.	8.11	5.46	C/C	10,272.1	58	50	1	Not free
El Salvador	n.a.	20.3	7.6	C/C	7,450.0	68.5	70	2	Free
Georgia	n.a.	41.9	20.5	C/C	5,654.9	n.a.	n.a.	2	Partly free
Ghana	36.3	29.7	25.3	C/C	3,445.8	59.4	65	2	Free
Guyana	14.7	48.9	41.8	PC/PC (2007)	5,344.4	48.4	40	1	Free
Honduras	n.a.	19.3	14.1 (2010)	C/C (2010)	4,345.2	58.6	60	2	Partly free
India	n.a.	6.01	5.1	C/C	4,883.1	54.5	35	2	Free
Indonesia	8.25	21.4	10.4	C/C (2004)	8,026.7	55.5	35	2	Partly free
Lao P.D.R.	116.2	193.7	315.1	PC/PC (2005)	3,667.5	50.4	30	0	Not free
Mauritania	23.5 (2012)	21.1 (2012)	26.8 (2012)	C/C (2010)	2,828.8	53	40	0	Not free

Moldova	n.a.	13.5	10.9	C/C	3,648.2	n.a.	n.a.	2	Partly free
Mongolia	38.1 (2006)	8.3 (2009)	6.04 (2009)	PC/C (2010)	5,918.9	60.3	60	2	Free
Morocco	n.a.	19.2 (2008)	10.6 (2008)	C/C	5,258.8	56.7	70	0	Partly free
Nicaragua	23.2	34.8	38.2	C/C	3,908.5	69.8	70	2	Partly free
Nigeria	337.4	674.8	890.4	C/C (2001)	4,714.7	55.1	30	2	Partly free
Pakistan	10.7 (2012)	13.3 (2012)	15.5 (2012)	C/C	4,360.3	54.7	35	2	Not free
Paraguay	60.0 (2006)	28.9 (2006)	25.1 (2008)	C/C	6,007.1	53.4	50	2	Partly free
Philippines	24.3	22.1	12.9	n.a./C (2002)	5,613.7	53.6	40	2	Free
Senegal	87.09	40.1	61.3	PC/C (1998)	2,161.7	58.3	50	2	Free
Sri Lanka	3.9 (2004)	23.2 (2008)	16.4 (2008)	PC/PC (2009)	5,343.7	62.5	50	2	Partly free
Swaziland	280.3	1,781.8	873.2	M/PC (1983)	6,526.8	58.4	50	0	Not free
Syria	n.a.	51.7 (2008)	34.9 (2008)	PC/PC	1,799.9	51.2	30	0	Not free
Ukraine	15.4	18.9	9.7	C/C	8,775.8	n.a.	n.a.	2	Partly free
Viet Nam	37.8	3.1	25.7 (2008)	C/C	3,484.8	48.1	30	0	Not free
Yemen	147.8	221.8	277.8	n.a./C (2004)	4,493.9	56.9	50	1	Partly free
Zambia	962.8	58	65	C/C (2011)	2,779.0	58	50	1	Partly free

Source: Author's elaboration. Note: for countries where the *'IU-technological take-off'* was not reported—data are for 2012 (reported if otherwise). The data for 'Democracy,' are for 2010. Data for Internet freedom (Obstacles to use, Limits on content and Violations of users rights) are for 2012

aIf necessary the period of analysis is extended

bInvestment Freedom Index is a subindex of Economic Freedom Index

Burkina Faso and 12.9 in Zimbabwe; while the average $crit\text{MCS}_{\text{low-income, y}}$ is 7.2[12] per 100 inhab.,[13] demonstrating that these countries inevitably head toward the '*MCS-technological take-off*' once the $\text{MCS}_{\text{i, y}}$ penetration rates reaches an average of 7.2 per 100 inhab.[14] Our empirical evidence also demonstrates that the average duration of the diffusion initial phase[15]—the length of time required for the '*MCS-technological take-off*' to emerge—in the low-income countries, was approximately 12 years; however, it varied significantly, ranging from 15 years in Bangladesh to 5 years in Comoros. Careful examination of the country-specific structural characteristics that are reported for the first year of the '*MCS-technological take-off*' interval leads to a few important conclusions. First, we consider the elements that may be described as *direct* stimuli for the '*MCS-technological take-off*', which are the following[16]: the price of a 1-min call, the price of sending one SMS, the cost of mobile-cellular prepaid connection, the mobile cellular sub-basket, per capita income and fixed telephony penetration rates, type of competition in the telecommunication market, economic freedom and investment freedom. Elements such as price of a 1-min call, price for sending an SMS, and mobile-cellular prepaid connection charges along with per capita income most directly affect the basic affordability of mobile cellular services. Overall affordability is also demonstrated through the mobile cellular sub-basket, which accounts for the percent of GNI per capita per month that must be spend to buy the standard basket of mobile cellular services; the influence of per capita income is thus demonstrated throughout this channel. The degree of competition (full competition, partial competition or monopoly) in the telecommunication market determines companies' possibilities of operating freely in a country. Economic freedom, as such, constitutes an essential element in shaping a country's economic environment, and investment freedom coherently measures country's market openness for inflows and outflows of goods and services; investment freedom also reflects possible constraints on and restrictions of investment capital flows. Fixed telephony penetration may, to a point, affect the adoption of mobile cellular telephony as a favourable alternative, if the mobile telephony is not freely accessible. The respective prices of 1-min calls varied significantly across countries. The highest prices are reported for Kenya (US$0.37),[17] and the lowest are for Nepal (US$0.02). In the lower-middle-income countries, the price of a 1-min call ranges from US$0.48 in Nicaragua, to US$0.02 in India. The differences in SMS prices are

---

[12] If the two extreme observations (Zimbabwe and Nepal) are eliminated, the average decreases until $crit\text{MCS}_{\text{i,y}} = 6.1$ per 100 inhab.

[13] Author's calculations.

[14] Obviously, the $\text{MCS}_{\text{i,y}} = 7.2$ (per 100 inhab.) stands for different absolute numbers of people in each country.

[15] The length of the initial diffusion phase we calculate as the number of years between the year when given ICT was first introduced until the first year of the '*technological take-off*'.

[16] For detailed description of variables—see Sect. 5.2.1.

[17] The prices of one-minute calls and SMS are expressed in United States dollars in PPP terms.

not so striking, although they are still essential across the examined countries in both income groups. Although the analysis of absolute mobile cellular service costs provides elementary information on the potential demand for these services, we argue that it would be far more informative to put mobile cellular service prices into an 'income perspective', which allows for assessing the overall affordability of ICT services. With this aim, we use the cost of the mobile cellular sub-basket expressed as a percentage of GNI per capita per month to draw conclusions on the affordability of mobile cellular services, which mirrors an individual's overall propensity to buy these services in a given country. The extensive analysis of cross-country mobile cellular sub-basket costs supports the supposition that—surprisingly—even *low* affordability does not inhibit the rapid expansion of mobile cellular networks. This is a far-reaching observation that reflects unusual tendencies in low-income countries. In both the low-income and the lower-middle-income groups, the '*MCS-technological take-off*' occurred under highly unfavourable conditions, while the affordability of mobile cellular services was low. According to the evidence summarised in Table 5.2, a few countries reflect extremely low $MCS_{i, y}$ affordability[18]: Togo (60 %) or Niger (59 %). The comparison between Niger and e.g. Bangladesh is striking; in Niger, the mobile cellular sub-basket accounts for approximately 59 % of GNI per capita per month, whereas in Bangladesh, the amount is only 3.38 %. Despite the vast differences in the values of mobile cellular sub-baskets, these two countries are primed for exponential growth in $critMCS_{i, y} = 6.3$ %; they both achieved similar $MCS_{i, y}$ penetration rates in the terminal year of our analysis (2012), approximately $MCS_{i,2012} = 60$ %. In the lower-middle-income economies, the cross-country disparities in the value of mobile cellular sub-baskets are less striking. The average mobile cellular sub-basket cost was estimated at roughly 7.18 %; the highest costs were reported in Zambia (18.5 %), and the lowest were in India (2.0 %). Although the results, especially in the case of the low-income countries, are at odds with basic intuition, they demonstrate that low affordability does not constitute a significant barrier for mobile cellular services acquisition and does not impede its rapid spread. This evidence also reflects individuals' astonishingly high propensity to acquire mobile cellular telephony even in the most economically backward countries. The cost of mobile-prepaid connection during the '*MCS-technological take-off*' varies extensively across countries, ranging from US$114.7 in Burkina Faso to US$3.03 in Zimbabwe. This evidence coincides with the previous findings and may suggest that the '*MCS-technological take-off*' is possible even if the one-time initial charge for mobile cellular telephony usage is relatively high and could potentially limit the rapid spread of mobile cellular services. Regarding the lower-middle-income countries, the variability in mobile-prepaid connection charges is far lower. The average cost of a mobile-prepaid connection was US$9.32, and there were no substantial differences across countries. As a reminder, the penetration rates for fixed telephony in both income groups remained extremely low over the examined

---

[18] Note that the first data on mobile cellular sub-basket prices are available in 2008 (ITU 2010).

period; that is, the majority of individuals and firms rarely accessed and used telephone landlines. Because the emergence of the '*MCS-technological take-off*' is a complex phenomenon, we additionally intend to focus on its deep determinants, mostly associated with institutional environments and political regimes (Rodrik et al. 2004).

Table 5.2 also summarizes the information on political regimes, political and economic freedom and types of competition in telecommunication markets across the countries in our scope. The first and very important thing to note is that in 12 (out of 15) low-income countries,[19] the telecommunication markets were fully liberalized during the '*technological take-off*'. The presence of full competition yields increasing telecommunication market efficiency, and provides a solid background for creating benefits for consumers owing to more balanced tariffs and growing geographic coverage. In only two countries, Ethiopia and Comoros, were the telecommunication markets fully monopolized; in another, Nepal, the telecom market was labelled partial competition[20] (World Bank Group 2014). In Ethiopia, in 2010 (the $Y_{crit,MCS}$), the telecommunication market was fully controlled by Ethio-Telecom (provider of fixed, mobile and Internet services), which significantly impeded tariff reductions and any increase in affordable and innovative services. Although the '*MCS-technology take-off*' was observed in Ethiopia in 2010–2011, the overall penetration remained relatively low (in 2012, $MCS_{ETH,2012} = 22.4$ per 100 inhab.). In turn, in Comoros, despite the fully monopolized telecommunication market (the mobile operator is Comoros Telecom/Huri), the relatively high prices of 1-min calls and sending SMSs, and the relatively low affordability; in 2012, the mobile cellular telephony penetration rate reached $MCS_{i, y} = 39.5$ per 100 inhab., although according to various sources, the of mobile cellular telephony network coverage was limited to urban areas. Meanwhile, in the lower-middle-income economies, in 22 countries (out of 29 where the '*MCS-technology take-off*' was reported), 'full competition' in the telecommunication markets was observed; 'partial competition' was observed in six countries; and 'full monopoly' was observed in one country (Swaziland). The lack of full competition, however, did not restrict either the '*MCS-technology take-off*' or the rapid expansion of mobile cellular networks. As a reminder, in 2012 (the terminal year of our analysis), the $MCS_{i, y}$ penetration rates were unexpectedly high in, e.g., Mongolia (120.7 per 100 inhab.) and Sri Lanka (91.6 per 100 inhab.); the costs of mobile cellular sub-baskets were, respectively, 2.2 % and 1.8 % of GNI per capita per month. The only country where the telecommunication market was *not* liberalized was

---

[19] In Malawi, although the telecommunication market is labelled 'full competition' (World Bank Group 2014), there are only two telecom operators, Airtel and Telecom Networks Malawi. In Zimbabwe, although from 2000 onward, the telecommunication market was labelled 'full competition', since 2009, it has been labelled 'partial competition'. In 2014 in Zimbabwe, there were three mobile operators, Econet Wireless, Telecell Zimbabwe Ltd., and TelOne.

[20] In Nepal, there are two mobile operators, Ncell and Nepal Telecom. Source: www.africantelecomsnews.com and www.nta.gov.np/en/; accessed: May 2014).

Swaziland. Notably, despite the existence of a fully monopolized telecommunication market,[21] the '*MCS-technology take-off*' took place in 2004–2005, fostering the rapid spread of cellular telephony, so that in 2012, $MCS_{SWZ,2012} = 65.4$ per 100 inhab. Not surprisingly, in Swaziland, because of the absence of liberalised telecommunication services, the prices of both a 1-min call and sending an SMS were comparably high, US$0.40 and US$0.12, respectively, among the highest rates in the lower-middle-income countries. However, despite the relatively high prices for basic mobile cellular services, the cost of acquiring mobile cellular sub-baskets was 5.6 % in 2008; the affordability of mobile cellular services was high in Swaziland. Therefore, high affordability may be recognized as a major driving force of exponential increases in the number of mobile cellular networks users in Swaziland during the period 2003–2012. Regarding the results on political regimes and countries' freedoms, the evidence is rather mixed and reveals little regularity. Using the Freedom House methodology, ten counties were classified as 'partly free', another four were 'not free', and only one country (*sic!*), Benin, attained 'free' status.[22] These results are striking. The remaining four countries labelled 'not free' are those where both political rights and civil liberties were heavily violated. In another ranking of broadly perceived political freedoms, provided in the Human Development Report 2010, seven countries scored[23] '2' and were claimed to be democracies; another seven scored '1' and were claimed to be democracies but with no alternation; and only one country, Bangladesh, scored '0' was labelled nondemocratic. The analogous comparison for the lower-middle-income group reveals that according to Freedom House, eight counties out of the considered were classified as 'not free', and another 13 economies were recognized as 'partly free', and the remaining eight were labelled 'free'. In the classification presented in the Human Development Report 2010, six countries attained a score of '0' and thus were classified as nondemocratic; another five scored '1', and the remaining 18 scored '2' and were considered democracies. Similar to the low-income countries, the lack of democracy and/or heavy violations of political rights and civil liberties did not preclude the emergence of the '*MCS-technology take-off*' and the broad expansion of mobile cellular networks in undemocratic and politically restricted countries. Addressing the results of countries' ratings regarding economic and investment freedoms (see the Heritage Foundation), the cross-country variation is high. Economic freedom is reflected in the freedom to choose to '*work, consume and produce*' (Heritage Foundation 2014) without being constrained '*beyond the extent necessary to protect and maintain the liberty itself*' (Heritage Foundation 2014). However, for the expansion of mobile networks, the level of investment freedom is arguably seminal, as shown in the degree of constrains that are arbitrarily imposed on flows of investment capital. Multiple restrictions on investments generally, depending on state policies and national

---

[21] The only mobile operator is MTN Swaziland.

[22] The meanings of 'country status' are provided in Sect. 5.2.1.

[23] The meaning of the 'scores' are provided in Sect. 5.2.1.

development strategies, promote or limit the effective investment actions undertaken by domestic and/or foreign companies. Across the low-income countries where the '*MCS-technological take-off*' took place, the average investment freedom index was 39.2, the best-performing country (with the weakest investment restrictions) was Madagascar at 70.0; the worst was Zimbabwe (10.0), where the investment process was highly restricted and state-regulated. The related disparities among the lower-income group are less striking. The average score for the investment freedom index was 49.3, with Bolivia the best performer at 90.0 (*sic!*) and with the worst being Honduras, Lao P.D.R, Nigeria, Pakistan, Syria and Viet Nam (30.0 in each case). The examples of Viet Nam and Swaziland appear to be the most interesting. In Viet Nam, despite the authoritarian regime, the lack of political rights and civil liberties, and the limited investment freedom,[24] the telecommunication market was fully liberalized.[25] For the rapid expansion of mobile cellular networks, the seminal factor was the approval, in 2001 (4 years before the '*MCS-technological take-off*'), of The Vietnam Post and Telecommunication Development Strategy to 2010; this legal document directly states a strong willingness to build, by 2020, a modern ICT infrastructure and, resultantly, an information society in Viet Nam[26] (Tuan 2011). The latter induced the '*MCS-technological take-off*' (in 2005–2006), which in a relatively short period dramatically shifted the mobile cellular penetration rates. The basic analysis of the degree of economic freedom (especially investment freedom) shows that there might be no single correct answer to the question: 'To what extent does economic freedom affect the '*MCS-technological take-off*'?'. The evidence might suggest that even under relatively unfavourable conditions for investment capital flows, the rapid expansion of mobile cellular services is not restricted. In contrast to what might have been expected, the combined evidence on countries' political regimes (democracies or dictatorships), freedom status (regarding violations of political rights and civil liberties) and, especially, investment freedom, has demonstrated that mobile cellular network expansion has relatively little to do with these three elements. The case of Swaziland is even more striking. In 2003 (the $Y_{crit,MCS}$), the country was classified as 'not free' and 'nondemocratic', with a fully monopolized telecommunication market. However, the numerical evidence demonstrates that even under extremely unfavourable conditions, the emergence of the '*MCS-technological take-off*' is still possible. Important to note is that in Swaziland, the cost of a standard mobile cellular sub-basket was relatively low (5.6 %, as mentioned previously), which was below the lower-middle-income group average and may be considered a

---

[24] Viet Nam has adopted a two-track approach to trade liberalization: By government decision, the country has been opened to foreign investment capital while at the same time providing high protection to multiple sectors (Tuan 2011).

[25] According to ITU data, in 2012 in Viet Nam, there were six active mobile operators, Viettel, Mobifone, Vinaphone, S-Telecom, Hanoi-Telecom, GTEL.

[26] In following years - 2005, 2006, 2008 and 2010, the government of Viet Nam adopted another four documents that enabled a national policy on broad ICT deployment. For details, see Broadband in Vietnam: Forging Its Own Path. Washington, D.C: infoDev/World Bank. 2011.

seminal driver of $MCS_{i, y}$ diffusion in Swaziland. It is also not insignificant that in 1995, the United Nation Economic Commission for Africa (ECA) released and adopted the first African Information Society Initiative (AISI), the primary target of which was to promote and assist actions that were designed to build information societies in African countries. In response, in 2000 (4 years before the *'MCS-technology take-off'*) in Swaziland, in cooperation with UNDP, UNESCO, ECA[27] and the Swaziland National Association of Journalists, the first national workshop where national ICT policy was discussed was organized (ECA 2003), which resulted in agreement on the future development of national ICT industries and media and telecommunication markets that contributed to the creation of ICT-enabling environments and increased empowerment stemming from the rapidly increasing ICT penetration rates. Eritrea and Myanmar are the only countries where through the final year of the analysis, 2012, the emergence of the *'MCS-technological take-off'* was not reported. Eritrea is a highly centralized authoritarian regime, classified by Freedom House (2014) as 'not free'. Although according to the Human Development Report 2010, the country is recognized as 'democratic with no alternation' (score '1'), from its independence from Ethiopia (1993) until 2011, no free elections were enforced. In 2012 in Eritrea, investment freedom was '0' (*sic!*); thus, the flows of investment capital were completely restricted. In 2010, the cost of a mobile cellular sub-basket accounted for 33 % of GNI per capita per month, which was slightly below the low-income group average. Although according to ITU data (ITU 2013), the telecommunication market was officially partially liberalised, in 2010, only one company, completely controlled by the government—Eritrea Telecommunications Services Corp. (Eritel)—was operating in the telecommunication market. In addition, Eritrea is recognized as one of the most censored countries in the world, where the freedom of expression and of the press is essentially violated. An authoritarian regime, heavy infrastructural underdevelopment, violations of human rights and censorship, and finally, the lack of a national 'e-strategy', all of these completely restricted the widespread deployment of mobile cellular telephony in Eritrea. According to our estimates, in Myanmar, the 'critical year' was found to be 2012. Because 2012 was the terminal year of our analysis, the *strict* identification of the emerging *'MCS-technological take-off'* was precluded. The country's environment is highly unfavourable: it is recognized as nondemocratic, it lacks basic political freedoms and basic investment freedoms were completely eliminated (the investment freedom index was reported '0' in 2012). In addition, the telecommunication market was monopolised. Moreover, the prices of mobile cellular services were extremely high; the cost of a mobile cellular sub-basket was 69.6 % of GNI per capita per month. All of these elements effectively restricted broad usage of mobile cellular networks in Myanmar. The government of Myanmar has adopted the Myanmar ICT Development Master Plan (2011–2015), the major objectives of which are, *inter alia,* the strong enhancement of broader countrywide ICT deployment, with the intent to achieve $MCS_{i, y} = 45$ per 100 inhab. by 2015 (ITU 2012). For the country of

---

[27] Economic Commission for Africa.

Myanmar, the plan brings prospects for the future in achieving gains from higher mobile cellular coverage, accessibility and usage. The picture arising from the $IU_{i,y}$ diffusion pattern analysis, is far less promising (see Table 5.3). Regarding the low-income countries, the '*IU-technological take-off*' was indentified in only seven (out of 17). An important observation is that among the countries listed above, Kenya is the only economy in which the '*IU-technological take-off*' interval may be undoubtedly reported for the time interval 2010–2011. In another two countries, Bangladesh and Cambodia, $Y_{crit,IU} = 2012$; as such, for the consecutive period 2013–2014, the '*IU-technological take-off*' is projected. In Nepal, Rwanda, Uganda and Zimbabwe, the $Y_{crit,IU}$ has been designated,[28] but the paths that reflect the changes in $\Omega_{IU,i,y}$ and $\Phi_{IU,i,y}$ are unstable; thus, the identification of a country-specific '*IU-technological take-off*' is marked by uncertainty. The time span when both $Y_{crit,IU}$ and '*IU-technological take-off*' were observed during the 4-year period (2008–2012), and the time required for the '*IU-technological take-off*' to emerge was, on average, 14.3 years.[29] According to our calculations, in the low-income countries, the average $critIU_{i,y} = 7.3$ %, which may be identified as the critical (threshold) level of Internet penetration rates that enhance the emergence of the '*IU-technological take-off*' leading to exponential growth of $IU_{i,y}$ penetration rates. The time span for the '*IU-technological take-off*' interval may be denoted for 2004–2012. The average length of the initial diffusion phase was 14.4 years; in India, it took 20 years for '*IU-technological take-off*' to emerge, whereas in Paraguay, it only took 10 years. Our evidence has also demonstrated that in the respective $Y_{crit,IU}$, the average $IU_{i,y}$ penetration rate was approximately 9.52 %; thus, we claim this to be the critical (threshold) Internet penetration rate, $critIU_{lower-middle,y} = 9.52$ %, in the lower-middle-income economies. However, the country-specific $critIU_{i,y}$ values vary significantly, ranging from $critIU_{LKA,y} = 1.4$ % in Sri Lanka to $critIU_{MLD,y} = 23.4$ % in Moldova. Examining the remaining country's specific conditions under which the '*IU-technological take-off*' occurred, a few conclusions of seminal interest arise. The first important observation is the average penetration rates of both fixed and wireless networks, enabling access to Internet connections. In the low-income group, the backbone infrastructure required to provide both fixed-narrowband and fixed-broadband networks was heavily underdeveloped. In consequence, the average fixed-narrowband penetration rate was $FIS_{aver,y} = 0.45$ per 100 inhab. and the fixed-broadband was a meagre $FBS_{aver,y} = 0.24$ per 100 inhab.; thus, the accessibility of fixed Internet connections was negligible. Regarding the spread of wireless-broadband infrastructure, the picture is somewhat more promising—average[30] $WBS_{aver,y} = 2.4$ %. Extremely limited access to fixed

---

[28] In Zimbabwe, because of rapid changes in $\Omega_{IU,i,y}$ and $\Phi_{IU,i,y}$, there emerged three potential $Y_{crit,IU}$.

[29] Author's calculations.

[30] Note that in the $Y_{crit,IU}$, wireless-broadband networks were reported in only three (out of seven) countries: Bangladesh (0.47 %), Cambodia (6.7 %) and Kenya (0.01 %).

and wireless infrastructure was an important hindrance to unbounded growth in the number of individuals who used the Internet.

The analogous exercise for the lower-middle-income countries finds that the penetration rates of fixed-narrowband and fixed-broadband networks, on average, reached $FIS_{aver, \; y} = 1.62$ per 100 inhab. and $FBS_{aver, \; y} = 0.69$ per 100 inhab., reflecting substantial shortages in access to the landline Internet infrastructure. The average performance in terms of wireless-broadband penetration rates was slightly better, $WBS_{aver, \; y} = 5.13$ per 100 inhab. Important to observe is that across the examined economies, wireless-broadband networks were available exclusively in seven (out of 26). Still, limited access to both fixed and wireless networks did not impede the emergence of the '*IU-technological take-off*', and a great majority of the lower-middle-income economies managed to enter the exponential growth phase along the $IU_{i, \; y}$ diffusion trajectory. Surprisingly, in the low-income countries, the reported prices of fixed-broadband connection and fixed-broadband monthly subscriptions were extremely high, which induced the indecently low affordability of Internet network access. The average fixed-broadband subscription charge was US$93.6 (if Zimbabwe, at US$64.7, is excluded); the average fixed-broadband monthly subscription charge was US$52.1 (again excluding Zimbabwe[31]). The lowest-cost fixed-broadband monthly subscription was reported in Bangladesh, US$4.2, and the highest was in Uganda,[32] US$131.2. The high costs of accessing Internet networks were mirrored by the critically low affordability. The cost of acquiring a standard fixed-broadband sub-basket was 166.1 %[33] of GNI per capita per month. Moreover, the observed cross-country disparities in Internet access affordability are enormous. For example, in Bangladesh, the price of a standard fixed-broadband sub-basket in 2012 was 7.3 % of GNI per capita per month; in Uganda it was 600 %, and in Rwanda, it was 344.3 %. Regarding the lower-middle-income group, the numerical evidence on the costs of a fixed-broadband connection and a fixed-broadband monthly subscription is even more striking. The average fixed-broadband connection charge[34] was reported to be US$131.5 (US$79.5 excluding Zambia[35]), and the average fixed-broadband monthly subscription charge[36] was US$133 (US$67.03 excluding Swaziland[37]). Shifting focus to the affordability of Internet network access, it is shown that although the cross-country

---

[31] According to ITU statistics, in 2006 in Zimbabwe, a fixed-broadband monthly subscription cost approximately US$2,673 (*sic!*).

[32] Excluding Zimbabwe from this comparison.

[33] Excluding Zimbabwe, where the price of a standard fixed-broadband sub-basket was 1,059 % (in 2010) of GNI per capita per month.

[34] The price of a fixed-broadband connection ranged from US$3.9 in Sri Lanka to US$337.4 in Nigeria.

[35] In Zambia, in 2010, the fixed-broadband connection charge was US$962.8.

[36] The price of a fixed-broadband monthly subscription ranged from US$3.1 in Viet Nam to US$674.8 in Nigeria.

[37] In Swaziland, in 2008, the fixed-broadband monthly subscription charge was US$1,781.8.

variability is tremendous, the average price of a fixed-broadband sub-basket was approximately 26 % of GNI per capita per month (24 % excluding Nigeria and Swaziland). This rate reflects the essentially higher affordability of accessing Internet connections and services compared with the low-income economies and is possibly the reason that the '*IU-technological take-off*' occurred in a great majority of the lower-middle-income countries while the great part of the low-income economies remained stuck in the low-level trap, unable to take off.

Demonstrably, in the vast majority of both the low-income and the lower-middle-income countries (in the 'critical years'), the telecommunication market (for fixed broadband connections and Internet services) was fully liberalized and free competition was introduced, allowing for the presence of multiple operators. In only four countries was the telecommunication market labelled 'partial competition' in both areas; meanwhile, only in Swaziland was there a telecommunication monopoly (in fixed broadband connections). This evidence sharply contrasts with the fact that according to the data provided by the Freedom House (House 2013),[38] none of the examined low-income countries was classified as 'free' (*sic!*) in terms of political rights and civil liberties; three countries were 'not free' and the remaining four were 'partly free'. Moving to the lower-middle-income group, the evidence shows that in the 'critical years', five countries were classified as 'not free', another 13—'partly free', and the remaining eight were labelled 'free' (for the specifications, see Table 5.3). Still, despite the significant lack of broadly defined freedoms, in a great number of the analysed economies, the emergence of '*IU-technological take-off*' was not restricted. This coincides with the conclusion derived from the analysis regarding the '*MCS-technological take-off*' (see the preceding paragraphs). Significant restrictions on political freedoms and civil liberties are mirrored in the limited digital media and Internet freedoms in the analysed countries. According to the Freedom House Freedom on the Net index, (see the reports Freedom on the Net 2011, 2012 and 2013), five[39] out of seven countries in our scope were classified as 'partly free'; that is, none was identified as free. The Freedom on the Net index comprehensively measures the level of Internet and ICT freedom (Freedom House 2013) in three major areas: Obstacles to use (refers to infrastructural and economic barriers to unbounded Internet and digital media access, legal control of Internet service providers and the independence of the relevant regulatory bodies); limits on content (refers to legal regulations on content, filtering or blocking websites, censorship, and the diversity of online media); and Violations of rights (refers to surveillance and repercussions for online activity, e.g., imprisonment or cyber attacks). Although in Bangladesh, Cambodia, Kenya, Uganda and Zimbabwe, the Internet network and other digital media access and use are nominally free from any governmental restrictions, there are still

---

[38] Officially, the data on Internet freedom are available beginning in 2009. However, for most low-income and lower-middle-income countries, data are available exclusively for 2013 and are reported as such.

[39] No data were available for either Nepal or Rwanda in 2010 and 2008, respectively.

violations in this area. The most prominent hindrance to unlimited access to and use of the Internet was still poorly developed backbone infrastructures (especially in rural regions), power shortages, low bandwidth for Internet connections and high pricing. Online media and Internet net were officially unfettered; however, in some cases (e.g., Bangladesh, Uganda and Cambodia), filtering and censorship were observed (Freedom House 2013). Internet users' rights were violated, especially in Bangladesh and Cambodia; a number of attacks on government websites were documented, mainly owing to their technical weaknesses and vulnerability. Additionally, the analogue evidence for lower-middle-income economies reveals that the degree of Internet freedoms regarding the obstacles to use and limits on content, is very close to that found among the low-income group. As reported by Freedom House (Freedom House 2013), an important obstacle to broader Internet us is poorly developed infrastructures, underserved rural areas, and the relatively high costs of acquiring Internet services (see, e.g., Georgia, Yemen and Lao P.D.R.). In 2012, in many countries, Internet users' rights, especially in terms of broad censorship and/or filtering content in digital media, were significantly violated. The worst-performing countries in this regard were Syria (35),[40] Viet Nam (31), Egypt (33) and Morocco[41] (24). Moreover, in 2012, Syria and Viet Nam faced extremely high obstacles to use and limits of contents arbitrary imposed by legal authorities. Finally, we consider the data that explain the degree of economic and investment freedoms in both income groups. Overall examination of the cross-country statistics shows that on average, these results do not differ significantly from those reported for the *MCS-technological take-off* study (to compare, see Table 5.2). In a small number of economies, we observe increasing values for various economic freedom measures. Slight improvements can be found in, e.g., Bangladesh, where investment freedom was at 55 in 2012 (as opposed to 30 in 2006), and Cambodia, where investment freedom increased from 50 (in 2006) to 60 (in 2012). Among the lower-middle-income economies, the sharpest changes were observed in Bolivia, where investment freedom decreased from 90 (in 2001) to 20 (in 2009).

Section 5.2 was intended to trace the country-specific *technological take-off* interval and the *critical mass* effects that are closely associated with ICT diffusion patterns. With this aim, we have indentified: *critical years*, *critical penetration rate of ICT* and the country-specific conditions during the *technological take-off* intervals. In the analysis outcomes regarding the mobile cellular telephony adoption, the important observation is that the *critical penetration rates* vary slightly between the low-income and lower-middle-income countries, accounting for 7.05 per 100 inhab. in the low-income group and 8.22 per 100 inhab. in the lower-middle-income group. The duration of the initial (early) phase of diffusion is roughly 12 years in both income groups. Deeper investigation into the issue reveals that both within and between income groups, the country-specific features vary widely and, countries share very few common conditions. These findings suggest

---

[40] Forty is the worst score.

[41] Data are for 2013 (earlier not available).

that there are no commonly recognized country conditions that predetermine leaving the early diffusion phase and the emergence of the '*MCS-technological take-off*'. In the low-income countries, an even more striking observation is that they experienced the '*MCS-technological take-off*' in extremely unfavourable environments. However, it is important to note that in a great majority of countries, the telecommunication markets were fully liberalised, which unquestionably facilitated the rapid expansion of mobile cellular service in even the most backward economies. Regarding Internet usage, the analysis of the 'critical conditions' yields similar conclusions to those in the previous case. Although the '*IU-technological take-off*' was identified in only 7 low-income and 26 lower-middle-income countries, the countries' individual conditions appeared to be highly unfavourable for any increases in Internet usage; there were high costs for fixed-broadband network access, low per capita incomes, and poor infrastructural development.

Bearing in mind that the analysis presented in Sect. 5.2.2 is unconventional and its results may be questionable, we have intended to complement and broaden the latter by providing additional empirical evidence, which can contribute to better understanding of the issues discussed, and shed more light on the considered relationships. To this aim, using the regressing analysis, the next Sect. 5.3 extends and enriches the evidence presented above, unveiling which factors have fostered—or conversely impeded—the $MCS_{i,y}$ and $IU_{i,y}$ diffusion across examined countries. Section 5.3.1 presents the data used, Sect. 5.3.2 displays the preliminary graphical evidence demonstrating the relationships between $MCS_{i,y}$ and $IU_{i,y}$ and their potential determinants, while Sect. 5.3.3 explains and discusses the regression results.

5.3   ICT Diffusion Determinants. A '*Traditional*' Approach

The following section provides additional evidence on $MCS_{i,y}$ and $IU_{i,y}$ diffusion determinants across low-income and lower-middle-income countries during the period of 1997–2012. Hence, the primary objective is to trace these variables empirically, which affected the most increases of $MCS_{i,y}$ and $IU_{i,y}$ penetration rates. To this target, we arbitrary select a bundle of various factors and investigate whether their impact on $MCS_{i,y}$ and $IU_{i,y}$ growth has been positive and strong, or conversely—negligible.

Estimating the relationships between ICTs diffusion and its factors is a challenging task, not only because countries in the scope of the analysis are highly heterogeneous but also because the examined relationships are complex and are influenced by multiple factors, which are often difficult to identify or quantify. Econometric modeling, by convention, is 'traditionally' used to report on the relationships between variables. However, it is important to mention that a country's individual features heavily pre-determine the nature of the investigated relationships, which are poorly captured through econometric models and statistics. Hence, to a point, the relationship between the process of ICTs diffusion and its determinants remains empirically intractable, and this should be borne in mind while reading this section. Although voluminous empirical literature has been published that attempts to provide adequate explanations for cross-country differences in new technology adoption, the evidence is mixed, lacks robustness, and yields different conclusions. The seminal contribution to identifying technology diffusion determinants was made by Comin and Hobijn (2004). They present a long-term analysis of technology adoption determinants across countries over the period 1788–2001, and they find that the most prominent determinants of the present adoption of technologies are factors such as human capita, government type, openness to international trade, and the degree of adoption of predecessor technologies (Comin and Hobijn 2004). These results are consistent with the evidence presented in another paper by Comin and Hobiijn (2006). This study (Comin and Hobiijn 2006), covering 19 different technologies across 21 countries over the period 1870–1998, demonstrates that democracy, quality of human capita and trade openness contribute significantly to technology diffusion. In another study (Comin and Hobijn 2009) that covered 23 countries over the last two centuries, they explore the similarities in the diffusion of 20 technologies. Their main finding is that quality of institutions and political lobbying play important roles in the growth of adoption of newly emerging technologies. The evidence presented in the study by Norris (2000) covering 179 countries and relied on multivariate regression, demonstrates that for Internet penetration neither literacy rate, level of education nor democratization showed a significant and positive influence. Internet diffusion, however, was strongly attributed to GDP per capita and R&D expenditures. Caselli and Coleman (2001) adopt random and fixed-effects regressions for the extensive study of Internet diffusion determinants, covering 89 countries between 1970 and 1990. Their major findings confirm the positive role of investment per worker, property right protection, and a small share of the agriculture sector in GDP in fostering Internet penetration. Kiiski and Pohjola (2002) demonstrate the evidence for cross-country determinants of Internet diffusion. They present evidence for OECD and non-OECD countries over the period 1995–2000. Using the Gompertz model, they find that neither the level of competition in the telecommunication market nor investments in education and mean years of schooling are statistically insignificant in explaining the differences in Internet penetration rates in OECD countries. However, the proxy for level of education became significant in the sample of developing countries. Factors that were significant in both OECD and non-OECD countries were GDP per capita and the costs of accessing Internet networks. These results contrast with the earlier findings provided by Hargittai (1999), who used OLS estimates and reported that across 18 OECD countries (1995–1998), both GDP per capita and regulation of telecommunication markets significantly affected Internet penetration rates. He also found that level of education and state policies positively affected Internet usage, whereas the price of access to the Internet showed negligible significance. Baliamoune-Lutz (2003), analysing developing countries, finds that Internet and mobile cellular penetration rates are positively affected by per capita incomes and government trade policies, whereas—contrary to expectations—freedom proxies and level of education were found to be statistically insignificant in explaining cross-country

ICT diffusion. Dasgupta et al. (2005), in their study of 44 economies over the period 1990–1997, found that among the factors that positively affected Internet penetration were per capita income, degree of urbanization, level of education and quality of institutions. Crenshaw and Robison (2006), concentrating exclusively on 80 developing countries during the period 1995–2000, underline the seminal impact of urbanization in enhancing network effects on Internet use. They also note the important role of government in ensuring property rights, which may induce an increase in Internet hosts and Internet penetration rates. In 2010, Chinn and Fairlie (2010) examined ICTs' (computer and Internet penetration rates) determinants in a panel of 161 countries over the period 1999–2001. They found that both the computer and Internet penetration rates were significantly attributed to income per capita, illiteracy rate, mean years of schooling, degree of urbanization, telecommunication market regulations and electricity consumption. Trade openness and prices on telecommunication markets were reported as insignificant for computer usage. Andrés et al. (2010), examining the Internet diffusion determinants across 214 countries (they divide the sample into two subsamples: low-income and high-income economies) during 1990–2004 and unveil the strong role of network effects in Internet diffusion that are very robust and were noted in both low-income and high-income economies. Bakay et al. (2011), examining the ICT diffusion factors in Latin American countries, affirm the seminal roles of per capita income, literacy and urbanization. They also find that social networks are essential in fostering ICT diffusion among individuals. In 1999, Ahn and Lee (1999), using observations for 64 countries, modelled the demand for mobile cellular telephony. Their major findings were that per capita income and fixed telephony penetration positively affected the increase in mobile cellular subscriptions, whereas pricing revealed little relevance. Madden et al. (2004), in their study of 56 countries during 1995–2000, show that network effects have great explanatory power in the increase in mobile cellular subscriptions, while Madden and Coble-Neal (2004) demonstrate similar results with respect to mobile cellular telephony determinants. These results, however, contradict the findings of Garbacz and Thomson (2007), who in a study of developing countries (time span 1996–2003) report high price elasticity of mobile telephony and note that pricing may be the seminal factor that spawns mobile cellular telephony diffusion. The results of Garbacz and Thomson (2007) coincide with those provided by Barrantes and Galperin (2008), who, based on their evidence for Latin American countries, argue that affordability is the main driver of or barrier to broad mobile cellular dissemination. Factors that determine the process of the spread of mobile cellular telephony were extensively studied by Rouvinen (2006). Using the Gompertz model and a broad array of economic and non-economic factors, he examined 200 developing and developed countries in the 1990s. He found that in developing countries, the total population variable was positively and statistically significantly associated with the increase in mobile telephony users, mainly owing to emerging network effects. Other variables that entered the regression with positive signs were degree of urbanization, development of fixed infrastructure, and trade openness. The overwhelming conclusion from Rouvinen's (2006) study is that in developing countries, the role of social and infrastructural factors are far more important compared with developed economies. Billon et al. (2009), in a study that covered 142 countries in total, reported that in low-income economies, the key determinants of ICT (mobile cellular telephony and Internet usage) diffusion were market regulations, competition in the telecommunication market, and relatively low prices. They also suggested that more urbanization may foster the spread of ICTs in less developed countries. More evidence regarding ICT diffusion's determinants may be found in, e.g., studies by Islam and Meade (1997), Michalakelis et al. (2008), Singh (2008), Jakopin and Klein (2011), Yates et al. (2011), Gupta and Jain (2012), Lee et al. (2011) and Liu et al. (2012).

5.3.1   The Data

To meet the main goals of this empirical analysis, we use a sample including 17 low-income and 29 lower-middle-income countries, which are examined for the period between 1997[42] and 2012. Depending on the data availability, 17 explanatory variables have been isolated, which are applied to provide complex and insightful explanation of the $MCS_{i,\,y}$ and $IU_{i,\,y}$ growth in the analyzed countries. Hence, the explanatory variables are as following[43]: Price of a 1-min call ($Call_{i,\,y}$), Price of one SMS ($SMS_{i,\,y}$), Fixed telephony penetration rate ($FTL_{i,\,y}$), Mobile Cellular Sub-Basket ($MCSIPB_{i,\,y}$), Number of 1-min calls per GNI per capita per month ($CallsMonth_{i,\,y}$), Number of SMSs per GNI per capita per month ($SMSMonth_{i,\,y}$), Number of mobile-cellular prepaid connection charges per GNI per capita per month ($MCSChargeMonth_{i,\,y}$), Fixed Internet Subscriptions ($FIS_{i,\,y}$), Fixed-Broadband Subscriptions ($FBS_{i,\,y}$), Wireless-Broadband Subscriptions ($WBS_{i,\,y}$), Fixed (wired)-broadband monthly subscription charge ($FBSCharge_{i,\,y}$), Fixed-Broadband Sub-Basket ($FBSIPB_{i,\,y}$), Number of fixed-broadband subscription charges per GNI per capita per month ($FBSChargeMonth_{i,\,y}$), Gross Domestic Product per capita ($GDPPPPpc_{i,\,y}$), School Enrollment ($School_{i,\,y}$), Population density ($PopDens_{i,\,y}$) and Urban population ($Urban_{i,\,y}$). The main data sets used in this study are the World Development Indicators 2013 and the World Telecommunication/ICT Indicators database 2013 (17th Edition). Additional information has been extracted from global reports—Measuring the Information Society 2010, 2011, 2012 and 2013, developed by the International Telecommunication Union. We presume that mobile cellular telephony penetration rates might be predominantly affected not only by per capita income but also by costs of adoption and the usage of mobile services, e.g., the cost of a 1-min call. Both per capita income and costs of usage, should strongly affect affordability for the adoption of mobile cellular telephony. We have also chosen the fixed telephony penetration rates as the determinant of the usage of mobile cellular services. We argue that poor

---

[42] In this case, to ensure the maximal reliability of estimates we have arbitrary extended the period of analysis so that it covers 1997–2012.

[43] Full description of the variables used in the analysis is presented in Sect. 5.2.1.

diffusion of fixed telephony should strongly enhance the acquisition of mobile telephony as a good alternative for the previous. As explained in Chap. 4, economically backward countries suffer significantly from lack of broad access to fixed telephony. In such cases, mobile services are an attractive, and often the sole, alternative for the traditional telephony. Additionally, we claim that primary school enrollment might be a factor determining the usage of cellular telephony as access to education, determining the level of a country's human capital, assures basic skills to use and benefit from this type of ICT. Finally, we argue that due to the effects of emerging networks, mobile cellular telephony spread should be favored in densely populated and highly urbanized areas, hence we argue that population density and the degree of urbanization might enhance the broader adoption of mobile cellular telephony. With respect to the penetration rates of Internet users, it is argued here that the level of usage of Internet connections is predominantly gauged by access to necessary infrastructure. Hence, we test the relationships between $IU_{i, y}$ against fixed Internet subscription rates, fixed-broadband subscription rates and wireless-broadband subscription rates. Similarly, as in the case of mobile cellular telephony, the usage of Internet by individuals hypothetically shall be fostered by the growth of per capita income and the decreasing costs of the usage of Internet connections. The reasoning lying behind recognizing school enrollment, population density and the degree of urbanization as potential determinants of Internet usage is similar to the case of mobile cellular telephony.

5.3.2 Graphical Evidence

Figures 5.5 and 5.7 graphically explain the relationship between the level of adoption of mobile cellular telephony ($MCS_{i, y}$) and Internet usage ($IU_{i, y}$) *versus* their selected determinants, in low-income economies over the period 1997–2012; while Figs. 5.6 and 5.8 present analogous relationships in the group of lower-middle-income countries. Visual inspection of the empirical findings reveals that certain regularities can be identified with regard to the examined relationships. Not surprising, all the evidence that is considered with respect to mobile cellular telephony determinants, both in low-income and lower-middle-income economies, reveals that the $MCS_{i, y}$ penetration rates are inversely correlated with the variables explaining the costs of acquiring and using mobile cellular services, which are: mobile cellular sub-basket, the price of a 1-min call,[44] price of SMSs,[45] and mobile-cellular prepaid[46] connection charges. The negative impact of the costs associated with the adoption and usage of mobile cellular telephony on respective penetration rates, seems to be relatively stronger in the group of low-income countries. During

---

[44] Peak and on-net.

[45] Peak and on-net.

[46] For analytical purposes, the prepaid tariffs have been chosen, because among low-income users they are usually the only available method of payment for mobile services.

the analyzed period 1997–2012, significant reduction in the prices of 1-min calls and/or of sending SMSs, as well as drops in mobile-cellular prepaid connection charges, fostered growth in the affordability of mobile services, which in turn boosted the use of mobile cellular telephony, even in the most economically backward countries. Interestingly, in three low-income and 14 (*sic!*) lower-middle-income countries, the value of a Mobile-Cellular Sub-Basket *increased* during the period 2008–2012.[47] Surprisingly, the unfavorable trends did not impede the spread of mobile telephony in some countries, despite that fact that mobile cellular services became *less* affordable. It is important to mention that regardless of the substantial increases of $MCSIPB_{i, y}$ in a few countries, still the prices of calls ($Call_{i, y}$) and SMSs ($SMS_{i, y}$) were gradually falling. Hence, the downward trends in the prices of basic mobile cellular telephony services was revealed to be a powerful stimulus for the rapid expansion of mobile cellular telephony across low-income and lower-middle-income countries. Referring back to Figs. 5.5 and 5.6, conversely to what was initially hypothesized, the variable showing the degree of development of fixed telephony ($FTL_{i, y}$) is positively correlated with $MCS_{i, y}$ penetration rates. Such results are valid both for low-income and lower-middle-income economies, which generally contradicts our preliminary expectations. However, detailed research of country-wise fixed telephony penetration rates demonstrates that during the period 1997–2012, the development of fixed telephony networks was extremely poor, especially in the group of low-income countries,[48] and any positive changes with this respect are negligible.[49] Henceforth, we claim that this result is inconclusive, and the variable $FTL_{i, y}$ has little explanatory power with respect to $MCS_{i.y}$ changes. The other two explanatory variables—per capita income ($GDPPPPpc_{i, y}$) and primary school enrollment ($School_{i, y}$)—seem to positively impact changes in mobile cellular penetration rates. The established relationships $GDPPPPpc_{i, y}$ *versus* $MCS_{i, y}$, and $School_{i, y}$ *versus* $MCS_{i, y}$, might suggest that growth of per capita income, along with the growth of human capital (approximated by primary school enrollment) translate into greater deployment of mobile cellular telephony, in both income groups. The impact of per capita income on mobile cellular telephony deployment seems to be unquestionable, mainly in terms of affordability. Meanwhile, it is interesting to observe how various countries that differ greatly with regard to $GDPPPPpc_{i, y}$, perform equally well in terms of $MCS_{i, y}$ penetration rates. The results displaying the connections between primary school enrollment and access to mobile cellular telephony reveal a positive relationship. It is clear that education matters, and shifts in human capital may profoundly reshape the way people act. In our case, providing basic education may be identified as an important driver of the increasing usage of mobile cellular telephony, even though significant

---

[47] The data on the value of Mobile-Cellular Sub-Basket are available only since 2008.

[48] In low-income countries, the average $FTL_{i,y}$ in 1997 and 2012 was respectively 0.52 and 1.43 (per 100 inhab.).

[49] For a detailed discussion of the relationship between the state of development of fixed telephony versus mobile telephony expansion—see Chap. 4.

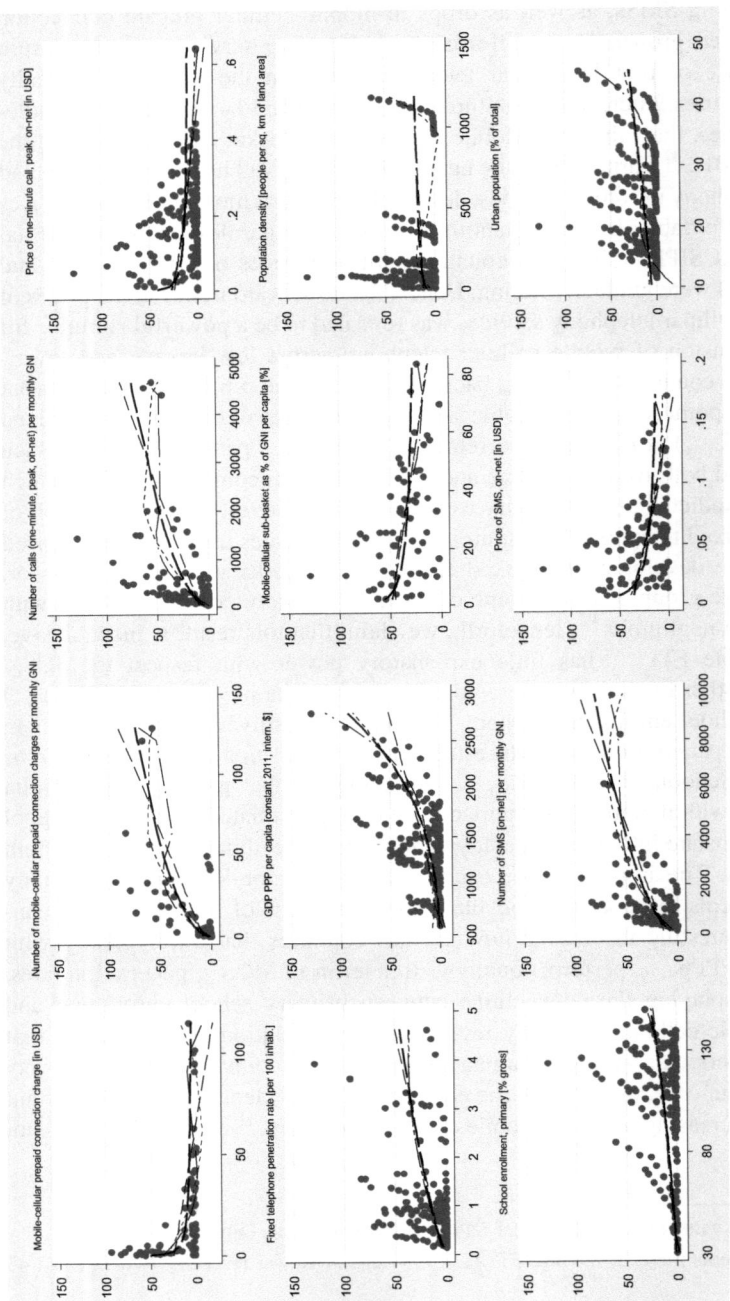

Fig. 5.5 Mobile cellular telephony penetration rates and its selected determinants. 16 low-income countries. 1997–2012. Note: Outlier (Zimbabwe)—excluded. On vertical axis—MCS₁,ᵧ penetration rates (per 100 inhab.). *Dash line*—linear prediction; *very short dash*—quadratic prediction; *long dash dot line*—cubic prediction; *long dash*—power prediction

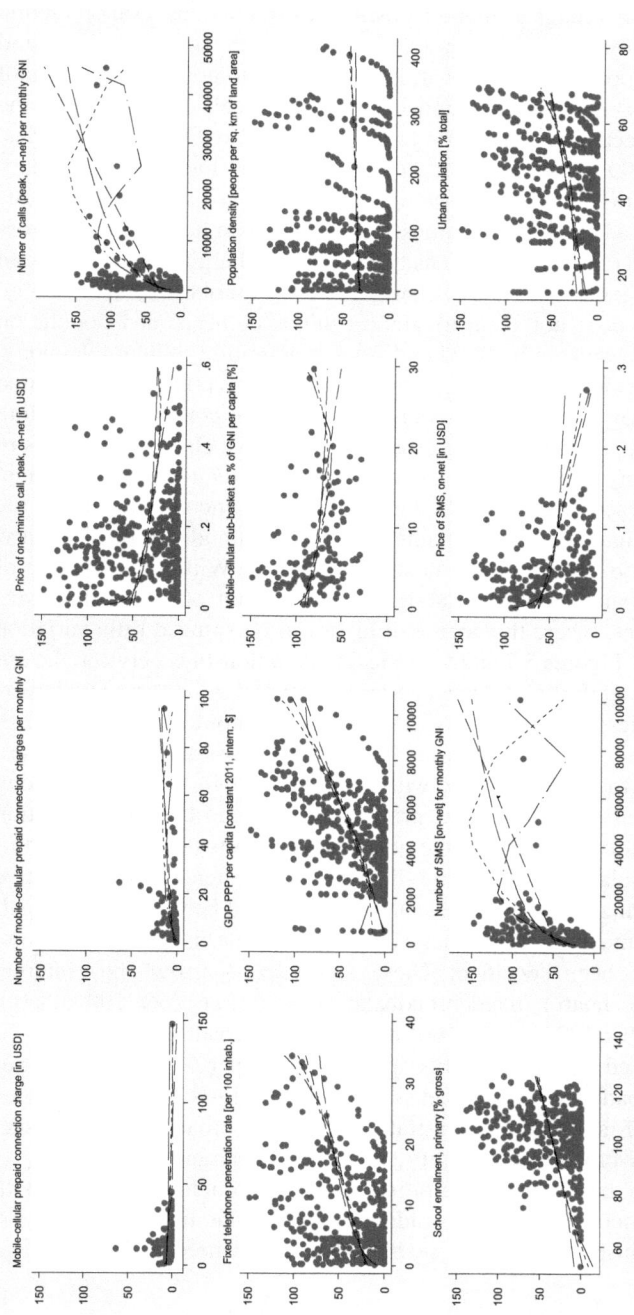

Fig. 5.6 Mobile cellular telephony penetration rates and its selected determinants. 29 lower-middle-income countries. 1997–2012. Note: On vertical axis— $MCS_{i,y}$ penetration rates (per 100 inhab.). *Dash line*—linear prediction; *very short dash*—quadratic prediction; *long dash dot line*—cubic prediction; *long dash*—power prediction

delays between the cause (growth in education) and effect (growth in $MCS_{i, y}$) may emerge. The evidence also suggests that the positive impact of education on mobile cellular telephony deployment is comparably strong in both the low-income and lower-middle-income economies. However, it is important to note that with regard to the relationship between education and use of mobile cellular telephony, the potentially stronger effects may be reported in the group of low-income countries, as during the period 1997–2012 these countries progressed the most in primary school enrollment. With regard to the variable, population density, the results obtained slightly contradict the predictions. We have hypothesized that across more densely populated regions the propensity of mobile cellar telephony to spread would be relatively higher, mostly due to emerging network effects. Unfortunately, the graphical evidence does not seem to support this hypothesis, and population density shows little relevance with regard to diffusion of mobile cellular telephony. Conversely, the variable denoting the degree of urbanization is positively correlated with $MCS_{i, y}$, both in low-income and lower-middle-income countries. According to the evidence, the impact of a growing urban population on changes in access to mobile cellular telephony seems to be relatively stronger in the low-income group. This is probably because, between 1997 and 2012 in low-income countries, the growth in urbanization has been more notable (see, e.g., Cambodia, Kenya, Malawi or Rwanda) compared to lower-middle-income economies. With the exception of Viet Nam or Yemen, such prominent shifts have not been observed in lower-middle-income countries, where the degree of urbanization showed little variation during analyzed period. Figures 5.7 and 5.8 reflect the relationships between the use of Internet connections against its selected determinants, in low-income and lower-middle-income countries over the period 1997–2012. Factors considered which hypothetically may affect the use of Internet connections across analyzed countries, are partially analogous to those discussed with respect to mobile cellular telephony and are as follows: per capita income, primary school enrollment, population density and degree of urbanization. As the quantitative results do not vary significantly from those displayed for the low-income group, hence the qualitative conclusions would be analogous, and thus, are not discussed here. However, apart from the factors just mentioned, another six potential determinants of Internet penetration rates have been specified. These are: fixed (narrowband) Internet subscriptions (per 100 inhab.), fixed broadband subscriptions (per 100 inhab.), wireless-broadband subscriptions (per 100 inhab.), fixed broadband subscriptions charges, number of fixed broadband subscription charges per GNI per capita per month, and fixed broadband sub-basket. Graphical analysis of the evidence displayed in Figs. 5.7 and 5.8 demonstrates that fixed-broadband sub-basket ($FBSIPB_{i, y}$) and fixed-broadband monthly subscription charges ($FBSCharge_{i, y}$) are inversely related to the Internet penetration rates. The conclusion is valid both for the group of low-income and lower-middle-income economies. Nevertheless, more detailed visual inspection of the respective charts where $FBSIPB_{i, y}$ and

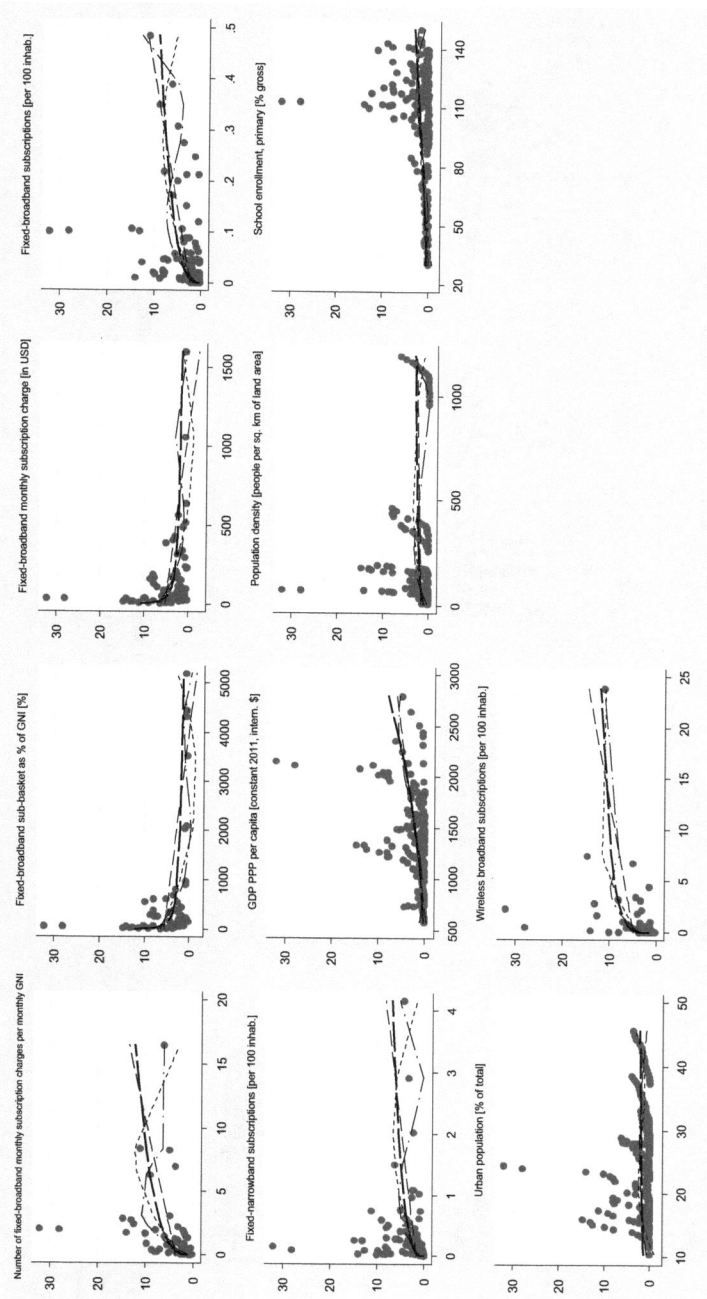

Fig. 5.7 Internet users penetration rates and its selected determinants. 16 low-income countries. 1997–2012. Note: Outlier (Zimbabwe)—excluded. On vertical axis—IU$_{i,y}$ penetration rates (% of individuals). *Dash line*—linear prediction; *very short dash*—quadratic prediction; *long dash dot line*—cubic prediction; *long dash*—power prediction

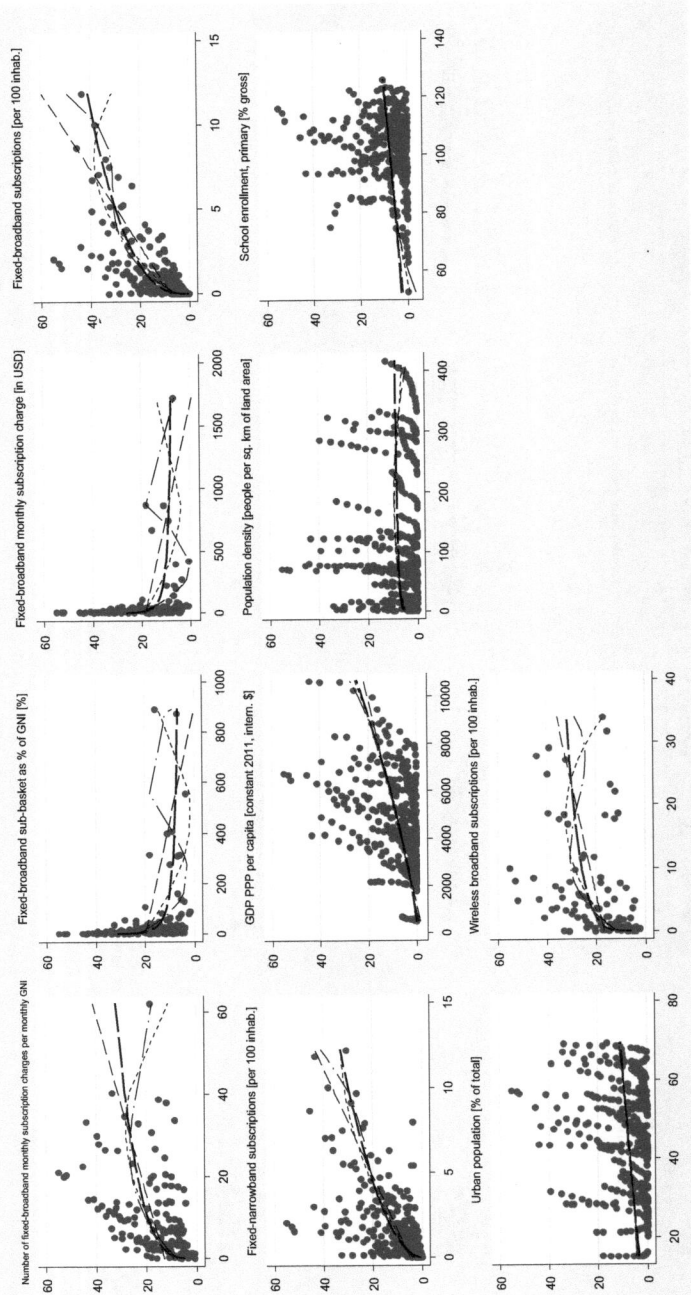

Fig. 5.8 Internet users penetration rates and its selected determinants. 1997–2012. Note: On vertical axis—IU$_{i,y}$ penetration rates (% of individuals). *Dash line*—linear prediction; *very short dash*—quadratic prediction; *long dash dot line*—cubic prediction; *long dash*—power prediction

FBSCharge$_{i, y}$ are plotted against IU$_{i, y}$, shows that the curves approximating the respective relationships are mostly flat. The latter suggests that the striking differences in FBSIPB$_{i, y}$ and FBSCharge$_{i, y}$, are poorly reflected by the differences in Internet penetration rates, which vary moderately across countries. The evidence suggests that, conversely to what was witnessed with regard to mobile cellular telephony, the impact of dramatically falling prices of access to an Internet network had a relatively weak impact on its broad deployment and usage. In most low-income countries (except Zimbabwe and Eritrea), during the period 2008–2012,[50] the cost of Fixed-broadband connection charges was rapidly decreasing; however, in only few countries has this price decrease generated significant increases in IU$_{i, y}$. In Kenya, the FBSCharge$_{i, y}$ dropped from US\$158.8 in 2008 to US\$35.3 in 2012, which enhanced growth of IU$_{i, y}$ from 8.6 % in 2008 to 32.1 % in 2012; in Uganda the analogous values were, respectively FBSCharge$_{UGA,2008}$ = US\$328.5, FBSCharge$_{UGA,2012}$ = US\$14.1, IU$_{UGA,2008}$ = 1.7 % and IU$_{UGA,2012}$ = 14.7 %. Conversely to what might have been expected, e.g., in Ethiopia drops in fixed-broadband connection charges from US\$635 (in 2008) to US\$22.5 (in 2012), or Malawi—from US\$1,057.4 (in 2008) to US\$30.2 (in 2012), the price decreased hardly impacted the shifts in access to and use of the Internet among individuals.[51] This suggests that in low-income countries the IU$_{i, y}$ variable revealed little sensitivity to essential decreases of costs of access to the Internet; while there might have been other factors that impeded the growth of individuals using Internet connections.[52] Closer analysis of the statistics on FBSIPB$_{i, y}$ seems to support the previously explained results, namely, that decreased charges for fixed-broadband connection have negligible impact on the growth of Internet penetration rates.

The variable FBSIPB$_{i, y}$ gives the representation of the price of a standard basket of fixed-broadband monthly usage and is expressed as a percentage of an average GNI per capita per month; hence, it sheds light on the affordability of fixed-broadband use. According to data collected in the Measuring the Information Society reports (ITU 2010, 2013), in the vast majority of low-income countries during the period 2008–2012, the reported values of FBSIPB$_{i, y}$ significantly exceed 100 %, which suggests that people in low-income countries can barely afford to buy a standard fixed-broadband basket. In only a few countries—Bangladesh, Cambodia, Nepal and Uganda—between 2008 and 2012, drops in FBSIPB$_{i, y}$ were enough,[53] to fairly increase the affordability of buying a standard fixed-broadband basket. Analysis of the analogous evidence for the group of lower-middle-income countries leads to similar conclusions as for the low-income group. Still, despite notable decreases in the prices of fixed-broadband connection charges and increasing affordability of the standard fixed-broadband basket, the use

---

[50] The data on Fixed-broadband connection charged are available only for the period 2008–2012.

[51] In Ethiopia in 2012 the IU$_{ETH,2012}$ = 1.5 %; in Malawi—IU$_{MWI,2012}$ = 4.3 %.

[52] For broader discussion—see Sect. 5.2.2.

[53] In 2012, the FBSIPB$_{i,y}$ in Bangladesh, Cambodia, Nepal and Uganda were respectively 7.3 %, 34 %, 17.8 % and 32.9 %.

of Internet connections by individuals remains relatively low. For example, in Sri Lanka in 2012 the $FBSIPB_{LKA,2012} = 2.1$ %, while the $IU_{LKA,2012} = 18.2$ %; while Senegal performed comparably well in terms of Internet penetration rates ($IU_{SEN,2010} = 19.2$ %), however at a significantly lower affordability—$FBSIPB_{SEN,2012} = 42.8$ %. Examples of this type abound in the group of lower-middle-income economies, hence the evidence explaining the relationships between $IU_{i, y}$ versus $FBSIPB_{i, y}$ and $FBSCharge_{i, y}$ is rather mixed and shows little robustness; thus, this evidence might suggest that the prices of access to, and use of, the Internet have a relatively weak impact on $IU_{i, y}$ growth, compared to the influence of prices of mobile cellular services on $MCS_{i, y}$ shifts. Finally, we exhibit the evidence regarding the relationships between $IU_{i, y}$ against the access indicators, namely: Fixed (narrowband) Internet subscriptions (per 100 inhab.), Fixed Broadband subscriptions (per 100 inhab.), and Wireless-broadband subscriptions (per 100 inhab.). For both graphical and numerical results, see Figs. 5.7 and 5.8. It is expected that gradually increasing access to infrastructure, which in this study is approximated by the number of subscriptions of fixed or wireless networks, should inevitably foster growth in the number of individuals using the Internet. Considering the group of lower-middle-income countries, the empirical results generally confirm our supposition that improvements in backbone infrastructure positively influence the Internet penetration rates. Plotting $IU_{i, y}$ *versus* $FIS_{i, y}$, $FBS_{i, y}$ and $WBS_{i, y}$ (see Fig. 5.8), it is discovered that rapid advances in the number of subscription to either fixed or wireless-networks brings considerable shifts in the broad use of the Internet connections. The results displayed in the correlation matrices in Appendix H reveal growing reliance on fixed-broadband technologies, compared to fixed-narrowband, across the countries covered in this analysis, and, at least up till now, wireless-broadband connections. The analogous evidence for low-income countries, gives few prospects for the future (see Fig. 5.7). It is important to note that, over the period 1997–2012, the average $FIS_{i, y}$ and $FBS_{i, y}$ remained at extremely low levels (in 2012, the respective averages[54] were 0.59 and 0.13), with the exception of Malawi, which significantly exceeded group average scores with respect to fixed-narrowband penetration rates. Analyzing plotted Internet penetration rates against wireless-broadband subscriptions per 100 inhabitants (see Fig. 5.7), evidence that is slightly more promising is emerging. Since 2009 onward, in a few low-income countries gradual expansion of wireless-broadband technologies is reported, which is mirrored by the growing number of individuals using the Internet.

The evidence provided earlier in this section yields to be confirmed by the statistical analysis which results are demonstrated in the consecutive Sect. 5.3.3.

---

[54] Author's calculations.

5.3.3 Panel Regression Results

The forthcoming Sect. 5.3.3 is fully subjected to present complementary evidence on the relationships between $MCS_{i, y}$ and $IU_{i, y}$ versus selected determinants. We do so by building two separate panels—for low-income and lower-middle-income countries—and re-examining the hypothesized relationships. Similarly, in the preceding sections, we separately consider low-income and lower-middle-income countries, which are analyzed between 1997 and 2012. The mobile cellular telephony ($MCS_{i, y}$) and Internet user ($IU_{i, y}$) penetration rates are denoted as response variables, while as predictors we consider all of the variables specified in Sect. 5.3.2, except the mobile-cellular prepaid connection charge. By doing so, we aim to draw inferences about the intensity of the influence of selected factors on $MCS_{i, y}$ and $IU_{i, y}$ in countries in our scope of study. Relying on the fixed effects regression,[55] which allows for heterogeneity across countries, we estimate the Eq. (5.1):

$$ICT_{i, y} = \alpha + \beta\left(x'_{i, y}\right) + u_{i, y}, \tag{5.1}$$

where α is the scalar, $\mathbf{ICT}_{i,y}$ denotes alternatively $MCS_{i, y}$ or $IU_{i, y}$; β is the $L \times 1$ and $x'_{i,y}$ stands for the iyth observation on L explanatory variables (Baltagi 2008). The subscripts $i = \{1, \ldots\ldots N\}$ stand for country and $y = \{1, \ldots\ldots T\}$ for the time period. In Eq. (5.1), the $u_{i, y} = \mu_i + \nu_{i, y}$, while the μ_i accounts for the unobservable and time-invariant country-specific effect, which is not captured in the model, and $\nu_{i, y}$ is the remainder disturbance (the observation-specific errors) (Greene 2003). To control for the possibly of emerging heteroskedasticity or within-panel serial correlations, robust standard errors are specified and reported (Arellano 1987; Hoechle 2007). In addition, to investigate the potential importance of the earlier technology adoption level in explaining current ICTs deployment, using one-step Arellano-Bond difference GMM estimator (Arellano and Bond 1991) we estimate the dynamic panel regression model, specified in Eq. (5.2):

$$\mathbf{ICT}_{i, y} = \left(\mathbf{ICT}_{i, y-1}\right) + \beta\left(x'_{i, y}\right) + u_{i, y}, \tag{5.2}$$

where $\mathbf{ICT}_{i, y-1}$ shows the lagged[56] value of $MCS_{i, y}$ or $IU_{i, y}$, the ξ stands for $ICT_{i, y-1}$ coefficient, and the remaining notations are as in Eq. (5.1). For the

---

[55] To select between the *fixed* or *random* effects regression, the authors have tested both to choose the most appropriate specification. Relying on the Hausman specification test (Hausman 1978; Maddala and Lahiri 1992), for the vast majority of estimates models, the *fixed* effects specification was reported as more appropriate to examine the relationship between covariates. In only few cases, was the *random* effects regression suggested as the superior specification compared to the *fixed* effects model.

[56] As demonstrated in Chap. 4, the yearly dynamic of $MCS_{i,y}$ and $IU_{i,y}$ diffusion is extremely high and, thus, it is important to explain its diffusion in consecutive periods; we argue that the most justifiable would be 1-year lagged values of $MCS_{i,y}$ and $IU_{i,y}$.

model specified in Eq. (5.2), as in the previous (see Eq. (5.1)), we assume the $u_{i,y}=\mu_i+\nu_{i,y}$, if $\mu_i \sim IID\left(0,\sigma_\mu^2\right)$ and $\nu_{i,y} \sim IID(0,\sigma_\nu^2)$ (Baltagi 2008). Analogously to the fixed effects regression, we estimate Eq. (5.2) using robust standard errors to obtain the errors consistent with panel-specific autocorrelations and heteroskedasticity. As the Sargan test of over identifying restrictions is not available after robust estimations, we calculate the Arellano-Bond test for second-order autocorrelation in the first-differenced errors (Arellano-Bond 1991). To control for possibly emerging multicollinearity among variables, we calculate bivariate correlation coefficients along with Variance Inflation Factors[57] between respective variables. The calculated correlation coefficients are summarized in respective tables in Appendix H. In addition, as the distributions of selected variables across the examined samples are heavily-tailed, to avoid strong violation of the regression analysis results, all extreme observation have been detected and excluded from the main data set.

The results of the panel regression analysis are displayed in respective tables summarized in Appendix I. Considering the low-income group, the results of random effects regressions estimations reporting on the $MCS_{i,y}$ determinants (see Tables I.1 and I.3), show that the final results differ with regard to various specifications. The only explanatory variable which reveals persistence in explaining the mobile cellular telephony penetration rates is population density ($PopDens_{i,y}$). In consecutive specifications (1), (2), (4), (5) and (11) in Table I.1, the variable $PopDens_{i,y}$ enters the regressions with the expected positive sign and is statistically significant at the 5 % level of significance. The β coefficients explaining the impact of growth of population density on $MCS_{i,y}$ increase vary from $\beta_{PopDens}=10.5$ in regression (2) to $\beta_{PopDens}=17.98$ in regression (11). The rationale behind these results is rather simple. In densely populated areas, the access to mobile cellular telephony is much easier mainly due to a better developed backbone infrastructure, as well as easier contacts between users and non-users of new technology (the 'word of mouth' effect), the network effects emerge, and hence the technology spread is highly facilitated. By contrast, in low-income countries, in poorly populated and often geographically isolated regions, the access to mobile cellular infrastructure is still restricted and contacts between people are rarer, which may impede diffusion of $MCS_{i,y}$. With respect to lower-middle-income countries, the impact of population density on mobile cellular telephony diffusion is equally strong and positive. In each estimated regression, the coefficients explaining the strength of $PopDens_{i,y}$ impact on $MCS_{i,y}$ are high (varying from 7.16 in specification (2) to 19.17 in specification (12)) and statistically significant. The rest of the estimated coefficients in the consecutive

---

[57] The Variation Inflation Factor (VIF) is the reciprocal of the Tolerance $(1-R_i^2)$, and determines how much of the variance of estimated regression coefficients are being *inflated* due to emerging collinearity between examined variables. Usually, we should be concerned about the multicollinearity once the VIF exceeds 10 (Mansfield and Helms 1982; O'Brien 2007; Dormann et al. 2013).

specifications suggest that this finding is robust and has a controlling effect for other variables. It also shows that in this lower-middle-income income group the positive networks effects are revealed, which fosters the dynamic spread of mobile cellular telephony among society members. Analyzing the impact of population density on mobile cellular telephony diffusion, however, it is important to note that a vast majority of examined countries carry one important characteristic. In the great majority of low-income and lower-middle-income countries, high fertility rates are reported, which translates into high natural growth rates, and finally contributes significantly to increases in population density. Thus, it shall be borne in mind that because both $PopDens_{i, y}$ and $MCS_{i, y}$ demonstrate relatively high annual growth rates across the analyzed countries during the period 1997–2012, it might have heavily determined the panel regression outcomes. Another factor that demonstrates a positive influence on increasing the number of mobile cellular telephony users, both in low-income and lower-middle-income countries, is per capita income ($GDPPPPpc_{i, y}$). In only two instances—(1) and (3) for the low-income group, the variable $GDPPPPpc_{i, y}$ is reported as statistically insignificant. In the remaining models, the impact of per capita income on $MCS_{i, y}$ penetration rates is found as intensive and positive, statistically significant and unaffected by inclusion or exclusion of various variables in the regressions. These findings suggest that economic growth may strongly shift the usage of mobile cellular telephony by individuals, mainly due to the increasing affordability of buying mobile services. Interestingly, the potential effect of economic growth on $MCS_{i, y}$ is relatively smaller compared to the intensity of impact of population density (sic!). In the group of low-income economies, the estimated impact of level of education and fixed telephony penetration rates is relatively unrobust and generally reported as statistically insignificant. Conversely, in lower-middle-income countries, both the $School_{i, y}$ and $FTL_{i, y}$ variables reveal positive associations with the increasing number of mobile cellular services users. However, earlier investigations and evidence show that these results might be misleading—see the discussion in preceding section (Sect. 5.2.2). According to our estimates, unexpectedly, the degree of urbanization ($Urban_{i, y}$) shows little relevance with the increasing number of mobile cellular telephony users. In both income-groups, the estimated coefficients are statistically insignificant, with the only exception being when the $Urban_{i, y}$ is the only explanatory variable included in the model. Further evidence, however suggests, that the results produced in models (12)'s[58] lack robustness and reveal strong justification for including other variables in the regression. Essential for understanding these 'strange' results is keeping in mind that in the countries examined in this study, a vast majority of people live in rural areas, while the degree of urbanization remains extremely low (for 2012, see, e.g., Cambodia—20 %, Ethiopia—17 % or Malawi—15 %), which arguably is not unimportant for the results. Conversely to what might be hypothesized, the two consecutive variables—$Call_{i, y}$ and $SMS_{i, y}$, which denote the basic costs of using

---

[58] Separately for low-income and lower-middle-income economies.

mobile cellular services, are identified as statistically insignificant in most of the specifications. Moreover, in model (1) for the low-income group, the variable $Call_{i, y}$ enters the regression with a 'wrong' positive sign. The same is reported in specification (2) in Table I.1—the same income group, with regard to the $SMS_{i, y}$ variable. These results seem surprising, however, Figs. 5.5 and 5.6, clearly demonstrate that in various countries, similar $MCS_{i, y}$ penetration rates are achieved at substantially different prices of 1-min calls and SMSs, and this is likely to have strongly affected the regression estimates. Turning to the analysis of the explored relationships presented in Table I.3 an important issue arises. The estimated coefficient for the respective variables $CallsMonth_{i, y}$, $SMSMonth_{i, y}$ and $MCSChargeMonth_{i, y}$ show that increasing affordability positively affects the growing number of mobile cellular telephony users in both country-income groups. The positive effects of the decreasing costs of mobile cellular services on the number of mobile telephony users is then explicitly, although indirectly, demonstrated through the growing availability of mobile cellular services to individuals. Therefore, removing a key factor such as 'low-affordability' enhances the spread of $MCS_{i, y}$, and accounts for a 'joint effect' of economic growth and drops of prices of mobile cellular telephony services. The empirical results summarized in Tables I.2 and I.4 (see Appendix I), illustrate the dynamic panel regression estimates with regard to $MCS_{i, y}$ in low-income and lower-middle-income countries. Including the lagged value of the $MCS_{i, y}$ variable in each of the models fundamentally reshapes the results. Nevertheless, when the $MCS_{i, y-1}$ is entered solely, or jointly, with other control variables, it remains positive and statistically significant. Moreover, most of regressors, except $GDPPPpc_{i, y}$ and $FTL_{i, y}$ in the selected specifications, lose their explanatory power; while the influence of 'epidemic mechanism' (Gray 1973; Sarkar 1998; Kumar and Krishnan 2002; Gomulka 2006) in the spread of $MCS_{i, y}$ is dominant over other determinants. Such evidence leads to a seminal conclusion on the existence of strong network effects with respect to the process of mobile cellular telephony diffusion. It might be claimed that once the critical conditions (see Sect. 5.2.2) are achieved, the process of diffusion is self-sustaining and predominantly conditioned by intensity and frequency of interpersonal contacts.[59] The results presented in Tables I.5 and I.7 (see Appendix I), help to explore the impact of selected factors of Internet usage in low-income and lower-middle-income countries over the period 1997–2012. First, we investigate the importance of the determinants of Internet penetration rates in both income groups. An important observation is that in low-income economies, specifications (1)–(3) (Table I.5) with multiple explanatory variables, although relatively high R^2 (within), report that the degree of urbanization ($Urban_{i, y}$) exclusively produced positive and statistically significant effect on the growth of Internet users penetration rates. In models (2) and (3), the inverse, and statistically significant, impact of fixed-broadband connection charges ($FBSCharge_{i, y}$) on

---

[59] For broader discussion—see Chaps. 3 (theoretical aspects of diffusion mechanism) and 4 (empirical evidence on ICTs diffusion).

$MCS_{i, y}$ is shown. The fixed-broadband connection charge, which presents the basic cost of acquiring Internet, is a seminal factor that may significantly encourage, or contrariwise, hinder, the possibility of paying for access and usage of the Internet by individuals. Importantly, the previous results coincide with the evidence presented in Table I.5, which confirms the importance of fixed-broadband connection charges on broad access to, and use of, an Internet network. It is worth noting that, despite that in the regressions (1) and (3), the $GDPPPpc_{i, y}$ is observed as statistically insignificant, the positive impact of economic growth on $IU_{i, y}$ is, however, indirectly captured by $FBSIPB_{i, y}$ and $FBSChargeMonth_{i, y}$, variables that explain the affordability of accessing the Internet network (see evidence in Table I.7). The impact of the remainder of the control variables on $IU_{i, y}$ changes, is found to be statistically insignificant.[60] Because the estimated models demonstrate little evidence on $IU_{i, y}$ seminal determinants in low-income economies, these results may be perceived as slightly disappointing. Yet, it is important to keep in mind that during the period 1997–2012, the average $IU_{i, y}$ in the low-income group persisted, with the exception of few prominent examples of Kenya, Uganda and Zimbabwe, at an extremely low level, which partially explains the lack of the robustness of the evidence in this regard. Concerning the lower-middle-income countries, we observe a marked positive effect of improving access to wireless-broadband networks on the share of individuals using the Internet. In each case (see respective models (1), (2), (3) and (8) in Table I.5), the coefficient going with the $WBS_{i, y}$ variable, is positive and statistically significant. These findings yield a straightforward conclusion regarding the increasing importance of wireless-broadband infrastructure in enabling broad usage of the Internet in lower-middle-income countries. Interestingly, this importance is reported neither for fixed-narrowband-, nor for fixed-broadband networks. Similarly, as in the case of low-income countries, the variable $FBSCharge_{i, y}$ turns out to be inversely correlated with $IU_{i, y}$ and statistically significant, suggesting that due to increasing competition and decreases in the price of access to fixed-broadband infrastructure, shifts in the number of individuals using the Internet network are observed. Moreover, as suggested by the evidence in Table I.5, the strong and positive effect of economic growth on $IU_{i, y}$ is demonstrated through the growing affordability of buying and using fixed-broadband networks by individuals. Because an important constraint such as 'low-affordability' is being gradually eradicated, there emerges an enormous potential of further expansion of Internet infrastructure, resulting in striking growths of Internet penetration rates. Contrary to what was reported for the low-income group, in lower-middle-income countries the population density arises as an important factor, positively contributing to the increasing number of individuals who use the Internet. The emphasized $IU_{i, y}$ determinant—population

---

[60] The consecutive models (4)–(11) with only one explanatory variable introduced demonstrate each of explanatory variables as statistically significant; but in some cases the overall fit of the model to the empirical data is poor (e.g. see regression (8) and (9)). For this reason, it is questionable to consider these results as valid and conclusive—see evidence from models (1), (2) and (3).

density, may play a role in enhancing the use of Internet connections because in more densely populated areas the access to fixed-, or wireless-networks is highly facilitated due to better developed backbone infrastructure, compared to remote and isolated regions. Hence, the population density may emerge as a country-specific feature conductive to $IU_{i, y}$ growth. Finally, the evidence summarized in Tables I.6 and I.8 in Appendix I, mirrors the results of the dynamic panel regression estimates of $IU_{i, y}$ determinants in low-, and lower-middle-income group. It provides support in favor of the supposition that, as in the case of $MCS_{i, y}$ determinants analysis, inserting the lagged value of $IU_{i, y}$ into the regression, reshapes the outcomes. The main finding is that regardless of the model and the regressors included, the coefficient for $IU_{i, y-1}$ (ζ) is always positive and statistically significant. This exercise yields a sharp conclusion that the current level of $IU_{i, y}$ penetration rates are highly pre-conditioned by the number of Internet users in the preceding period, which confirms the hypothesis that an existing strong network affects the underlying mechanism of technology diffusion. Interestingly, according to the dynamic panel regression estimates for lower-middle-income countries, the $WBS_{i, y}$ is reported as significant in each case and hence may be considered as valid explanatory factor of $IU_{i, y}$, changing in scope over time and across countries. In turn, the variable standing for population density ($PopDens_{i, y}$) has 'lost' its explanatory power, which shows that population density does not play an essential role in enhancing $IU_{i, y}$ growth, as was previously suggested by the estimates reported from the respective fixed effects regressions. Additionally, contrary to what might have been expected, the degree of urbanization remains insignificant. The rationale behind this is that in the examined countries, a vast majority of people still live in rural regions that persistently suffer from underdevelopment of the backbone infrastructure that enables Internet connections. This finding is also supported by the fact that in the majority of backward countries, the urban-rural divide with regard to Internet penetration rates is substantial and persistent. According to the data provided in the report Measuring the Information Society 2011 (ITU 2011), in developing countries fundamental differences still exist between urban and rural areas in access to and use of Internet networks. The Internet penetration rates differ remarkably between urban and rural areas; people living in rural regions are still heavily deprived of the opportunity of using the Internet.

In the final part of Chap. 5, we have investigated the factors, which might potentially influence mobile cellular telephony and Internet penetration rates in low-income and lower-middle-income countries during the period 1997–2012. First we have estimated the fixed effects regressions to test which variables might be considered as important determinants of $MCS_{i, y}$ and $IU_{i, y}$ diffusion. Our estimates suggest that in the examined countries (in both income groups), $MCS_{i, y}$ was positively attributed to GDP per capita, level of education ($School_{i, y}$) and population density, and although these results are not fully robust, they reveal little sensitivity to the inclusion or exclusion of other control variables in the model. We may also conclude that the overall affordability explains changes in $MCS_{i, y}$ growth in both income-groups relatively well. The population density variable has been shown to be statistically significant, and these effects are robust. Somewhat unexpectedly, the price of a 1-min call, and of SMSs, in most instances did not demonstrate any statistical significance to explain the variability in cross-country $MCS_{i, y}$. Our estimates of $IU_{i, y}$ diffusion determinants show that, in low-income countries, GDP per capita and the price of fixed-broadband connection revealed statistical significance and may be considered factors positively influencing a growing number of Internet users. In the group of lower-middle-income countries, the variables GDP per capita, the prices of fixed-broadband connection and wireless broadband penetration rates are reported as having positive impact on increasing Internet penetration rates. However, if the fixed effects models, both for $MCS_{i, y}$ and $IU_{i, y}$, are refined by including the lagged values of response variables, the overall picture changes dramatically. Relying on the dynamic panel regressions, we have revealed the existence of strong network effects with regard to mobile cellular telephony and Internet user growth. The coefficients going with the lagged values of $MCS_{i, y}$ and $IU_{i, y}$ are positive and statistically significant regardless of the specification and are insensitive to the inclusion/exclusion of various control variables. Hence, it is justified to claim that the network effects are fully robust and reveal great explanatory power in cross-country ICTs diffusion.

5.4   Summary

The main targets of Chap. 5 were twofold. First, adopting the newly developed methodological approach, it aimed to trace the 'critical mass' effects. Henceforth, we have identified the '*critical year*', '*critical penetration rate*', and the '*technological take-off*' and explored country's individual conditions during the specific '*technological take-off*' interval. Regarding the mobile cellular telephony the important observation is that the '*critical penetration rates*' barely vary between the low-income and lower-middle-income countries—7.05 (per 100 inhab.) in low-income and 8.22 in lower-middle-income group. The country-wise analysis revealed that in both within and between income groups, the country-specific features vary widely and countries share *very few* common conditions that predetermine leaving the early diffusion phase and the emergence of the '*MCS-technological take-off*'. Regarding Internet network diffusion, the analysis of the 'critical conditions' yields similar conclusions to those in the previous case. However, importantly to note that the overall Internet penetration rates in many of the examined countries in 2012 were still very low, which indicates that access to Internet connections was still a 'luxury' good and could not be unboundedly afforded in a vast majority of economically backward countries. The latter implies that the analysis results regarding $IU_{i,y}$ and detecting country-specific conditions during the '*IU-technological take-off*' are—to a point—violated, and thus shall be interpreted carefully. Second, we targeted to trace those factors which have had positive impact of ICTs diffusion across analyzed countries. Regarding $MCS_{i,y}$ diffusion we have found that GDP per capita, level of education and population density impact positively the latter. Contrary, factors like price of a 1-min call and of SMSs are reported as statistically insignificant. Across analyzed countries, the

$IU_{i,y}$ was mostly enhanced by GDP per capita, changes in price of fixed-broadband connection and (in lower-middle-income group) by growing access to wireless broadband solutions. In addition, the analysis has demonstrated that both in case of $MCS_{i,y}$ and $IU_{i,y}$ ICT diffusion is predominantly conditioned and enhanced by the 'word of mouth', which give rise to the emergence of strong network effects. Finally, a few important issues should be mentioned with regard to the evidence provided earlier in this chapter. Due to short data time series in the case of some variables and limited data availability, this may heavily violate analysis outcomes and conclusions. This is a serious limitation, which may cause lack of robustness of our results. Moreover, the analysis predominantly explains statistical relationships between variables. Hence, the question arises: Are the explanatory variables causes of, or simply correlates of, $MCS_{i,\ y}$ and $IU_{i,\ y}$? Considering the type of selected explanatory variables, it might be justified to argue that these are factors driving profound changes in access to and use of basic ICTs, although these relationships may not be straightforward, and severe time lags may emerge between the cause and the outcome.

References

Ahn, H., & Lee, M. H. (1999). An econometric analysis of the demand for access to mobile telephone networks. *Information Economics and Policy, 11*(3), 297–305.

Andrés, L., Cuberes, D., Diouf, M., & Serebrisky, T. (2010). The diffusion of the internet: A cross-country analysis. *Telecommunications Policy, 34*(5), 323–340.

Arellano, M. (1987). Practitioners' corner: Computing robust standard errors for within-groups estimators. *Oxford Bulletin of Economics and Statistics, 49*(4), 431–434.

Arellano, M., & Bond, S. (1991). Some tests of specification for panel data: Monte Carlo evidence and an application to employment equations. *The Review of Economic Studies, 58*(2), 277–297.

Bakay, A., Okafor, C. E., & Ujah, N. U. (2011). Factors explaining ICT diffusion: Case study of selected Latin American countries. *International Journal on Advances in ICT for Emerging Regions (ICTer), 3*(2), 25–33.

Baliamoune-Lutz, M. (2003). An analysis of the determinants and effects of ICT diffusion in developing countries. *Information Technology for Development, 10*(3), 151–169.

Baltagi, B. (2008). *Econometric analysis of panel data* (Vol. 1). Chichester: Wiley.

Barrantes, R., & Galperin, H. (2008). Can the poor afford mobile telephony? Evidence from Latin America. *Telecommunications Policy, 32*(8), 521–530.

Billon, M., Marco, R., & Lera-Lopez, F. (2009). Disparities in ICT adoption: A multidimensional approach to study the cross-country digital divide. *Telecommunications Policy, 33*(10), 596–610.

Caselli, F., & Coleman, W. J., II. (2001). *Cross-country technology diffusion: The case of computers* (No. w8130). National Bureau of Economic Research.

Cheibub, J. A., Gandhi, J., & Vreeland, J. R. (2010). Democracy and dictatorship revisited. *Public Choice, 143*(1–2), 67–101.

Chinn, M. D., & Fairlie, R. W. (2010). ICT use in the developing world: An analysis of differences in computer and internet penetration. *Review of International Economics, 18*(1), 153–167.

Comin, D., & Hobiijn, B. (2006). *An exploration of technology diffusion* (No. w12314). National Bureau of Economic Research.

Comin, D., & Hobijn, B. (2004). Cross-country technology adoption: Making the theories face the facts. *Journal of Monetary Economics, 51*(1), 39–83.

Comin, D., & Hobijn, B. (2009). Lobbies and technology diffusion. *The Review of Economics and Statistics, 91*(2), 229–244.

Crenshaw, E. M., & Robison, K. K. (2006). Globalization and the digital divide: The roles of structural conduciveness and global connection in internet diffusion. *Social Science Quarterly, 87*(1), 190–207.

Dasgupta, S., Lall, S., & Wheeler, D. (2005). Policy reform, economic growth and the digital divide. *Oxford Development Studies, 33*(2), 229–243.

Dormann, C. F., Elith, J., Bacher, S., Buchmann, C., Carl, G., Carré, G., et al. (2013). Collinearity: A review of methods to deal with it and a simulation study evaluating their performance. *Ecography, 36*(1), 027–046.

Economic Commission for Africa. (2003). E-strategies. National, sectoral and regional ICT policies, plans and strategies. E/ECA/DISD/CODI.3/3.

Freedom House. (2011). Freedom on the Net. A global assessment of Internet and digital media freedom. Freedom House.

Freedom House. (2012). Freedom on the Net. A global assessment of Internet and digital media. Freedom House.

Freedom House. (2013). Freedom on the Net. A global assessment of Internet and digital media. Freedom House.

Freedom House. (2014). https://freedomhouse.org

Garbacz, C., & Thompson, H. G. (2007). Demand for telecommunication services in developing countries. *Telecommunications Policy, 31*(5), 276–289.

Gomulka, S. (2006). *The theory of technological change and economic growth.* London: Routledge.

Gray, V. (1973). Innovation in the states: A diffusion study. *The American Political Science Review, 67*, 1174–1185.

Greene, W. H. (2003). *Econometric analysis.* New Delhi: Pearson.

Gupta, R., & Jain, K. (2012). Diffusion of mobile telephony in India: An empirical study. *Technological Forecasting and Social Change, 79*(4), 709–715.

Hargittai, E. (1999). Weaving the Western Web: Explaining differences in Internet connectivity among OECD countries. *Telecommunications Policy, 23*(10), 701–718.

Hausman, J. A. (1978). Specification tests in econometrics. *Econometrica: Journal of the Econometric Society, 46*, 1251–1271.

Heritage Foundation. (2013, 2014). www.heritage.org

Hoechle, D. (2007). Robust standard errors for panel regressions with cross-sectional dependence. *Stata Journal, 7*(3), 281.

House, F. (2013). *Nations in transit 2013: Democratization from Central Europe to Eurasia.* New York: Rowman & Littlefield.

Human Development Report. (2010). *The real wealth of nations, pathways to human development,* 20th Anniversary Edition.

Islam, T., & Meade, N. (1997). The diffusion of successive generations of a technology: A more general model. *Technological Forecasting and Social Change, 56*(1), 49–60.

ITU. (2010). *Measuring the information society 2011.* Geneva: ITU.

ITU. (2011). *Measuring the information society 2011.* Geneva: ITU.

ITU. (2012). *Wireless broadband masterplan for the Union of Myanmar.* Geneva: ITU.

ITU. (2013). *Measuring the information society 2011.* Geneva: ITU.

Jakopin, N. M., & Klein, A. (2011). Determinants of broadband internet access take-up: Country level drivers. *info, 13*(5), 29–47.

Kiiski, S., & Pohjola, M. (2002). Cross-country diffusion of the Internet. *Information Economics and Policy, 14*(2), 297–310.

Kumar, V., & Krishnan, T. V. (2002). Multinational diffusion models: An alternative framework. *Marketing Science, 21*(3), 318–330.

Lee, S., Marcu, M., & Lee, S. (2011). An empirical analysis of fixed and mobile broadband diffusion. *Information Economics and Policy, 23*(3), 227–233.

Liu, X., Wu, F. S., & Chu, W. L. (2012). Diffusion of mobile telephony in China: Drivers and forecasts. *IEEE Transactions on Engineering Management, 59*(2), 299–309.

Maddala, G. S., & Lahiri, K. (1992). *Introduction to econometrics* (Vol. 2). New York: Macmillan.

Madden, G., & Coble-Neal, G. (2004). Economic determinants of global mobile telephony growth. *Information Economics and Policy, 16*(4), 519–534.

Madden, G., Coble-Neal, G., & Dalzell, B. (2004). A dynamic model of mobile telephony subscription incorporating a network effect. *Telecommunications Policy, 28*(2), 133–144.

Mansfield, E. (1986). Microeconomics of technological innovation. In R. Landau & N. Rosenberg (Eds.), *The positive sum strategy* (pp. 307–325). Washington, DC: National Academies Press.

Mansfield, E. R., & Helms, B. P. (1982). Detecting multicollinearity. *The American Statistician, 36*(3a), 158–160.

Michalakelis, C., Varoutas, D., & Sphicopoulos, T. (2008). Diffusion models of mobile telephony in Greece. *Telecommunications Policy, 32*(3), 234–245.

Norris, P. (2000, August). The global divide: Information poverty and Internet access worldwide. In *Internet conference at the international political science world congress in Quebec city* (pp. 1–6).

O'Brien, R. M. (2007). A caution regarding rules of thumb for variance inflation factors. *Quality & Quantity, 41*(5), 673–690.

Rodrik, D., Subramanian, A., & Trebbi, F. (2004). Institutions rule: The primacy of institutions over geography and integration in economic development. *Journal of Economic Growth, 9*(2), 131–165.

Rostow, W. W. (1990). *The stages of economic growth: A non-communist manifesto.* Cambridge: Cambridge University Press.

Rouvinen, P. (2006). Diffusion of digital mobile telephony: Are developing countries different? *Telecommunications Policy, 30*(1), 46–63.

Sarkar, J. (1998). Technological diffusion: Alternative theories and historical evidence. *Journal of Economic Surveys, 12*(2), 131–176.

Singh, S. K. (2008). The diffusion of mobile phones in India. *Telecommunications Policy, 32*(9), 642–651.

Tuan, T. M. (2011). *Broadband in Vietnam: Forging its own path.* Washington, DC: infoDev/ World Bank.

UNDP. (2010). Available at: http://hdr.undp.org/en/content/human-development-report-2010

WDI. (2013). *World Development Indicators 2012 database.* Available online at: http://data. worldbank.org/data-catalog/worlddevelopment-indicators

World Bank Group. (2014). *The little data book on information and communication technology, 2014.* Washington, DC: World Bank.

Yates, D. J., Gulati, G. J., & Weiss, J. W. (2011, January). Different paths to broadband access: The impact of governance and policy on broadband diffusion in the developed and developing worlds. In *2011 44th Hawaii international conference on system sciences (HICSS)* (pp. 1–10). IEEE.

Conclusion, Recommendations and Implications

6

Abstract
The main purpose of this chapter is to present a comprehensive overview of empirical findings regarding ICT diffusion in 17 low-income and 29 lower-middle-income countries over the period 2000–2012. It shows major ICT diffusion trends, demonstrates the main features of the technological substitution process, and shows technological convergence dynamics. It also provides insight into seminal factors that accelerate—or, conversely, hinder—rapid ICT diffusion in developing economies. Moreover, it briefly discusses ICT policies that aim to foster ICT deployment in economically backward countries. Finally, it sheds light on the potential role of ICT in boosting growth and development in economically backward countries.

Keywords
ICT diffusion • Developing counties • ICT determinants • ICT policies

6.1 Introduction

The last chapter comprehensively summarizes the seminal contributions to the present state of knowledge regarding the process of ICT diffusion in economically backward countries; which predominantly consists in:

- Developing new methodological framework designed to trace the 'critical mass' effects with respect to ICT diffusion process. Along with the latter it proposes new terms 'technological take-off', 'critical year' and 'critical penetration rate', which are consistent of the novel methodological approach. Moreover, the notion of 'critical mass' has been put into broad social, economical and institutional perspective which constitutes a novelty on this field of study.

© Springer International Publishing Switzerland 2015
E. Lechman, *ICT Diffusion in Developing Countries*,
DOI 10.1007/978-3-319-18254-4_6

- Elaborating detailed country-specific ICT diffusion trajectories allowing for in-depth analysis of the process and detection of countries' unique features with this respect;
- Tracing the technological substitution effects and explaining why this process is rather illusive when regarding the group of developing countries;
- Examining the process of technology convergence technology club convergence, showing the group of developing countries in the worldwide perspective;
- Dedicating the analysis to economically backward countries, which so far have received relatively little attention in empirical literature and the evidence on these countries, is highly fragmented and limited.

Moreover, this final chapter is to provide the reader with a comprehensive summary of the empirical findings, which are presented throughout the book, shedding light on the seminal issues associated with the progress in ICTs growth that took place in a great majority of economically backward countries over the analyzed period, 2000–2012. It also provides a brief overview of ICT policies that aim to foster the deployment of new technologies even in the least developed countries. Finally, it introduces some highlights on the role of ICTs in boosting growth and development in economically underdeveloped countries.

6.2   Underlying Conclusions

This study of ICTs diffusion has covered 17 low-income and 29 lower-middle-income countries[1], for the period 2000–2012. We have examined the process of diffusion for five different ICT indicators, namely: mobile cellular subscriptions ($MCS_{i, y}$), fixed[2] Internet subscriptions ($FIS_{i, y}$), fixed-broadband subscriptions ($FBS_{i, y}$), wireless-broadband subscriptions ($WBS_{i, y}$) and Internet users ($IU_{i, y}$).
    The analysis was designed to consider the ICTs diffusion from four perspectives:

- Explaining the ICTs diffusion patterns and the dynamics of the process itself;
- Detecting technological substitution;
- Examining technology convergence;
- Identifying the 'critical conditions' that enhanced the emergence of the 'technological take-off.'

The major findings are summarized in the following of this section.

---

[1] According to World Bank classification.
[2] Narrowband.

6.2.1 The First Perspective. Moving Ahead or Lagging Behind?

Undoubtedly, over the period 2000–2012, both low-income and lower-middle-income countries have experienced significant increases in ICTs deployment[3]. We have contributed to the better understanding of the process by developing country-specific ICT diffusion trajectories, which allowed capturing various unique countries characteristics that the process is attributed to. Our main findings strongly confirm that overall the mobile market and the total number of users of mobile cellular technologies have been steadily growing during the examined period, albeit with significant variations among countries both in the speed and the achieved saturation in 2012. From 1984 onward, in lower-middle-income economies the number of users of mobile cellular networks has been continuously increasing at an estimated average annual rate[4] 39.2 %, which resulted in the growth of the average penetration rate from .001 in 1984, to 3.28 in 2000, and finally in 2012 up to 95.35 per 100 inhab. In lower-middle-income counties the mobile cellular services have been almost equally available for all individuals, although cross-country disparities still persisted at the end of 2012. Analogous calculations, however, for low-income economies demonstrate that the process of diffusion of mobile cellular telephony, even in the most economically backward countries, has been dynamically proceeding by approximately 47.7 % per year, on average, over the period 1992–2012. The strong annual growth of mobile cellular users resulted in increases in total penetration rates from 0.001 in 1992, to .45 in 2000, and 51.76 per 100 inhab. in 2012. Although there have been tremendous changes in the uptake of mobile cellular telephony in this income group, large parts of these societies still have not been reached by this digital technology, and some income-groups are lagging behind regarding the uptake of mobile cellular telephony. Notwithstanding, the fast and broad diffusion of mobile telephony in most economically underdeveloped countries was facilitated by multiple factors, like, for example, the growing accessibility and affordability has made uptake of mobile cellular telephony relatively easy for the majority of societies in the low-, and lower-middle-income economies (ITU 2011c); the universal establishment of the prepaid systems (UNCTAD 2007), which allowed the barriers of insufficiently developed infrastructure to be overcome. The growth of the fixed-narrowband networks ($FIS_{i, y}$) and fixed-broadband networks ($FBS_{i, y}$), albeit relatively stable, has been notably slower in the countries in both income groups between 2000 and 2012. Despite the relatively high estimated annual growth rates of fixed-narrowband and fixed-broadband subscriptions, in 2012 the overall access to fixed infrastructure facilitating Internet connections has been extremely restricted, especially in low-income economies. In the lower-middle-income countries, the growth of $FIS_{i, y}$ and $FBS_{i, y}$ is only marginally higher (in 2012 stand at approximately 2.67 and 2.45 per 100 inhab., respectively). Heavy underdevelopment of fixed-networks has been, at least

---

[3] See evidence in Chap. 4.

[4] Author's calculations.

partially, compensated by the relatively dynamic expansion of wireless-broadband infrastructure (since 2007 onward), which is broadly recognized as an attractive alternative in countries with poorly developed wired infrastructure. On average, low-income countries, in 2012, reached higher $WBS_{i,\,y}$ penetration rates—4.9 per 100 inhab., compared to lower-middle-income ones—$WBS_{i,2012} = 9.58$ per 100 inhab. It is important to highlight that these numbers are promising and suggest prospects for the future. Although the picture arising from the analysis of the respective trends in lower-middle-income economies is slightly more promising, still huge cross-country disparities exist, as evidenced by the identifiably limited access to wireless-broadband infrastructure in most of these countries. The emergence of wireless-broadband networks is reported in both country groups; however, poor penetration of fixed-, and wireless-networks is the reality of the vast of majority of countries and is recognized as the main barrier for unrestricted access to, and use of, Internet connections. In the examined countries, still quite a share of individuals is permanently 'unconnected' and suffers from digital deprivation.

6.2.2   The Second Perspective. Technological Substitution: Illusion or Fact?

The data on ICTs collected from ITU (2013) suggest that the rapid expansion of mobile telephony and wireless broadband connections, might have resulted in the number of subscriptions to the 'new technologies' surpassing the number of subscriptions to the 'old technology'. Thus, our analysis we have intended to uncover major tendencies in the process of technological substitution in developing countries, which has received relatively little attention in empirical literature. Henceforth, we have broadly documented the emergence of technological substitution effects, referring to 'fixed-to-mobile telephony substitution' and 'fixed-to-wireless Internet connections substitution'. The empirical evidence on the process of fixed-to-wireless Internet connections substitution' is so far very limited and thus constitutes a great value added of this analysis. Our evidence strongly confirms the emergence of fixed-to-mobile substitutions and fixed-to-wireless Internet connections substitution effects. Regarding the fixed-to-mobile substitution, our estimates report that the average number of years required for the 'take over'[5] was approximately 10 years, in both income groups; while, with respect to fixed-to-wireless Internet connections, the average 'take over' time was approximately 5.8 years and 5.2 years in the low-income and lower-middle-income countries, respectively. This demonstrates that economically backward countries have undergone the dynamic process of switching from the 'old' to 'new' technologies, which exhibits newly emerging trends and patterns in the scope of development of mobile

---

[5] The 'take over' time is the estimated number of years necessary so that the 'new' technology passes from level of saturation 10 % until 90 %. Put another way, it shows the time required for the new technology to achieve 90 % share on telecommunication market.

cellular telephony and wireless-broadband Internet connections. The in-depth analysis of the fixed-to-mobile technological substitution process reveals several critical issues, which are essential for interpretation of the examined technological substitution effects. First, in most of economically backward countries the vast share of society has never before owned a fixed telephone mainline and went straight to cellular technologies instead; which suggests that the mobile networks are diffusing '**in place**' of traditional fixed mainlines and not alongside them. Thus, the observed fixed-to-mobile substitution is rather an '**illusion**' than the fact, as the vast majority of people living in developing world has never before adopted any type of 'fixed' technology, but directly adopted 'mobile' solutions. The explanations for this situation are various. For example, the relative ease of infrastructure deployment (compared to fixed infrastructure) to provide mobile services makes the mobile alternative highly attractive, and mobile phones are deployed instead of fixed ones (UNDP 2014); implementation of a regulatory framework that fostered competition among mobile telephony operators from the first introduction[6] of mobile telephony into the market, which attracted new operators and totally broke the monopoly of fixed telephone companies. Finally, unlike the fixed telephone mainline services, the mobile telephony operators started to offer pre-paid subscriptions schemes, which generated an unprecedented boom in mobile cellular telephony subscriptions. Turning to the fixed-to-wireless Internet connection substitution, we have observed analogous trends and regularities and the reported substitution is rather illusive. Significantly, the low penetration of fixed-narrowband or fixed-broadband infrastructure resulted in people tending to go straight to wireless broadband networks, rather than using both simultaneously or gradually switching from the fixed to wireless technologies. The **illusive** technological substitution effects that have been reported on here, may also be associated with technological leapfrogging; this happens when countries do not follow the 'classical' development patterns but instead 'jump' directly to more advanced stages of development. The technological leapfrogging phenomenon perfectly explains the process of the successful deployment of 'new' and more advanced technologies in economically backward countries without ever adopting the 'old' (prior) version of technological solutions.

6.2.3   The Third Perspective. Digital Gaps Closing or Growing?

Arguing that a great majority of developing countries have made enormous progress in ICT deployment, we contribute by verifying the hypothesis of technology convergence and technology club convergence. To this aim, the original empirical sample has been extended by inclusion another 25 upper-middle-income and

---

[6] The year when mobile telephony was introduced to the market varies significantly even in low-, and lower-middle-income countries (see Appendix B).

42 high-income countries[7]; so it encompassed 113 countries. We have tested our hypothesis on technology convergence regarding mobile cellular telephony, fixed-narrowband and fixed-broadband Internet connections and Internet penetration rates; while due to very limited data availability and the short time series, we have excluded the wireless-broadband technology from the 'standard' analysis of technology convergence. Our major findings with respect to all four analyzed ICT indicators confirm the existence of strong unconditional technology-beta-convergence in the case of each ICT indicator. The most dynamic process of technology-beta-convergence has been reported in the case of mobile cellular telephony, as—according to our estimates—the MCS-beta-convergence proceeds at the rate of 17.6 % annually, which demonstrates that the cross-country disparities may be halved within 3.92 years (*sic!*). For the remaining ICT indicators, the technology-beta-convergence is still confirmed, while being identifiably the slowest in the case of the fixed-narrowband networks. The results of technology-sigma-convergence are ambiguous. Relying on the 'standard deviation' we have found no evidence in support of technology-sigma-convergence; however, when using the coefficient of variation as a measure, the technology-sigma-convergence hypothesis is positively verified. It shows that, although in relative terms the cross-country disparities are gradually diminishing, the absolute digital gaps have grown enormously between 2000 and 2012. Additionally, to test whether all countries have been included in the technology-convergence process, we have examined the existence of the technology-convergence clubs, which has not been tested so far. We have used two distinct empirical approaches to technology-convergence clubs identification (Baumol and Wolff 1988; Chatterji and Dewhurts 1996); however, *no* technology clubs have been detected. This shows that none of analyzed countries was left outside the 'exclusive' technology convergence clubs, which again supports the supposition of the worldwide catching-up process that, in the long-term horizon, shall inevitably lead to gradual eradication of digital gaps (at least in relative terms).

6.2.4 The Fourth Perspective. Ready for the 'ICT Revolution'?

Finally, deploying newly developed methodological framework, we aimed to trace the country-specific '*technological take-off*' intervals and the '*critical mass*' closely associated with the ICTs diffusion patterns. The issues associated with the identification of the 'critical mass' regarding ICT diffusion in developing countries, have been relatively rarely discussed both in theoretical and empirical literature. Henceforth, we have filled this gap in knowledge, and proposed totally new methodological framework designed to trace the 'critical mass' effects, however putting it into wide social, economical and institutional perspective. Such conceptualization of the 'critical mass' has so far never been discussed in economic

---

[7] Based on the World Bank 2013 country classifications (see: http://data.worldbank.org/news/new-country-classifications, accessed: May 2014).

literature, and thus constitutes an important contribution to the presents state of the art.

To meet the major goals of this analysis of 'critical mass' effects, we have identified the critical year and the critical penetration rates, along with the country-specific conditions that potentially might have fostered entering the exponential growth phase on country-specific ICTs diffusion trajectories. Additionally, to gather additional evidence, using panel analysis we have tested what most determined the ICTs diffusion in both income groups. Bear in mind that over the period 2000–2012, considerable growth was observed with respect to exclusively two ICT indicators—$MCS_{i, y}$ and $IU_{i, y}$. Tables in Appendix J summarize our major findings, which give rise to several key conclusions. First, regarding the mobile cellular telephony adoption determinants, the 'critical' penetration rates vary slightly between the low-income and lower-middle-income countries, accounting for 7.05 per 100 inhab. in low-income group, and 8.22 per 100 inhab. in the lower-middle-income group. The duration of the initial (early) phase of diffusion is approximately 12 years in both income groups. The further analysis shows that the country-specific features vary greatly and countries' **common** socio-economic conditions are barely detectable (see Tables 5.2 and 5.3 in Chap. 5). This suggests that there are **no common** country 'conditions' that would predetermine leaving the early diffusion phase and the emergence of the 'MCS-technological take-off'. The striking observation is that, the '*MCS-technological take-off*' interval was reported in countries characterized by the following dynamics: on average, barely 18.7 % of the population had access to electrification, only 67 % has access to improved drinking water, almost 44 % of adult population was illiterate, 52.1 % experienced extreme poverty, and more than ¾ of countries' total population lived in rural areas. Moreover, only one country (out of the low-income group) was classified as politically 'free,' ten were classified as 'partly 'free', and four as 'not free,' while the average GDP PPP per capita was only US\$1,434.6. In contrast to what might be expected, all the 'characteristics' listed above, which are traditionally considered as obstacles for development, did not appear to be the insurmountable barriers for mobile cellular telephony adoption. Our analysis, however, yields another seminal conclusion. The ICTs—especially mobile cellular telephony—are broadly recognized as 'suitable for all,' and are claimed to be adequate technologies for underdeveloped countries. The overall empirical evidence provided in this book seems to confirm (at least partially) the supposition that mobile cellular services *are* technologies that may be easily and rapidly adopted even in the poorest countries and in widely differing environments.

Regarding Internet usage (see Appendix J), the analysis of the 'critical conditions' yields similar conclusions as in the previous case. Although the '*IU-technological take-off*' was identified in only 7 low-income and 26 lower-middle-income countries, the countries individual 'conditions' seemed highly unfavorable for the growth of Internet usage based on the following characteristics: high cost of access to fixed-broadband networks, low per capita income, and poor infrastructural development. However, conversely to what was reported in the case of mobile cellular telephony, in 2012, the overall Internet penetration rates in many of the examined countries were still very low, which indicates that access to Internet connections is still a 'luxury' good and cannot be afforded in a significant number of economically underdeveloped countries. The previous evidence was complemented by regression analysis, to trace the factors that might *potentially* influence mobile cellular telephony and Internet penetration rates in low-income and lower-middle-income countries. Our estimates suggest that in the examined countries (in both income groups), $MCS_{i, y}$ was positively attributed to GDP per capita, level of education and population density, and although these results are not fully robust, they reveal little sensitivity if we include other control variables in the model. 'Affordability' explains changes in $MCS_{i, y}$ growth relatively well, and in addition, the population density has been shown to be statistically significant, and these effects are robust. Surprisingly, the prices of a 1-min call and SMSs did not demonstrate any statistical significance to explain the variability in cross-country mobile cellular telephony penetration rates. Our estimates of $IU_{i, y}$ diffusion determinants show that, in low-income countries, GDP per capita and the price of fixed-broadband connection showed statistical significance and may be considered as factors positively influencing the growing number of Internet users. In the group of lower-middle-income countries, the variables GDP per capita, price of fixed-broadband connection and wireless broadband penetration rates are reported as having positive impact on increasing Internet penetration rates. However, if the fixed effects models, both for $MCS_{i, y}$ and $IU_{i, y}$, are refined by including the lagged values of response variables, the overall picture changes dramatically. Relying on the dynamic panel regressions, we have unveiled the existence of strong network effects with regard to mobile cellular telephony and Internet users growth. The coefficients associated with the lagged values of $MCS_{i, y}$ and $IU_{i, y}$ are positive and statistically significant regardless of the specification; they are also insensitive to the inclusion/exclusion of various control variables. Hence, it is justified to claim the network effects are fully robust and reveal great explanatory power in cross-country ICTs diffusion.

Before drawing final conclusions, it is important to bear in mind that each of the countries examined has different socio-economic conditions that may generate different causal relations with regard to the tested relationships. The deployed methodological framework may be contested, and some crucial questions still may not be satisfactory answered. Using econometric modeling to indicating causal order has significant limitations, the results often reveal high sensitivity to the inclusion/exclusion of outliers, hence this method is not always very conclusive and interpretive. Still, the influence of the factors 'left outside the model' is not captured; they are omitted and 'stay in the shadows' once we concentrate exclusively on arbitrarily selected factors (Mäki 2002). The overwhelming majority of cross-country variability seems to be unexplained by the factors considered, as most of the detected relationships are highly unstable. None of conclusions is entirely persuasive and convincing, and probably other factors, not identified in here, are an important part the whole story.

6.3   A Brief Look at ICT Policies in Developing Countries

6.3.1   What Needs to Be Addressed? Some Recommendations

Our study has demonstrated that during the period 2000–2012, many low-income and lower-middle-income countries managed to escape from the low-penetration trap and achieve broader ICT deployment. The diffusion of mobile cellular telephony was phenomenally rapid, which resulted in dramatic shifts in mobile cellular telephony penetration rates in nearly all of the analysed countries. However, in 2012, with fixed and wireless network penetration rates and share of individuals who use the Internet, the numbers are less optimistic, reflecting that although average annual growth rates were impressively high, in many countries (especially the low-income ones), ICT adoption remains low. This suggests that ensuring unbound access to ICTs is still restricted and impeded by multiple constrains.

Mobile telephony diffusion was mainly facilitated by the rapid development of wireless networks, which made the mobile services broadly accessible even in geographically remote rural areas (UNCTAD 2007). Rapid infrastructure development was also critically affected by increasing market liberalization, which enhances competition and contributed to lowering the cost of mobile telephony usage (World Bank 2006). These factors fostered importing innovative telecommunication solutions and enabled the expansion of access to and use of mobile telephony. Many developing countries, until 2010, also made significant progress in mobile service coverage (ITU 2011b), which improved access to mobile networks in geographically remote and rural areas (according to ITU estimates (ITU 2011b), in 2010, more than 62 % of the populations in low-income countries were covered by a mobile cellular signal). Although many countries have been very successful in the rapid expansion of mobile cellular service, there remain a few, such as Eritrea, Ethiopia and Myanmar, where mobile cellular network access in 2012 was still limited. Moreover, according to ITU data, the prices of mobile cellular sub-baskets were still indecently high in 2011 (e.g., in Zimbabwe, the cost was 53.7 % of GNI per capita per month; in Togo, it was 48 %, and in Eritrea, it was 42.8 %). High prices and low affordability of mobile cellular services are still perceived as the major barrier for the further deployment of mobile cellular telephony, and these must be eliminated if there are to be increases in the numbers mobile telephony adopters. Important to note is that in many countries, especially in Africa, although the telecommunication markets have been officially fully liberalized, there are still very few (two or three; see, e.g., Zambia, Uganda and Rwanda) operators, which restricts any additional price lowering for mobile services. It is likely that increased liberalisation of telecommunication markets and allowance for multiple operators would ease the low affordability and foster additional dynamic diffusion of cellular telephony. The mobile sector in developing countries demonstrated significant increases during the period 2000–2012; however, the deployment of fixed- and wireless-broadband networks remained relatively low, which resulted in very limited individual Internet use. In developing countries, the predominant barrier to the rapid and broad spread of the Internet is poorly developed backbone infrastructures—often recognized as the *Achilles heel* of these economies—which impedes the diffusion of wired (fixed) solutions that would provide access to data transfer. The vast majority of developing countries experience permanent shortages in the deployment of the basic infrastructure that preconditions the widespread development of cable-based Internet connectivity; meanwhile, the scarcely existent ICT infrastructure is often a direct cause of the permanent lack of financial resources, institutional weakness and political instability (ITU 2011b). Infrastructural shortages are especially awkward in rural, underserved, physically isolated and poorly populated areas, which additionally generates rural-urban asymmetries in Internet network access. Moreover, the low affordability of accessing and using fixed networks—note that the prices of fixed-broadband sub-baskets generally significantly exceed the average GNI per capita per month (see Table 5.3 in Chap. 5)—greatly hinders its broad usage. The problem of extremely low affordability is especially striking in the low-income countries and, together with infrastructural underdevelopment, may be recognized as the major barrier to the extensive use of Internet networks (Proenza 2006). The latter leads to the unbounded access to the Internet in low- and lower-middle-income countries is still perceived as a 'luxury good' that is not accessible and affordable to a wide audience.

In backward countries, the vast majority of populations live in rural regions rather than urban. These rural and often geographically isolated areas significantly lag in ICT deployment because they are *'difficult to serve because of rugged terrain, dispersion of costumers, and low-income and limited ability to pay for services'* (Proenza 2006, p. 22). Moreover, as was already stated, in rural regions, there is generally no fixed infrastructure available; thus, to enhance the expansion of Internet network coverage and usage, ensuring ubiquitous access to wireless broadband networks is considered an important target (Warren 2007). Wireless technologies appear to be an attractive panacea for fixed-infrastructure shortages; thus, their universalizing warrants special attention in developing countries. It is also suggested that wireless technologies can be easily deployed and accommodated even under rough rural conditions, and thus they have an obvious advantage over wire lines (Galperin 2005; Proenza 2006; Gunasekaran and Harmantzis 2007; Puumalainen et al. 2011; Gourhant et al. 2014).

Nevertheless, in a few countries, spectacular achievements in wireless broadband deployment were reported (see Chaps. 4 and 5) by 2012, most of the examined economies were heavily deprived of this type of network; access to wireless technologies was still highly constrained, mainly owing to remarkably high prices for leasing lines, which hinders the broad access and use of this ICT. There are also multiple country-specific barriers, including unfavourable locations, poorly developed hard infrastructures, or permanent problems with power supplies, which significantly impede the broader introduction of wireless broadband technologies. Keeping in mind the latter, special attention should be dedicated to eliminating the major bottlenecks and constraints in the adoption of wireless solutions and—in tandem—to building solid backgrounds for creating ICT-friendly environments (e.g., political stability, full liberalisation of telecommunication markets, ensuing legal regulations).

To avoid stagnation in further ICT deployment and to ensure equal and unrestricted access to ICTs, special assistance and efforts should be focused on the following (ITU 2006, 2011a, b) areas:

- Establishing autonomous regulatory authorities (if there are none) to provide legal frameworks for developing telecommunication markets, to lower the risk for operators who aim to invest in high-risk markets with low initial penetration rates (e.g., Addison and Heshmati 2003; Guerrieri et al. 2011);
- Complete the transition to fully liberalised and competitive markets (eliminate monopolies) to encourage maximum participation by private companies (e.g., Wresch and Fraser 2012);
- Implement and support pricing policies to ensure continuing price reductions (for both mobile cellular services and Internet access) for greater affordability (Baliamoune-Lutz 2003; ITU 2011a)
- Promote further development of backbone infrastructures to, *inter alia*, ensure greater access to electrification (regions permanently lacking power shortages should receive special attention) (e.g., Eberhard et al. 2008; ITU 2006, 2011b; Romijn and Caniëls 2011);
- Pay special attention to ICT deployment in rural, geographically isolated and poorly populated regions to gradually close the rural-urban digital divide (e.g., Chen and Wellman 2004; Robison and Crenshaw 2010; White et al. 2011; Nakamura and Chow-White 2013);
- Focus on developing wireless networks (e.g., Proenza 2006; Thapa 2011; Hanson and Narula 2013; Gourhant et al. 2014) to ensure better connectivity, especially in underserved and remote regions.

A final remark: the countries investigated in this study are economically and institutionally weak. These countries permanently suffer from heavy infrastructural underdevelopment and shortages, face political instabilities and lack good governance, and rules of law. The vast majority of the populations live in rural areas that are underserved and remote, with little contact with the 'outside world'; moreover, a number of these countries' populations live in extreme poverty and are poorly educated and illiterate. All of these obstacles hinder the further growth and effective use of ICTs, and they must be addressed through adequate ICT policies developing countries are to fully benefit from the potential of ICTs rather than being digitally isolated (Lucas and Sylla 2003).

6.3.2   A Few Words on ICT Policies and *e-Strategies* Implementation in Developing Countries

The message behind our analysis is that the deployment of ICTs in developing countries yields strong support. By convention, ICT policies may be designed to promote the development of national infrastructure, address specific sectorial needs, such as education or government, or to regulate the telecommunications market (e.g., pricing and tariffs, freedom of information). Developing countries are different from developed ones and therefore require specially tailored strategies and actions that target the constraints and barriers to ICT deployment, without which they will suffer from digital exclusion (Hanna 2010).

The vast majority of countries, regardless of their economic performance, political regimes or infrastructural shortages, have adopted national ICT policies that seek to create a favourable environment for nationwide ICT deployment. ICT polices determine the legal frameworks (specifying regulatory bodies, financial and reporting regimes, and level of telecommunication market competition) and tele-communication market regulations (e.g., access programmes or tariff policies) governing ICT deployment and are predominantly designed to overcome barriers of access to and use of ICTs. They also seek to support the development of backbone infrastructure and incorporate ICT deployment into national economic development goals. The regulatory environment in developing countries has undergone a tremendous transition between 1984 and 2012 (ITU 2011c). Since the 1980s, multiple countries have broken the hegemony of state monopolies on telecommunication markets, heading toward the liberalisation and privatisation of the telecommunication sector. The result was the inception of autonomous regu-latory bodies that design and implement national ICT strategies to facilitate the roll-out of telecommunication infrastructure through private capital engagement. The rapid transition from monopoly to competition, the emergence of politically inde-pendent agencies and the separation of telecommunications operations from other government tasks resulted in creation of more transparent and politically free telecommunication markets. According to ITU data (ITU 2011c), by the end of 2009, only 8 of the 49 least developed countries so classified by ITU lacked a separate regulatory body for the telecommunications market. Moreover, by 2010, the majority of the 49 least developed countries had broken the monopoly of state owned telecommunication companies (usually on the fixed-line markets) and fully liberalised both fixed and mobile markets. Similar tendencies are reported for the market for Internet providers. According to ITU data (ITU 2014) from the ITU World Telecommunication Regulatory Database, in 2010, 30 out of 46 analysed low-income and lower-middle-income countries had fully liberalised their mobile markets,[8] and 33 had fully liberalised the Internet services market.[9]

The liberalisation of telecommunication markets due to the growing accessibil-ity and affordability of ICT services has undoubtedly boosted ICT penetration rates in developing countries. From a long-term perspective, however, it is highly desirable to establish solid backgrounds for the deployment of broadband networks

---

[8] In 12 countries there was partial competition, and in 4—monopoly (Comoros, Ethiopia, Myanmar and Swaziland).

[9] In 9 countries there was partial competition, and in 4—monopoly (Comoros, Ethiopia, Myanmar and Niger).

(ITU 2013), as they ensure universal access and enable the fast and effective flow of information. To this end, governments in developing countries attempt to encourage private investment in broadband infrastructure by delivering specific broadband policies, legal frameworks and regulations that aim to accelerate the countrywide implementation of broadband connectivity (Ngwenyama and Morawczynski 2009; Unwin 2009; Kozma and Vota 2014). These strategies, widely recognised as national broadband plans, constitute an important part of national ICT policies, seek to facilitate the development of broadband connectivity and overcome major barriers for broadband network adoption in developing countries. Country-specific targets declared in national broadband plans are consistent with those defined in *'Broadband Targets for 2015'*[10] (Broadband Commission 2011), which was issued in 2011 by the Broadband Commission for Digital Development. Three out of four targets directly refer to developing countries and specify that by 2015, regulations should ensure that broadband services should be affordable for ordinary citizens (*Target 2*); 40 % of households should have Internet access (*Target 3*); and Internet penetration rates should reach at least 50 % (and 15 % in the least developed countries) (*Target 4*). By the end of 2012, 28 out of 46 countries examined in our study had implemented the national broadband plans. In 14 countries—Bolivia, El Salvador, Eritrea, Georgia, Lao P.D.R., Madagascar, Mauritania, Myanmar, Nepal,[11] Nicaragua, Swaziland, Syria, Ukraine and Yemen—national broadband plans were not developed or adopted until 2013, while in another four—Benin, Comoros, Senegal and Togo—they were still under development. In 2003, Malawi, one of the world's poorest economies, was the first country to adopt its national broadband plan, *'An Integrated ICT-led Socio Economic Development Policy for Malawi'* (Government of the Republic of Malawi 2003). The national broadband plan for Malawi is a comprehensive long-term strategy defining how, through the deployment and exploitation of ICTs, the country shall move toward the realisation of national socio-economic development goals (Mbvundula 2003).

Broadly defined, national ICT policies are designed to serve as a framework for achieving long-term socio-economic development goals and their objectives are thus twofold: stimulate wider uptake of ICTs, and support ICT deployment. This will contribute to a fundamental reshaping of the environment and thereby promote long-term economic development. Regarding developing countries, the unique combination of ICTs with socio-economic development goals deserves special attention (Hanna 2003). Many claim (see, e.g., Khakhar et al. 2007; Manyozo 2008; Hilty and Hercheui 2010; Hanson and Narula 2013) that harnessing ICTs' potential is key for strengthening a country's economy and makes a positive contribution to overall welfare. Introducing ICT policies in developing countries garners assistance and support (including financial support) from international bodies. As a response to these needs, during the Third World Telecommunication

---

[10] The full text of the *'Broadband Targets to 2015'* is available at: http://www.broadbandcommission.org/Documents/Broadband_Targets.pdf.

[11] The in 2012, the national broadband plan for Nepal was still under draft.

Development Conference (2002), ITU established the *'Special Programme for the Least Developed Countries'*, which aims to provide robust support and assistance in achieving the development targets listed in Brussels Declaration and Programme of Action for the Least Developed Countries for the decade 2001–2010 (BPoA).[12] Through this Special Programme for the Least Developed Countries, ITU (with the support of United Nations Regional Commissions) has assisted a number of developing countries in realising their national ICT policies in line with the commitments set out in the *'Declarations of Principles'* and *'Geneva Plan of Action'*, agreed upon during the World Summit on the Information Society[13] in 2003 in Geneva .[14] To stay in compliance with the *'Geneva Plan of Action'* (ITU 2003), which states that *'development of national e-strategies, (...), should be encouraged by all countries (...), taking into account different national circumstances'*, national *e-strategies* generally aim to create favourable conditions for further ICT development to enhance deployment and bridge the digital divide both within and between countries (ITU 2011d). They are of particular importance in low-income countries, where the development of an *e-strategy* is often a central interest of the national government, and they are frequently incorporated into long-term socio-economic development strategies. As of the year 2010, 43 of 46 countries analysed in our study had formulated comprehensive national and sectorial *e-strategies* that have been officially released and adopted (ITU 2011d). Only three countries had not developed national and/or sectorial *e-strategies* at that time: Cambodia, Eritrea and Yemen[15] (see ITU 2011d).

Examples of the adoption of national *e-strategies* abound. In Asia, national governments seeking to comply with the recommendations of the Regional Action Plan towards the Information Society in Asia and the Pacific (ESCAP 2009), have placed special emphasis on establishing an enabling regulatory environment to break national monopoles and move toward telecommunication market

---

[12] The Brussels Declaration and Programme of Action for the Least Developed Countries for the decade 2001–2010 (BPoA) was worked out during the Third United Nations Conference for the Least Developed Countries (2001). Seven major commitments to challenge the poverty reduction problem and sustain long-term economic development (see UNCTAD 2011) in developing backward countries were approved.

[13] The first World Summit on the Information Society was held in two phases, in Geneva in 2003 and in Tunis in 2005. Four major documents were agreed upon during those meetings: *'Declaration of Principles'* and *'Geneva Plan of Action'* in Geneva; *'Commitments'* and *'Agenda for the Information Society'* in Tunis. Together, they shape the road toward building information societies. The documents place special emphasis on creating mechanisms (including financial support) to bridge the digital divide between developed and developing countries, and to make decisive efforts in assisting the least developed counties to achieve their economic development and ICT deployment goals.

[14] Over the decade 2000–2010, ITU through the ITU LDC Programme has assisted in the implementation of 60 projects in developing countries (ITU 2011b) that were mainly focused on ICT deployment (especially in rural areas), establishing Multipurpose Community Centres, and capacity building.

[15] In Cambodia and Yemen, the national *e-strategies* were still under development in 2010.

liberalisation and privatisation. In national and sectorial *e-strategies* formulated for African and Arab countries, particular attention is devoted to overcoming insufficient infrastructure, weak institutions and a lack of regulatory environments and financing mechanisms. Moreover, few countries have addressed the employment opportunities that are generated due to ICT deployment. Similarly, some countries' national *e-strategies* have formulated goals that address country-specific problems and weaknesses. For example, Malawi's national *e-strategy*[16] targets the promotion of e-commerce as a complementary action for poverty reduction[17] plans, as well as to enhance the preparation of Malawian society to participate in the information economy. Moldova[18] has implemented the e-Moldova plan to encourage electronic payments systems to facilitate exports. Kenyan authorities consider the development of local ICT content to preserve the heritage of local communities a priority,[19] and in Nigeria,[20] ensuring media independence and freedom of expression is emphasised. Sectorial *e-strategies* are targeted for multiple, sector-specific applications of ICTs, as for example,, e-government (e.g., India[21]), e-business (e.g., Guyana[22]), e-employment (e.g., Bangladesh[23]), or other sectors where ICTs incorporation may contribute significantly to its development and effectiveness.

6.4   Toward the Great Escape...?

The desire to escape is always there. Yet the desire is not always fulfilled. New knowledge, new inventions, and new ways of doing things are the key to progress
Angus Deaton (2013)

Gerschenkron (1962) argues that developing countries mainly operate below the world technology frontier; however, by imitating developed technologies, they gain the opportunity to converge (catch-up) with developed countries in terms of economic development. 'Technological congruence', meaning a lack of appropriate technology to enter development path, has also been stressed in the works of Abramovitz (1986, 1994). Gerschenkron (1962) writes that *'borrowed technology,*

---

[16] Malawi, the Republic of. (2003). An integrated ICT-led socio-economic development policy for Malawi: A Policy Statement for the Realization of the Aspirations of the Vision 2020 through the Development, Deployment and Exploitation of ICTs within the Economy and Society.

[17] In Malawi in 2010, approximately 85 % of the population experienced extreme poverty.

[18] Moldova. (2005). National Strategy on Building Information Society—"e-Moldova". Chisinau, Moldova.

[19] Kenya, Ministry of Information and Communications. (2006). National Information and Communications Technology (ICT) Policy.

[20] Nigeria. (n.d.). Nigerian National Policy for Information Technology (IT), Use It.

[21] India, Government of, Department of Information Technology, Ministry of Communications & Information Technology. (n.d.). The National e-Governance Plan (NeGP): e-Governance initiatives across the country.

[22] Guyana, Ministry of Tourism, Industry and Commerce. (2005). Draft E-Commerce Bill.

[23] ITU. (n.d.). WSIS Stocktaking Database, Project ID 1103043762.

so much and rightly stressed by Veblen, was one of the primary factors assuring a high speed of development in a backward country'. In the same vein Castellacci (2006, 2008, 2011) claims that technology can foster the catch-up of developing countries mainly by enabling improvements in education, the diffusion of knowledge and shifts in labour productivity.

However, technology acquisition does not happen unconditionally. Baumol (1986), Perez and Soete (1988), and Verspagen (1991) argue that a country's ability to adopt new technologies is preconditioned by multiple specific features. While societies assess and assimilate technological novelties relying upon 'intellectual' capital (Soete and Verspagen 1993) as well as institutional, governmental and cultural conditions, some empirical evidence shows that the most prominent factor in a country's ability to adopt and effectively benefit from technological progress are the education level and skills of the labour force (Baumol 1989). Countries with a largely uneducated and/or unskilled labour pool are unlikely to ever be able to harness the full potential of technological change and they lose the opportunity to catch up with richer countries, instead remaining economically disadvantaged. Similar arguments have also been raised by Gregory Clark who, in 1987, wrote that *'poor countries have remained poor because they cannot absorb the technologies of the advanced countries'* (Clark 1987, p. 141). The explanation of this idea may be traced in works of, *inter alia*, Hirschman (1958), Rosenberg (1976) and Easterlin (1981), who argued that poor countries are simply not prepared to absorb technologies from developed countries because high-quality institutions are scarce and their labour pool is poorly educated and lacking in managerial skills.

The arguments raised above are undoubtedly true. Our empirical evidence, however, has clearly demonstrated that even in the most backward economies, new information and communication technologies—especially mobile cellular telephony—may spread, overcoming the classical barriers to adoption such as low literacy, underdeveloped infrastructure and poor access to electricity, extreme poverty, authoritarian regimes and weak institutions. Although unrestricted Internet access remains limited in some of examined countries (especially in low-income economies), many of them are rapidly improving Internet accessibility, which is mainly facilitated by development of wireless networks (see the examples of Nepal, Uganda and Zimbabwe). These trends are promising and reveal prospects for the future because new information and communication technologies are widely perceived as a solution for developing countries, providing them with the opportunity to embark on a stable path of socio-economic development (Hanna 2003; Torero and von Braun 2006; Unwin 2009; Gruber and Koutroumpis 2011). The rapid diffusion of ICTs in developing countries has given rise to questions about their role in promoting economic development and whether poor countries can harness ICTs' full potential (Heeks 1999; Elliott 2012), to catalyze economic growth and development. These issues are receiving growing attention worldwide (see, e.g., Heeks 2010; Hanson and Narula 2013; Khavul and Bruton 2013), as the near-ubiquitous spread of information and communication technologies offers unprecedented opportunities to take off on the development path (Desai and Potter 2013).

Undeniably, ICTs can play a critical role in the development process[24] by broadening access to information and all types of knowledge, which results in improvements in people's empowerment and their participation in socio-economic life (Mansell 1999, 2001; Wilson 2004; Mansell et al. 2009). The significant impact of ICTs on the economy of developing countries may be further enabled by the creation of positive links between market agents, providing opportunities for more flexible work and providing new contacts, which—in effect—results in increased economic activity with potential increases in productivity, firm efficiency and cost reduction. The latter is also closely associated with information asymmetries, which generate transaction costs, uncertainty, and hence market failures (Wolf 2001). ICTs can help to eradicate information asymmetries and therefore they enhance the efficiency of resource allocation. Reducing these asymmetries improves access to economic activities for a multitude of agents, fostering participation, *inter alia,* in the labour market of previously disadvantaged societal groups. Above all ICTs offer connectivity, the transfer of knowledge and information, regardless of physical distance. Thus, the '*death of distance*' (Cairncross 2001; Redding and Venables 2004) becomes a fact, rendering face-to-face contacts no longer necessary. Moreover, new information and communication technologies bring developing countries opportunities to fight rural and urban poverty (Forestier et al. 2002; Graham 2002; Cecchini and Scott 2003) by improving economic performance and the ability to compete on global markets; they also provide a means by which to exploit an unused labour force and increase social capital (Chong 2011). Most importantly, the impact of ICTs reaches far beyond the ICT sector itself (ITU 2012) to deeply affect and transform social and economic life, playing an enabling and unlocking role for economic growth and development (ITU 2012). There is a causal chain between ICT adoption and a country's ability to enter the pattern of long-term economic development, which finally *should* allow backward countries to catch up with the best performing economies.

Hence, the question becomes do ICTs foster economic development and help backward countries climb the development ladder? In attempting to provide a final answer, we can follow David Landes, who argued that '*A definite answer is impossible. We are dealing here with the most complex type of problem, one that involves numerous factors of variable weights working in changing combinations*' (Landes 2003, p. 14). Regardless, when attempting to identify and assess the effects

---

[24] Many international agencies like for example United Nations Development Programme, International Development Research Centre, International Telecommunication Union, ICT4D (Information and Communication Technologies for Development) or *info*Dev (a program in the World Bank group), offer a wide spectrum of evidence demonstrating the effects of ICTs deployment in developing countries. Although the evidence is still relatively scattered and focuses rather on explaining the 'success stories' of ICTs implementation for achieving various development goals; it shows that ICTs *are* adequate tools to fight poverty and economic isolation (Thompson and Garbacz 2007; Cortés and Navarro 2011), enhance communication, information and knowledge flows, promote education and skills improvements, foster trade and other economic activities which generate increase of per capita income, and above all forces countries to head toward social and economic progress.

of introduction of ICTs in developing countries, we face multiple difficulties that are *'(. . .) only partially technical. To a far greater extent, they are economic, social and political'*. As already mentioned, poor infrastructure, illiteracy, low affordability (as effect of low per capita income), investment restrictions and limitations *may* hinder the effective adoption of ICTs. Most importantly, the remarkable impact of ICTs on developing countries can only be confirmed when ICT is converted into human development and progress; a precise answer on how ICTs promote socio-economic development remains elusive. ICTs can be a double-edged sword for developing countries in that there are two distinct possible scenarios. The first pessimistic scenario posits that developing countries will be unable to harness the potential that ICTs offer due to multiple constraints and weak socioeconomic and institutional conditions, with the result that that gap between developed and developing countries will widen, leading to greater disparities. Following Hanna (2003), David (2001) and Perez (2002), we can state that countries that are permanently inactive in adopting new information and communication technologies and make little efforts to improve their situation may lose the opportunities and benefits offered by ICTs, which would result in their digital and economic marginalisation. The second optimistic scenario is that ICTs will offer a way towards development and growth that will encourage developing countries climb the ladder and enter a stable development pattern. Finding a way to close the gap enables developing countries to escape the underdevelopment trap and forge ahead economically.

Yet it is important to note here that the evidence demonstrating the channels through which ICTs impact social and economic life in developing countries is limited, predominantly due to short time series and the restricted availability of complete and reliable data. The available time span for empirical studies of the effect of new information and communication technologies on development in low-income and lower-middle-income countries is not longer than 20 years and is as brief as 10 years in some cases[25]. This is definitely too short a time to reveal whether changes caused by ICTs are profound enough to help countries escape the underdevelopment trap and head toward higher stages of development, or vice versa, if the impact of ICT deployment is superficial, spurious and short-term. Following David (1990), we may say that ICTs as general purpose technologies do not induce the productivity and growth gains right after its arrival. Put simply, it takes time to unleash the ICT potential and transform this potential into socio-economic development. The last argument coincided with what we have argued in Sect. 2.3 that productivity gains and increases of per capita incomes are revealed with substantial time lags compared to the technological progress; and 'Solow's productivity paradox', may possibly emerge regarding the information and communication technologies. Helpman and Trajtenberg (1994), and Helpman (1998) suggest that full incorporation and deployment of general purpose technologies (thus ICTs) takes time, and their role in fostering economic growth and

---

[25] See availability of data on ICTs deployment in examined countries (Appendix B).

development may be revealed in a long-perspective. They call the early phase of general purpose technologies diffusion as 'time to sow', as this is a time period when resources need to be *'diverted to the development of complementary inputs to take advantage of the new GPT'* (Helpman and Trajtenberg 1994, p. 85). The second stage of general purpose technologies diffusion they call a 'time to reap' when rises in total productivity and per capita output are unveiled. To discover whether ICTs have engendered in-depth structural transformations of national economies and created solid backgrounds for escaping the underdevelopment trap is empirically intractable as of yet; a longer time perspective is needed. ICTs unquestionably offer the opportunity to escape, but whether these opportunities are realised is a different question that still needs an answer. As claimed by Hanna (2010), *'economic history, the cumulative learning and transformation process involved in using ICT, and the pace of this wave of technological change suggest that a 'wait and see attitude' would keep many developing countries out of a technological revolution no less profound than the last industrial revolution'* (Hanna 2010, p. 29). And finally, it is hard to disagree with Rosenberg who said the *'Perhaps the reason why we do so poorly at predicting the impact of technological change is that we are dealing with an extraordinarily complex and interdependent set of relationships'* (Rosenberg 1986, p. 17).

References

Abramovitz, M. (1986). Catching up, forging ahead, and falling behind. *The Journal of Economic History, 46*(02), 385–406.

Abramovitz, M. (1994). Catch-up and convergence in the postwar growth boom and after. In W. Baumol, R. Nelson, & E. Wolff (Eds.), *Convergence and productivity: Cross-national studies and historical evidence* (pp. 86–125). New York: Oxford University Press.

Addison, T., & Heshmati, A. (2003). *The new global determinants of FDI flows to developing countries: The importance of ICT and democratization* (No. 2003/45). WIDER Discussion Papers//World Institute for Development Economics (UNU-WIDER).

Baliamoune-Lutz, M. (2003). An analysis of the determinants and effects of ICT diffusion in developing countries. *Information Technology for development, 10*(3), 151–169.

Baumol, W. (1986). Productivity growth, convergence, and welfare: What the long-run data show. *American Economic Review, 76*, 1072–1084.

Baumol, W. J., & Wolff, E. N. (1988). Productivity growth, convergence, and welfare: Reply. *The American Economic Review, 78*, 1155–1159.

Baumol, W. J. (1989). Reflections on modern economics: Review. *Cambridge Journal of Economics, 13*(2), 353–358. Oxford University Press.

Broadband Commission. (2011). Broadband targets for 2015. Broadband Commission for Digital Development/ITU/UNESCO.

Cairncross, F. (2001). *The death of distance: How the communications revolution is changing our lives.* Boston: Harvard Business Press.

Castellacci F. (2006). *Convergence and divergence among technology clubs* (DRUID Working Paper No. 06-21).

Castellacci, F. (2008). Technology clubs, technology gaps and growth trajectories. *Structural Changes and Economic Dynamics, 19*(2008), 301–314.

Castellacci, F. (2011). Closing the technology gap? *Review of Development Economics, 15*(1), 189–197.

Cecchini, S., & Scott, C. (2003). Can information and communications technology applications contribute to poverty reduction? Lessons from rural India. *Information Technology for Development, 10*(2), 73–84.

Chatterji, M., & Dewhurst, J. L. (1996). Convergence clubs and relative economic performance in Great Britain: 1977–1991. *Regional Studies, 30*(1), 31–39.

Chen, W., & Wellman, B. (2004). The global digital divide—Within and between countries. *IT & Society, 1*(7), 39–45.

Chong, A. (Ed.). (2011). *Development connections: Unveiling the impact of new information technologies.* New York: Macmillan.

Clark, G. (1987). Why isn't the whole world developed? Lessons from the cotton mills. *The Journal of Economic History, 47*(01), 141–173.

Cortés, E. A., & Navarro, J. L. A. (2011). Do ICT influence economic growth and human development in European Union countries? *International Advances in Economic Research, 17*(1), 28–44.

David, P. (1990, May). The dynamo and the computer: An historical perspective on the modern productivity paradox. *The American Economic Review*, Papers and proceedings of the hundred and second annual meeting of the American Economic Association, Vol. 80, No. 2, pp. 355–361.

David, P. A. (2001). Path dependence, its critics and the quest for 'historical economics'. In P. Garrouste & S. Ioannides (Eds.), *Evolution and path dependence in economic ideas: Past and present* (pp. 15–40). Cheltenham: Edward Elgar.

Deaton, A. (2013). *The great escape: Health, wealth, and the origins of inequality.* Princeton, NJ: Princeton University Press.

Desai, V., & Potter, R. B. (2013). *The companion to development studies.* New York: Routledge.

Easterlin, R. A. (1981). Why isn't the whole world developed? *The Journal of Economic History, 41*(01), 1–17.

Eberhard, A., Foster, V., Briceño-Garmendia, C., Ouedraogo, F., Camos, D., & Shkaratan, M. (2008). Underpowered: The state of the power sector in Sub-Saharan Africa.

Elliott, J. (2012). *An introduction to sustainable development.* New York: Routledge.

ESCAP. (2009). *Regional progress and strategies towards building the information society in Asia and the Pacific.* Bangkok: ESCAP.

Forestier, E., Grace, J., & Kenny, C. (2002). Can information and communication technologies be pro-poor? *Telecommunications Policy, 26*(11), 623–646.

Galperin, H. (2005). Wireless networks and rural development: Opportunities for Latin America. *Information Technologies and International Development, 2*(3), 47–56.

Gerschenkron, A. (1962). *Economic backwardness in economic perspective.* Cambridge: Belknap Press.

Gourhant, Y., Lukashova, E., Reddy Sama, M., Abdel Wahed, S., Meddour, D. E., & Venmani, D. P. (2014, June). Low-cost wireless network architecture for developing countries. In *2014 14th International conference on innovations for community services (I4CS)* (pp. 16–21). IEEE.

Government of the Republic of Malawi. (2003). An integrated ICT led socio-economic development policy for Malawi, Malawi.

Graham, S. (2002). Bridging urban digital divides? Urban polarisation and information and communications technologies (ICTs). *Urban Studies, 39*(1), 33–56.

Gruber, H., & Koutroumpis, P. (2011). Mobile telecommunications and the impact on economic development. *Economic Policy, 26*(67), 387–426.

Guerrieri, P., Luciani, M., & Meliciani, V. (2011). The determinants of investment in information and communication technologies. *Economics of Innovation and New Technology, 20*(4), 387–403.

Gunasekaran, V., & Harmantzis, F. C. (2007). Emerging wireless technologies for developing countries. *Technology in Society, 29*(1), 23–42.

Guyana, Ministry of Tourism, Industry and Commerce. (2005). Draft E-Commerce Bill.

Hanna, N. K. (2003). Why national strategies are needed for ICT-enabled development. *World Bank Staff Paper*. Washington, DC: World Bank.

Hanna, N. (2010). *e-Transformation: Enabling new development strategies*. New York: Springer.

Hanson, J., & Narula, U. (2013). *New communication technologies in developing countries*. New York: Routledge.

Heeks, R. (1999). *Information and communication technologies, poverty and development*. Manchester, UK: Institute for Development Policy and Management, University of Manchester.

Heeks, R. (2010). Do information and communication technologies (ICTs) contribute to development? *Journal of International Development, 22*(5), 625–640.

Helpman, E. (Ed.). (1998). *General purpose technologies and economic growth*. Cambridge, MA: MIT Press.

Helpman, E., & Trajtenberg, M. (1994). *A time to sow and a time to reap: Growth based on general purpose technologies* (No. w4854). National Bureau of Economic Research.

Hilty, L. M., & Hercheui, M. D. (2010). ICT and sustainable development. In *What kind of information society? Governance, virtuality, surveillance, sustainability, resilience* (pp. 227–235). Berlin: Springer.

Hirschman, A. O. (1958). *The strategy of economic development* (Vol. 58). New Haven, CT Yale University Press.

ITU. (2003). *Geneva plan of action*. Geneva: ITU.

ITU. (2006). *ICT and telecommunications in least developed countries. The mid-term review for the decade 2001-2010*. Geneva: ITU.

ITU. (2011a). *Measuring information the society*. Geneva: ITU.

ITU. (2011b). *ICT and telecommunications in least developed countries. Review of progress made during the decade 2000-2010*. Geneva: ITU.

ITU. (2011c). *The role of ICT in advancing growth in least developed countries. Trends, challenges and opportunities*. Geneva: ITU.

ITU. (2011d). *National e-strategies for development. Global status and perspectives 2010*. Geneva: ITU.

ITU. (2012). *Measuring information the society*. Geneva: ITU.

ITU. (2013). *The state of broadband 2013: Universalizing broadband*. Geneva: ITU.

ITU. (2014). *ITU world telecommunication regulatory database*. Geneva: ITU.

ITU. (n.d.). WSIS Stocktaking Database, Project ID 1103043762.

Kenya, Ministry of Information and Communications. (2006). *National information and communications technology (ICT) policy*.

Khakhar, D., Cornu, B., Wibe, J., & Brunello, P. (2007). ICT and development. In *Past, present and future of research in the information society* (pp. 63–73). Springer.

Khavul, S., & Bruton, G. D. (2013). Harnessing innovation for change: Sustainability and poverty in developing countries. *Journal of Management Studies, 50*(2), 285–306.

Kozma, R. B., & Vota, W. S. (2014). ICT in developing countries: Policies, implementation, and impact. In J. M. Spector, M. D. Merrill, J. Elen, & M. J. Bishop (Eds.), *Handbook of research on educational communications and technology* (pp. 885–894). New York: Springer.

Landes, D. S. (2003). *The unbound Prometheus: Technological change and industrial development in Western Europe from 1750 to the present*. New York: Cambridge University Press.

Lucas, H., & Sylla, R. (2003). The global impact of the internet: Widening the economic gap between wealthy and poor nations? *Prometheus, 21*(1), 1–22.

Mäki, U. (Ed.). (2002). *Fact and fiction in economics: Models, realism and social construction*. Cambridge: Cambridge University Press.

Mansell, R. (1999). Information and communication technologies for development: Assessing the potential and the risks. *Telecommunications Policy, 23*(1), 35–50.

Mansell, R. (2001). Digital opportunities and the missing link for developing countries. *Oxford Review of Economic Policy, 17*(2), 282–295.

Mansell, R., Avgerou, C., Quah, D., & Silverstone, R. (Eds.). (2009). *Information and communication technologies*. Oxford: Oxford University Press.

Manyozo, L. (2008). Communication for development: An historical overview. *UNESCO, Media, Communication, Information: Celebrating, 50*, 31–53.

Mbvundula, T. P. (2003). An integrated socio-economic and ICT policy and plan development framework for Malawi. UNECA/UNDP.

Moldova. (2005). National strategy on building information society—"e-Moldova". Chisinau, Moldova.

Nakamura, L., & Chow-White, P. (Eds.). (2013). *Race after the internet*. New York: Routledge.

Ngwenyama, O., & Morawczynski, O. (2009). Factors affecting ICT expansion in emerging economies: An analysis of ICT infrastructure expansion in five Latin American countries. *Information technology for development, 15*(4), 237–258.

Nigeria. (n.d.). Nigerian National Policy for Information Technology (IT).

Perez, C., & Soete, L. (1988). Catching-up in technology, entry barriers and windows of opportunity. In G. Dosi, et al. (Orgs.), *Technical change and economic theory*. London: Pinter.

Perez, C. (2002). *Technological revolutions and financial capital: The dynamics of bubbles and golden ages*. Cheltenham, UK: Edward Elgar.

Proenza, F. J. (2006). The road to broadband development in developing countries is through competition driven by wireless and internet telephony. *Information Technologies & International Development, 3*(2), 21–39.

Puumalainen, K., Frank, L., Sundqvist, S., & Tappura, A. (2011). The critical mass of wireless communications: Differences between developing and developed economies. *Mobile information communication technologies adoption in developing countries: Effects and implications* (pp. 1–17).

Redding, S., & Venables, A. J. (2004). Economic geography and international inequality. *Journal of International Economics, 62*(1), 53–82.

Robison, K. K., & Crenshaw, E. M. (2010). Reevaluating the global digital divide: Socio-demographic and conflict barriers to the internet revolution. *Sociological Inquiry, 80*(1), 34–62.

Romijn, H. A., & Caniëls, M. C. (2011). Pathways of technological change in developing countries: Review and new agenda. *Development Policy Review, 29*(3), 359–380.

Rosenberg, N. (1976). *Perspectives on technology*. CUP Archive.

Rosenberg, N. (1986). The impact of technological innovation: A historical view. In R. Landua & N. Rosenberg (Eds.), *The positive sum strategy: Harnessing technology for economic growth* (p. 17). Washington, DC: National Academy Press.

Soete, L., & Verspagen, B. (1993). Technology and growth: The complex dynamics of catching-up, falling behind and taking over. In A. Szirmai (Ed.), *Explaining economic growth*. Amsterdam: Elsevier.

Thapa, D. (2011). The role of ICT actors and networks in development: The case study of a wireless project in Nepal. *The Electronic Journal of Information Systems in Developing Countries, 49*(1), 1–16.

Thompson, H. G., & Garbacz, C. (2007). Mobile, fixed line and internet service effects on global productive efficiency. *Information Economics and Policy, 19*(2), 189–214.

Torero, M., & Von Braun, J. (Eds.). (2006). *Information and communication technologies for development and poverty reduction: The potential of telecommunications*. Washington, DC: Intl Food Policy Res Inst.

UNCTAD. (2007). Information economy report 2007-2008. Science and technology for development: The new paradigm of ICT. Geneva.

UNCTAD. (2011). Programme of action for the least developed countries. UN General Assembly, Brussels.

UNDP. (2014). Beyond geography. Unlocking human potential. Human Development Report 2014. Government of Nepal and United Nations, Nepal.

Unwin, P. T. H. (Ed.). (2009). *ICT4D: Information and communication technology for development*. Cambridge: Cambridge University Press.

Verspagen, B. (1991). A new empirical approach to catching up or falling behind. *Structural Change and Economic Dynamics, 2*(2), 359–380.

Warren, M. (2007). The digital vicious cycle: Links between social disadvantage and digital exclusion in rural areas. *Telecommunications Policy, 31*(6), 374–388.

White, D. S., Gunasekaran, A., Shea, T. P., & Ariguzo, G. C. (2011). Mapping the global digital divide. *International Journal of Business Information Systems, 7*(2), 207–219.

Wilson, E. J., III. (2004). *The information revolution and developing countries*. Cambridge, MA: MIT Press.

Wolf, S. (2001, September). Determinants and impact of ICT use for African SMEs: Implications for rural South Africa. In *Center for Development Research (ZEF Bonn). Paper prepared for TIPS Forum*.

World Bank. (2006). *Information and communication for development: Global trends and policies*. Washington, DC: World Bank.

Wresch, W., & Fraser, S. (2012). ICT-enabled market freedoms and their impacts in developing countries: Opportunities, frustrations, and surprises. *Information Technology for Development, 18*(1), 76–86.

Appendix A. ICT Core Indicators: Definitions

ICT indicator	Definition
Fixed telephone subscriptions per 100 inhabitants	Fixed-telephone subscriptions refers to the sum of active number of analogue fixed-telephone lines. voice-over-IP (VoIP) subscriptions, fixed wireless local loop (WLL) subscriptions, ISDN voice-channel equivalents and fixed public payphones
Mobile cellular telephone subscriptions per 100 inhabitants	Mobile-cellular telephone subscriptions refers to the number of subscriptions to a public mobile-telephone service that provide access to the PSTN using cellular technology. The indicator includes (and is split into) the number of postpaid subscriptions, and the number of active prepaid accounts (i.e. that have been used during the last three months). The indicator applies to all mobile-cellular subscriptions that offer voice communications. It excludes subscriptions via data cards or USB modems, subscriptions to public mobile data services, private trunked mobile radio, telepoint, radio paging and telemetry services
Fixed (wired) Internet subscriptions per 100 inhabitants	Fixed (wired) Internet subscriptions refers to the number of active fixed (wired) Internet subscriptions at speeds less than 256 kbit/s (such as dial-up and other fixed non-broadband subscriptions) and total fixed (wired)-broadband subscriptions
Fixed (wired)-broadband Internet subscriptions per 100 inhabitants	Refers to subscriptions to high-speed access to the public Internet (a TCP/IP connection), at downstream speeds equal to, or greater than, 256 kbit/s. This includes cable modem, DSL, fibre-to-the-home/building and other fixed (wired)-broadband subscriptions. This total is measured irrespective of the method of payment. It excludes subscriptions that have access to data communications (including the Internet) via mobile-cellular networks. It should exclude technologies listed under the wireless-broadband category

(continued)

© Springer International Publishing Switzerland 2015
E. Lechman, *ICT Diffusion in Developing Countries*,
DOI 10.1007/978-3-319-18254-4

ICT indicator	Definition
Wireless-broadband subscriptions per 100 inhabitants	Wireless-broadband subscriptions refers to the sum of satellite broadband, terrestrial fixed wireless broadband and active mobile-broadband subscriptions to the public Internet. The indicator refers to total active wireless-broadband Internet subscriptions using satellite, terrestrial fixed wireless or terrestrial mobile connections. Broadband subscriptions are those with an advertised download speed of at least 256 kbit/s. In the case of mobile-broadband, only active subscriptions are included (those with at least one access to the Internet in the last three months or with a dedicated data plan). The service can be standalone with a data card, or an add-on service to a voice plan. The indicator does not cover fixed (wired)-broadband or Wi-Fi subscriptions. Both residential and business subscriptions should be included. This is calculated by dividing wireless broadband subscriptions by the population and multiplying by 100
Proportion of individuals using Internet	Refers to the proportion of individuals that used the Internet in the last 12 months. Data are based on surveys generally carried out by national statistical offices or estimated based on the number of Internet subscriptions

Source: Derived directly from World Telecommunication/ICT Indicators database 2013 (17th Edition)
Note: Cited indicators were agreed as core ICT indicators during WSIS Thematic Meeting on Measuring the Information Society in Geneva (Switzerland) 7–9 February 2005. The list was discussed and adjusted during the meeting of Expert Group on Telecommunication/ICT Indicators that was held in Mexico City, on 2–3 December 2013

Appendix B. Core ICT Indicators. Low-Income and Lower-Middle-Income Economies. Years 1975 and 2012

© Springer International Publishing Switzerland 2015
E. Lechman, *ICT Diffusion in Developing Countries*,
DOI 10.1007/978-3-319-18254-4

Country	Fixed telephone subscriptions[a] (100 per inhab.)		Mobile cellular subscriptions[b] (100 per inhab.)		Fixed internet subscriptions[c] (100 per inhab.)		Fixed broadband subscriptions[d] (100 per inhab.)		Wireless broadband subscriptions[e] (100 per inhab.)		Internet users[f] (% of individuals)	
	1975[g]	2012	In brackets—first year of data availability	2012	In brackets—first year of data availability	2012	In brackets—first year of data availability	2012	In brackets—first year of data availability	2012	In brackets—first year of data availability	2012
Low income-economies												
Bangladesh	0.08	0.62	0.0002 (1992)	62.8	0.0008 (1997)	1.50	0.029 (2007)	0.39	0.042 (2009)	0.47	0.0007 (1997)	6.30
Benin	0.16	1.56	0.017 (1995)	83.7	0.029 (1999)	0.42	0.0002 (2002)	0.05	0.180 (2009)	0.37	0.0016 (1996)	3.80
Burkina Faso	0.04	0.86	0.005 (1996)	60.6	0.008 (1997)	0.18	0.0004 (2002)	0.09	0.004 (2009)	0.004	0.0009 (1996)	3.73
Cambodia	n.a.	3.93	0.047 (1993)	128.5	0.02 (1999)	0.41[h]	0.0003 (2002)	0.20	1.04 (2010)	6.73	0.0058 (1997)	4.94
Comoros	0.17	3.34	0.35 (2003)	39.5	0.06 (1999)	0.28	0.00017 (2004)	0.17	0.49 (2011)	0.85	0.037 (1998)	5.98
Eritrea	n.a.	0.98	0.42 (2004)	5.0	0.04 (2000)	0.02	0.0008 (2008)	0.002	0.001 (2012)	0.001	0.0090 (1997)	0.80
Ethiopia	0.14	0.87	0.01 (1999)	22.4	0.002 (1997)	0.23	0.00007 (2003)	0.01	0.094 (2009)	4.45	0.0017 (1996)	1.48
Kenya	0.39	0.58	0.0004 (1992)	71.2	0.12 (2000)	0.16	0.01 (2005)	0.10	0.010 (2009)	2.22	0.0088 (1996)	32.10
Madagascar	n.a.	1.09	0.002 (1994)	39.4	0.02 (1998)	0.06	0.007 (2006)	0.04	0.046 (2010)	0.32	0.0036 (1996)	2.05
Malawi	0.16	1.43	0.003 (1995)	29.2	0.003 (1997)	4.16[h]	0.0005 (2003)	0.01	0.57 (2010)	3.46	0.0046 (1997)	4.35
Myanmar	0.08	1.05	0.001 (1993)	10.3	0.00004 (1999)	0.03[h]	0.0004 (2005)	0.01	0.006 (2009)	0.03	0.00015 (1999)	1.07
Nepal	0.06	3.03	0.02 (1999)	59.6	0.04 (1999)	0.75	0.003 (2006)	0.48	23.7 (2012)	23.79	0.0045 (1996)	11.15
Niger	0.07	0.59	0.0009 (1997)	59.6	0.0009 (1999)	0.37[h]	0.0006 (2004)	0.02	0.38 (2011)	0.57	0.001 (1996)	1.41
Rwanda	0.05	0.39	0.06 (1998)	31.4	0.012 (2000)	0.03	0.012 (2004)	0.02	0.03 (2010)	3.22	0.0008 (1996)	8.02
Togo	n.a.	0.93	0.06 (1997)	49.7	0.033 (1997)	0.63	0.0205 (2007)	0.11	0.40 (2010)	1.81	0.01 (1996)	4.00
Uganda	0.19	0.87	0.008 (1995)	49.9	0.017 (1999)	0.26	0.002 (2005)	0.11	1.6 (2010)	7.41	0.004 (1996)	14.69
Zimbabwe	1.32	2.20	0.04 (1997)	45.0	0.08 (1998)	0.52	0.0061 (2001)	0.52	0.29 (2009)	28.14	0.0167 (1996)	17.09

Lower-middle-income economies

Armenia	6.72	19.7	0.009 (1996)	111.9	0.401 (2000)	7.07	0.0001 (2001)	6.75	6.55 (2009)	29.09	0.009 (1994)	39.16
Bolivia	2.51 (1980)	8.4	0.004 (1991)	90.4	0.033 (1996)	1.30	0.03 (2002)	1.05	1.033 (2010)	6.61	0.06 (1995)	34.19
Congo (Rep.)	0.36	0.30	0.03 (1996)	98.8	0.003 (2000)	0.05	0.003 (2009)	0.01	0.52 (2011)	2.11	0.0035 (1996)	6.11
Egypt	0.854	10.6	0.004 (1987)	119.9	0.069 (1999)	3.02	0.074 (2002)	2.83	11.69 (2009)	27.93	0.0009 (1993)	44.07
El Salvador	1.13	16.8	0.029 (1993)	137.3	0.045 (1996)	3.99	0.32 (2003)	3.84	2.31 (2010)	5.49	0.086 (1996)	25.50
Georgia	4.48	29.3	0.002 (1995)	107.8	0.078 (2002)	8.67	0.009 (2001)	8.67	5.69 (2010)	8.55	0.011 (1995)	45.50
Ghana	0.317	1.1	0.002 (1992)	101.0	0.054 (2000)	0.26	0.004 (2004)	0.26	0.177 (2009)	33.92	0.0003 (1995)	17.11
Guyana	1.89	19.4	0.11 (1992)	68.8	0.94 (1999)	2.17	0.26 (2005)	3.67	0.067 (2001)	0.10	0.065 (1996)	33.00
Honduras	0.547	7.7	0.04 (1996)	92.9	0.06 (1997)	0.95	0.01 (2010)	0.77	4.25 (2011)	8.56	0.036 (1995)	18.12
India	0.23	2.5	0.008 (1995)	69.9	0.008 (1997)	2.05	0.0047 (2001)	1.21	0.002 (2007)	4.99	0.0001 (1992)	12.58
Indonesia	0.16	15.4	0.001 (1984)	114.2	0.015 (1996)	0.87	0.0019 (2000)	1.21	0.70 (2009)	31.59	0.001 (1994)	15.36
Lao PDR	0.177	1.8	0.0064 (1992)	64.7	0.009 (1999)	0.35	0.00045 (2003)	0.11	0.087 (2010)	2.06	0.0096 (1998)	10.75
Mauritania	0.10	1.7	0.56 (2000)	106.0	0.004 (1997)	0.18	0.005 (2005)	0.18	0.37 (2009)	3.64	0.0041 (1997)	5.37
Moldova	3.55	34.3	0.0003 (1995)	102.0	0.07 (1998)	11.87	0.0059 (2001)	11.87	0.007 (2007)	5.12	0.0008 (1994)	43.37
Mongolia	2.41 (1982)	6.3	0.03 (1996)	120.7	0.11 (1997)	3.84	0.002 (2001)	3.75	0.11 (2009)	18.37	0.008 (1995)	16.40
Morocco	0.62	10.1	0.0002 (1987)	120.0	0.011 (1999)	2.10	0.006 (2002)	2.10	2.27 (2009)	10.09	0.003 (1995)	55.00
Nicaragua	0.78	5.0	0.007 (1993)	86.1	0.3 (2000)	1.91	0.013 (2000)	1.65	0.36 (2009)	1.01	0.01 (1994)	13.50
Nigeria	0.21 (1981)	0.20	0.008 (1993)	66.8	0.018 (2000)	0.19	0.0003 (2005)	0.01	0.59 (2010)	18.44	0.008 (1996)	32.88
Pakistan	0.30	3.2	0.0018 (1990)	67.1	0.022 (1997)	1.70	0.0092 (2005)	0.52	0.011 (2008)	0.66	0.0001 (1995)	9.96
Paraguay	1.06	6.1	0.03 (1992)	101.6	0.54 (2000)	1.21	0.0028 (2000)	1.19	1.186 (2008)	5.98	0.02 (1996)	27.08
Philippines	0.68	4.1	0.058 (1991)	106.5	0.14 (1996)	6.33	0.012 (2001)	2.22	1.85 (2009)	3.83	0.005 (1994)	36.24
Senegal	0.294	2.5	0.001 (1994)	83.6	0.017 (1997)	0.70	0.01 (2002)	0.70	0.18 (2010)	3.58	0.0006 (1995)	19.20
Sri Lanka	0.308	16.3	0.005 (1990)	91.6	0.0009 (1994)	2.01	0.0017 (2001)	1.68	1.418 (2010)	7.75	0.0027 (1994)	18.29
Swaziland	0.64	3.7	0.456 (1998)	65.4	0.38 (1999)	0.31	0.066 (2008)	0.28	0.39 (2010)	11.90	0.001 (1995)	20.78
Syria	1.69	20.2	0.025 (1999)	59.3	0.0021 (1998)	1.83	0.003 (2004)	1.11	0.19 (2009)	1.66	0.032 (1997)	24.30
Ukraine	5.74	26.8	0.0001 (1993)	130.3	0.47 (2002)	3.48	0.27 (2005)	8.00	4.12 (2010)	5.43	0.0007 (1993)	33.70

(continued)

Country	Fixed telephone subscriptions[a] (100 per inhab.)		Mobile cellular subscriptions[b] (100 per inhab.)		Fixed internet subscriptions[c] (100 per inhab.)		Fixed broadband subscriptions[d] (100 per inhab.)		Wireless broadband subscriptions[e] (100 per inhab.)		Internet users[f] (% of individuals)	
	1975[g]	2012	In brackets—first year of data availability	2012	In brackets—first year of data availability	2012	In brackets—first year of data availability	2012	In brackets—first year of data availability	2012	In brackets—first year of data availability	2012
Viet Nam	0.113 (1982)	11.2	0.0011 (1992)	147.7	0.0014 (1997)	1.93	0.001 (2002)	4.90	7.89 (2010)	18.77	0.0001 (1996)	39.49
Yemen	0.214 (1980)	4.6	0.011 (1992)	58.3	0.009 (1997)	3.93	0.0074 (2005)	0.70	0.0901 (2011)	0.20	0.0006 (1996)	17.45
Zambia	0.57	0.6	0.017 (1995)	74.8	0.0016 (1996)	0.11	0.0002 (2000)	0.11	0.0836 (2009)	0.65	0.006 (1994)	13.47

Source: Author's compilation

[a]Fixed-telephone subscriptions refers to the sum of active number of analogue fixed-telephone lines

[b]Mobile-cellular telephone subscriptions refers to the number of subscriptions to a public mobile-telephone service that provide access to the PSTN using cellular technology

[c]Fixed (wired) Internet subscriptions refers to the number of active fixed (wired) Internet subscriptions at speeds less than 256 kbit/s and total fixed (wired)-broadband subscriptions

[d]Refers to subscriptions to high-speed access to the public Internet (a TCP/IP connection), at downstream speeds equal to, or greater than, 256 kbit/s. This includes cable modem, DSL, fibre-to-the-home/building and other fixed (wired)-broadband subscriptions

[e]Wireless-broadband subscriptions refers to the sum of satellite broadband, terrestrial fixed wireless broadband and active mobile-broadband subscriptions to the public Internet

[f]Refers to the proportion of individuals that used the Internet in the last 12 months

[g]In few countries the data on $FTL_{i,y}$ is available since 1965, however for cross-country data comparability—reported since 1975

[h]Estimates based on time-trend

Appendix C. Countries Included in the Technology Convergence Analysis. Core ICT Indicators[a]. Period 2000–2012

© Springer International Publishing Switzerland 2015
E. Lechman, *ICT Diffusion in Developing Countries*,
DOI 10.1007/978-3-319-18254-4

Country	Mobile cellular subscriptions in 2000	Mobile cellular subscriptions in 2012	Mobile cellular subscriptions average annual growth (%)	Fixed internet subscriptions in 2000	Fixed internet subscriptions in 2012	Fixed internet subscriptions average annual growth (%)	Fixed broadband subscriptions in 2000	Fixed broadband subscriptions in 2012	Fixed broadband subscriptions average annual growth (%)	Internet users in 2000	Internet users in 2012	Internet users average annual growth rate (%)	Wireless broadband subscriptions in 2009	Wireless broadband subscriptions in 2010	Wireless broadband subscriptions in 2011	Wireless broadband subscriptions in 2012
Low-income economies																
Bangladesh	0.20	62.8	36.4	0.00	1.5	29.2	n.a.	n.a.	n.a.	0.10	6.3	37.4	0.00	0.10	0.40	0.50
Benin	0.80	83.7	49.6	0.00	0.40	19.1	0.00	0.10	43.4	0.20	3.8	23.5	0.20	0.20	0.30	0.40
Burkina Faso	0.20	60.6	33.7	0.00	0.20	15.8	0.00	0.10	48.8	0.10	3.7	32.3	0.00	0.00	0.00	0.00
Cambodia	1.1	128.5	70.6	0.00	0.40	20.6	0.00	0.20	46.9	0.00	4.9	38.8	n.a.	1.0	2.2	6.7
Comoros	n.a.	n.a.	n.a.	0.10	0.30	9.3	0.00	n.a.	79.3	0.30	6.0	25.8	n.a.	n.a.	0.50	0.90
Eritrea	n.a.	n.a.	n.a.	0.00	0.00	−8.9	0.00	n.a.	n.a.	0.10	0.80	14.7	n.a.	n.a.	n.a.	0.00
Ethiopia	0.00	22.4	33.4	0.00	0.20	34.3	0.00	0.00	64.6	0.00	1.5	38.1	0.10	0.10	0.30	4.4
Kenya	0.40	71.2	43.2	0.10	0.20	1.9	0.00	0.10	26.9	0.30	32.1	38.5	0.00	0.20	0.50	2.2
Madagascar	0.40	39.4	37.6	0.10	0.10	−0.6	0.00	0.00	24.0	0.20	2.1	19.6	n.a.	0.00	0.10	0.30
Malawi	0.40	29.2	58.0	0.00	4.2	36.9	0.00	0.00	12.6	0.10	4.4	29.5	n.a.	0.60	3.1	3.5
Myanmar	0.00	10.3	45.4	0.00	0.00	28.1	0.00	0.00	43.6	0.00	1.1	72.0	0.00	n.a.	0.00	0.00
Nepal	0.00	59.6	67.2	0.10	0.70	22.2	0.00	0.50	69.4	0.20	11.1	33.3	n.a.	n.a.	n.a.	23.8
Niger	0.00	31.4	35.1	0.00	0.40	25.8	0.00	0.00	37.0	0.00	1.4	30.5	n.a.	n.a.	0.40	0.60
Rwanda	0.50	49.7	32.30	0.00	0.00	8.6	0.00	0.00	9.10	0.10	8.0	40.4	n.a.	0.00	1.0	3.2
Togo	1.0	49.9	38.0	0.10	0.60	13.5	n.a.	n.a.	n.a.	0.80	4.0	13.4	n.a.	0.40	0.90	1.8
Uganda	0.50	45.0	25.4	0.00	0.30	20.3	0.00	0.10	51.1	0.20	14.7	37.5	n.a.	1.6	2.8	7.4
Zimbabwe	2.1	91.9	31.4	0.20	0.50	6.5	0.10	0.50	26.7	0.40	17.1	31.3	0.30	4.7	14.2	28.1
Lower-middle-income economies																
Armenia	0.60	111.9	23.3	0.40	7.1	23.9	0.10	6.7	66.1	1.3	39.2	28.4	6.6	12.2	27.3	29.1
Bolivia	6.9	90.4	30.8	0.50	1.3	8.5	0.10	1.1	29.0	1.4	34.2	26.4	n.a.	1.0	2.8	6.6
Congo	2.2	98.8	32.3	0.00	0.10	22.9	n.a.	n.a.	n.a.	0.00	6.1	45.4	n.a.	n.a.	0.50	2.1
Egypt	2.1	119.9	18.9	0.10	3.0	30.6	0.20	2.8	38.1	0.60	44.1	35.3	11.7	17.0	25.0	27.9
El Salvador	12.5	137.3	29.3	0.90	4.0	12.3	0.70	3.8	24.4	1.2	25.5	25.6	n.a.	2.3	3.6	5.5
Georgia	4.1	107.8	42.1	0.10	8.7	39.2	0.10	8.7	72.7	0.50	45.5	37.8	n.a.	5.7	7.2	8.5
Ghana	0.70	101.0	24.5	0.10	0.30	13.0	0.00	0.30	48.1	0.20	17.1	39.3	0.20	6.9	23.2	33.9
Guyana	5.4	68.8	27.7	1.7	2.2	1.8	0.30	3.7	37.7	6.6	33.0	13.4	n.a.	n.a.	0.10	0.10
Honduras	2.5	92.9	46.7	0.30	1.0	9.5	n.a.	n.a.	n.a.	1.2	18.1	22.6	n.a.	n.a.	4.3	8.6
India	0.30	69.9	30.7	0.30	2.0	16.4	0.10	1.2	33.1	0.50	12.6	26.4	0.00	0.90	1.9	5.0
Indonesia	1.8	114.2	51.5	0.20	0.90	13.0	0.00	1.2	46.0	0.90	15.4	23.4	0.70	18.6	22.1	31.6

Lao P.D.R.	0.20	39.5	64.7	0.00	0.40	20.9	0.00	0.10	43.5	0.10	10.7	38.1	n.a.	0.10	0.70	2.1
Mauritania	0.60	28.7	106.0	0.00	0.20	13.8	0.00	0.20	50.3	0.20	5.4	27.8	0.40	0.70	4.9	3.6
Moldova	3.4	23.0	102.0	0.30	11.9	30.4	0.30	11.9	53.7	1.3	43.4	29.3	2.2	3.4	3.6	5.1
Mongolia	6.4	22.5	120.7	0.30	3.8	20.4	0.10	3.7	56.6	1.3	16.4	21.4	0.10	7.6	17.8	18.4
Morocco	8.2	35.1	120.0	0.10	2.1	23.3	0.80	2.1	13.3	0.70	55.0	36.4	2.3	5.0	8.1	10.1
Nicaragua	1.8	68.1	86.1	0.30	1.9	15.3	0.20	1.6	30.6	1.0	13.5	21.9	0.40	0.70	0.80	1.0
Nigeria	0.00	47.9	66.8	0.00	0.20	19.5	0.00	0.00	45.2	0.10	32.9	52.0	n.a.	0.60	9.9	18.4
Pakistan	0.20	12.3	67.1	0.10	1.7	24.2	0.00	0.50	57.5	0.70	10.0	22.5	0.10	0.20	0.40	0.70
Paraguay	15.3	20.9	101.6	0.50	1.2	6.7	0.10	1.2	36.1	0.70	27.1	29.9	1.9	3.7	5.3	6.0
Philippines	8.3	31.1	106.5	0.50	6.3	20.9	0.10	2.2	39.1	2.0	36.2	24.2	1.9	2.3	3.4	3.8
Senegal	2.5	30.0	83.6	0.10	0.70	21.1	0.20	0.70	21.0	0.40	19.2	32.2	n.a.	0.20	1.5	3.6
Sri Lanka	2.3	28.2	91.6	0.20	2.0	18.6	0.10	1.7	39.6	0.60	18.3	27.8	n.a.	1.4	2.3	7.8
Swaziland	3.1	49.0	65.4	0.50	0.30	-3.5	n.a.	n.a.	n.a.	0.90	20.8	25.9	n.a.	0.40	1.1	11.9
Syria	0.20	29.8	59.3	0.10	1.8	28.3	0.00	1.1	61.6	0.20	24.3	40.8	0.20	0.50	0.90	1.7
Ukraine	1.7	40.8	130.3	0.50	3.5	16.6	0.30	8.0	48.1	0.70	33.7	32.1	n.a.	4.1	4.4	5.4
Viet Nam	1.0	55.8	147.7	0.10	1.9	22.7	0.20	4.9	42.7	0.30	39.5	42.0	n.a.	7.9	17.8	18.8
Yemen	0.20	34.1	58.3	0.00	3.9	39.0	0.00	0.70	64.9	0.10	17.4	44.6	0.00	0.00	0.10	0.20
Zambia	1.0	48.9	74.8	0.10	0.10	5.2	0.00	0.10	55.4	0.20	13.5	35.5	0.10	0.30	0.20	0.60
Upper-middle-income economies																
Argentina	17.5	28.1	151.9	3.2	11.4	10.4	2.4	10.9	21.6	7.0	55.8	17.3	n.a.	3.3	12.4	20.9
Azerbaijan	5.1	44.9	108.7	0.03	18.3	51.7	0.00	14.1	90.3	0.10	54.2	49.2	0.20	5.2	24.6	34.8
Belarus	0.49	23.1	113.5	0.04	28.0	53.3	0.00	26.9	105.9	1.9	46.9	26.9	29.5	12.7	19.1	33.3
Belize	7.04	25.7	53.2	1.7	4.3	7.2	1.8	3.1	7.3	6.0	25.0	11.9	n.a.	0.10	0.30	0.50
Bosnia and Herzegovina	2.4	16.1	87.5	0.30	12.7	29.1	0.40	10.6	48.6	1.1	65.4	34.2	3.1	7.5	10.7	12.2
Botswana	12.6	20.4	153.7	0.80	0.90	0.80	0.10	0.90	34.3	2.9	11.5	11.5	n.a.	1.5	12.0	74.9
Brazil	13.2	21.7	125.0	1.2	11.3	18.1	1.7	9.2	23.7	2.9	49.8	23.8	4.7	11.0	21.4	33.7
Bulgaria	9.2	25.8	148.1	0.08	18.0	44.2	2.2	17.9	30.3	5.4	55.1	19.4	2.5	35.4	40.5	48.5
China	6.6	22.1	80.7	0.70	13.1	24.4	2.8	12.7	21.5	1.8	42.3	26.4	0.90	3.5	9.4	16.9
Colombia	5.6	24.5	102.8	0.60	8.2	21.7	0.70	8.2	34.3	2.2	49.0	25.8	2.1	2.5	3.8	5.0
Costa Rica	5.3	21.8	111.9	0.90	9.3	19.4	1.0	9.3	31.3	5.8	47.5	17.5	0.10	7.2	10.2	20.3
Dominican Rep.	8.1	25.9	86.9	0.60	4.8	17.2	0.60	4.3	27.1	3.7	45.0	20.8	1.2	2.4	7.8	15.6
Ecuador	3.8	10.5	106.2	0.40	5.6	20.8	0.20	5.3	47.2	1.5	35.1	26.5	n.a.	9.0	10.2	21.6
Hungary	30.08	22.1	116.07	2.1	23.0	19.7	6.5	22.9	18.1	7.0	72.0	19.4	7.1	8.8	18.2	24.2
Jordan	8.1	31.8	128.1	0.60	2.8	12.0	0.40	2.8	26.2	2.6	41.0	22.9	0.60	1.6	6.1	11.3

(continued)

Country	Mobile cellular subscriptions in 2000	Mobile cellular subscriptions in 2012	Mobile cellular subscriptions average annual growth (%)	Fixed internet subscriptions in 2000	Fixed internet subscriptions in 2012	Fixed internet subscriptions average annual growth (%)	Fixed broadband subscriptions in 2000	Fixed broadband subscriptions in 2012	Fixed broadband subscriptions average annual growth (%)	Internet users in 2000	Internet users in 2012	Internet users average annual growth rate (%)	Wireless broadband subscriptions in 2009	Wireless broadband subscriptions in 2010	Wireless broadband subscriptions in 2011	Wireless broadband subscriptions in 2012
Maldives	2.8	165.6	19.9	0.38	5.3	21.7	1.1	5.3	22.5	2.2	38.9	23.9	0.00	6.6	16.8	25.3
Mauritius	15.1	119.8	18.1	2.9	11.7	11.4	0.40	11.2	46.1	7.3	41.4	14.5	14.6	14.6	13.2	22.9
Mexico	13.5	83.3	15.2	1.09	10.7	19.0	1.7	10.5	25.7	5.1	38.4	16.9	0.40	2.7	6.8	9.8
Panama	13.4	178.02	29.9	1.4	8.0	14.4	0.50	7.8	38.7	6.6	45.2	16.1	n.a.	3.2	13.9	14.3
Peru	4.8	98.0	18.1	0.50	4.8	18.4	1.3	4.7	18.8	3.1	38.2	21.0	0.40	1.0	1.4	2.9
Romania	11.1	104.9	25.5	1.6	16.2	19.1	1.7	16.2	32.1	3.6	50.0	21.9	6.9	9.4	13.9	27.0
Thailand	4.9	127.2	38.5	0.90	8.6	18.3	0.80	8.2	32.3	3.7	26.5	16.4	n.a.	n.a.	0.10	0.10
Tunisia	1.2	118.08	12.7	0.30	5.0	21.5	0.20	4.9	47.5	2.8	41.4	22.6	n.a.	0.90	2.4	8.9
Turkey	25.5	91.4	11.7	2.3	10.7	12.5	2.3	10.6	21.6	3.8	45.1	20.7	0.60	2.0	8.8	16.5
Venezuela	22.3	101.8	43.2	1.1	7.5	15.9	1.3	6.7	23.3	3.4	44.0	21.4	2.2	3.2	4.3	4.8
High-income economies																
Australia	44.4	105.5	2.7	20.3	26.2	2.1	9.8	24.3	13.0	46.8	82.3	4.7	n.a.	56.1	80.2	96.2
Austria	76.2	160.5	13.7	13.0	29.3	6.7	14.2	25.0	8.0	33.7	81.0	7.3	29.7	33.1	46.0	56.3
Bahrain	30.7	161.1	8.9	3.2	13.2	11.6	2.4	13.2	24.1	6.2	88.0	22.2	6.9	10.6	18.2	91.2
Belgium	54.8	111.3	11.3	11.2	33.4	9.1	19.1	33.3	7.9	29.4	82.0	8.5	5.8	9.7	19.0	33.0
Brunei Darussalam	28.6	113.9	11.5	6.7	6.7	0.0	2.2	4.8	11.1	9.0	60.3	15.9	5.0	5.5	6.3	7.6
Canada	28.4	80.05	10.7	14.08	33.2	7.1	21.7	32.5	5.7	51.3	86.8	4.4	n.a.	30.4	39.2	42.1
Chile	22.01	138.1	14.9	3.7	11.7	9.4	4.3	12.4	15.0	16.6	61.4	10.9	n.a.	8.5	18.0	28.0
Croatia	23.08	115.4	13.3	4.1	21.6	13.7	2.6	20.7	29.4	6.6	63.0	18.7	6.5	8.4	47.7	53.9
Cyprus	23.1	98.4	7.01	5.5	19.9	10.7	3.1	19.2	26.1	15.3	61.0	11.5	n.a.	29.6	31.0	34.1
Czech Republic	42.3	126.8	5.8	4.0	16.4	11.6	6.9	16.4	12.3	9.8	75.0	17.0	25.9	40.8	51.0	52.1
Denmark	63.0	117.5	8.8	31.5	40.3	2.0	24.8	38.8	6.4	39.2	93.0	7.2	30.1	64.7	84.3	97.2
Estonia	40.7	160.4	9.8	6.0	26.6	12.4	13.5	25.5	9.0	28.6	79.0	8.5	4.6	27.6	45.9	76.9
France	49.06	97.0	4.9	9.1	37.8	11.8	15.4	37.5	12.7	14.3	83.0	14.7	28.1	36.2	43.6	51.8
Greece	53.9	120.03	3.7	2.4	24.5	19.1	1.5	24.1	40.2	9.1	56.0	15.1	12.4	25.1	37.2	45.7
Iceland	76.4	108.05	4.2	16.7	35.7	6.3	26.3	34.3	3.8	44.5	96.2	6.4	31.6	46.3	57.9	70.9
Ireland	64.6	107.2	3.1	14.4	24.4	4.4	7.8	22.7	15.4	17.9	79.0	12.4	53.6	50.9	61.1	65.9
Israel	73.1	120.6	4.06	13.3	25.3	5.3	18.6	25.3	4.4	20.9	73.4	10.5	n.a.	49.9	40.7	53.0

Italy	74.1	159.7	9.1	10.1	22.1	6.5	11.6	22.1	9.2	23.1	58.0	7.7	17.1	37.8	44.6	52.2
Japan	53.1	110.9	5.3	14.4	30.9	6.3	18.4	27.7	5.9	30.0	79.1	8.1	78.7	87.6	106.1	115.1
Korea (Rep. of)	58.3	109.4	16.5	10.6	37.2	10.4	25.9	37.2	5.2	44.7	84.1	5.3	88.6	97.7	104.3	105.1
Lithuania	14.9	165.05	7.2	1.5	21.3	22.0	7.1	21.1	15.5	6.4	68.0	19.7	7.4	12.0	14.4	18.8
Luxembourg	69.5	145.3	13.7	5.6	32.4	14.6	15.3	32.4	10.7	22.9	92.0	11.6	n.a.	50.0	66.6	80.6
Malta	28.08	126.9	5.2	8.4	32.0	11.1	12.4	32.0	13.6	13.1	70.0	14.0	17.5	20.5	33.0	35.3
Netherlands	67.8	117.9	9.01	37.	39.8	0.5	25.1	39.8	6.6	44.0	93.0	6.2	n.a.	38.0	52.6	61.3
New Zealand	39.9	110.3	3.5	12.9	30.7	7.2	7.8	27.8	18.2	47.4	89.5	5.3	n.a.	39.6	53.9	65.9
Norway	71.7	116.6	22.9	26.1	36.3	2.7	21.4	36.3	7.5	52.0	95.0	5.0	69.2	75.1	77.0	84.8
Oman	7.3	159.2	18.3	1.0	2.3	6.3	0.50	2.1	20.1	3.5	60.0	23.6	16.0	26.4	36.4	50.9
Poland	17.5	140.3	6.4	2.4	15.7	15.6	2.5	15.5	26.3	7.3	65.0	18.2	44.8	52.5	53.3	63.5
Portugal	64.6	116.1	14.5	6.4	22.8	10.6	11.1	22.5	10.1	16.4	64.0	11.3	20.7	24.4	27.6	32.8
Qatar	20.3	126.8	33.6	1.7	10.7	15.0	3.1	10.5	17.3	4.9	88.1	24.1	12.1	51.7	56.8	61.8
Russia	2.2	182.9	27.3	0.30	14.7	31.5	1.1	14.5	36.8	2.0	53.3	27.4	0.60	n.a.	47.8	52.8
Saudi Arabia	6.8	187.4	8.1	0.90	6.9	16.2	0.30	6.9	46.2	2.2	54.0	26.6	27.7	43.8	42.1	45.4
Singapore	70.1	152.1	15.5	21.7	25.9	1.4	14.6	25.4	7.9	36.0	74.2	6.0	70.1	99.7	115.5	126.1
Slovak Republic	23.08	111.9	5.0	1.2	15.2	20.7	3.4	14.7	21.0	9.4	80.0	17.8	18.0	24.1	35.9	39.7
Slovenia	61.09	108.6	4.9	7.0	24.4	10.4	9.8	24.3	12.9	15.1	70.0	12.8	n.a.	24.3	29.2	37.0
Spain	60.2	108.3	3.4	7.9	24.5	9.3	11.6	24.4	10.6	13.6	72.0	13.9	10.3	24.1	37.4	53.6
Sweden	71.8	124.5	5.4	25.3	33.4	2.3	27.9	32.3	2.1	45.7	94.0	6.0	69.9	83.8	97.3	104.9
Switzerland	64.7	130.2	19.3	23.2	40.3	4.6	22.5	39.9	8.2	47.1	85.2	4.9	24.9	30.1	34.7	39.6
Trinidad and Tobago	12.7	140.8	9.1	2.0	13.9	15.8	0.80	13.8	40.1	7.7	59.5	17.0	0.30	1.9	2.5	8.4
United Arab Emirates	47.1	149.6	5.9	6.9	10.4	3.4	3.1	10.3	17.1	23.6	85.0	10.7	n.a.	13.4	19.2	44.8
United Kingdom	73.7	135.2	10.4	14.2	34.5	7.4	16.4	34.0	10.4	26.8	87.0	9.8	37.0	43.2	52.6	72.1
United States	38.4	95.4	14.0	19.8	28.3	3.0	17.2	28.3	7.2	43.1	81.0	5.3	40.5	60.6	76.4	88.2

Source: Author's compilation and calculations

[a]For detailed definitions of respective ICT indicators—refer back to Appendix A

Appendix D. ICTs Inequalities. 113 World Countries. Period 2000–2012

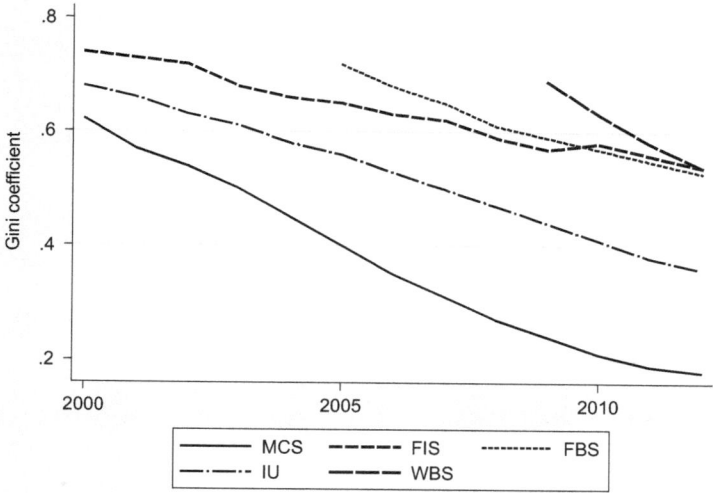

Fig. D.1 Gini coefficients time trends—$MCS_{i,y}$, $FIS_{i,y}$, $FBS_{i,y}$, $WBS_{i,y}$ and $IU_{i,y}$. 113 countries. Period 2000–2012. *Source*: Author's elaboration

© Springer International Publishing Switzerland 2015
E. Lechman, *ICT Diffusion in Developing Countries*,
DOI 10.1007/978-3-319-18254-4

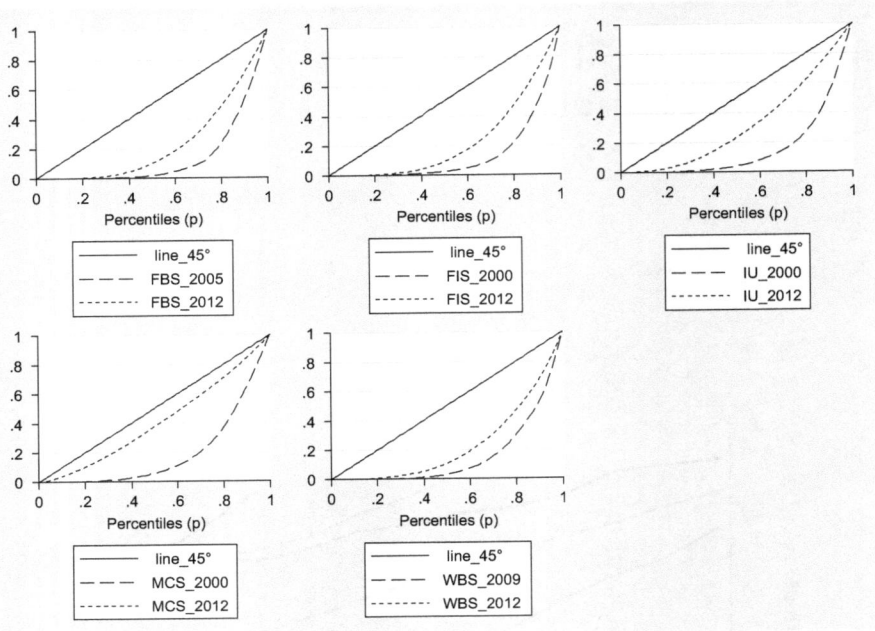

Fig. D.2 Lorenz curves—MCS$_{i,y}$, FIS$_{i,y}$, FBS$_{i,y}$, WBS$_{i,y}$ and IU$_{i,y}$. 113 countries. Years 2000 and 2012. *Source*: Author's elaboration

Appendix E. ICTs Distributions. 113 World Countries. Period 2000–2012

© Springer International Publishing Switzerland 2015
E. Lechman, *ICT Diffusion in Developing Countries*,
DOI 10.1007/978-3-319-18254-4

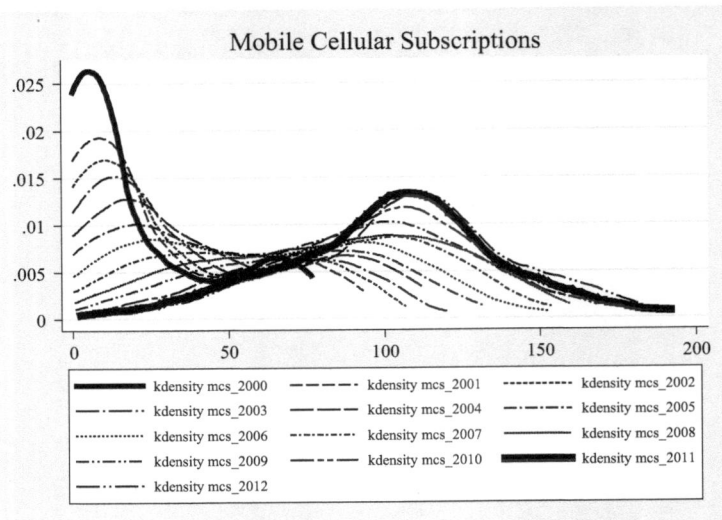

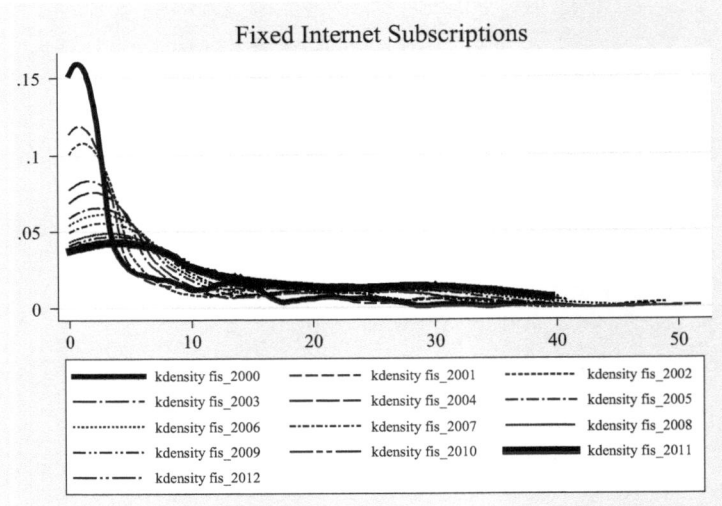

Fig. E.1 Changes in core ICT indicators distribution. 113 world countries. Period 2000–2012. *Source*: Author's elaboration. *Note*: for FBS—Kernel densities estimates since 2005. For MCS—111 countries (Comoros and Eritrea excluded—significant lacks in data time series). For FBS—107 countries (Bangladesh, Congo, Honduras, Eritrea, Swaziland and Togo excluded—significant lacks in data time series); for Madagascar and Nepal—initial values for 2006. For WBS—Kernel densities estimates since 2009. For WBS—111 countries (Eritrea and Nepal—excluded—data available only for 2012). On X-axis—absolute data

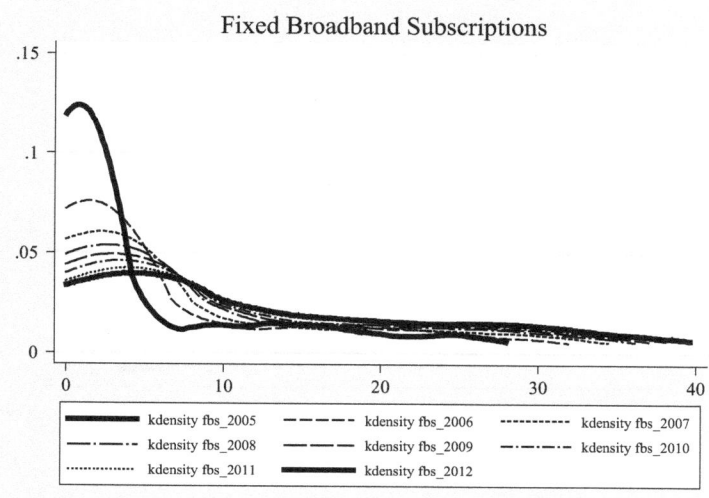

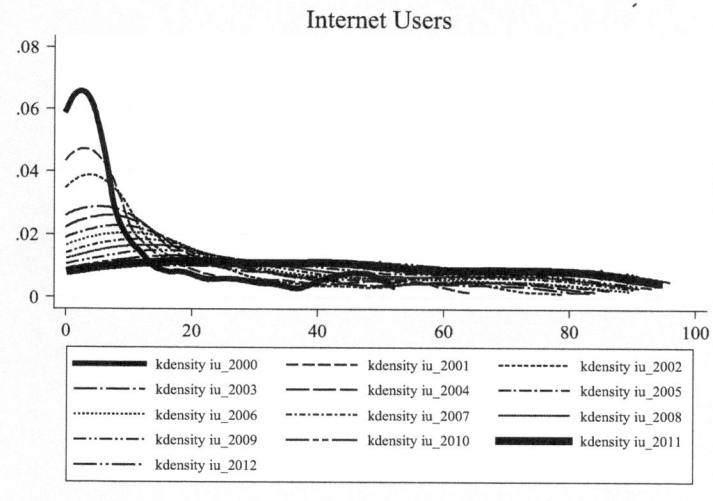

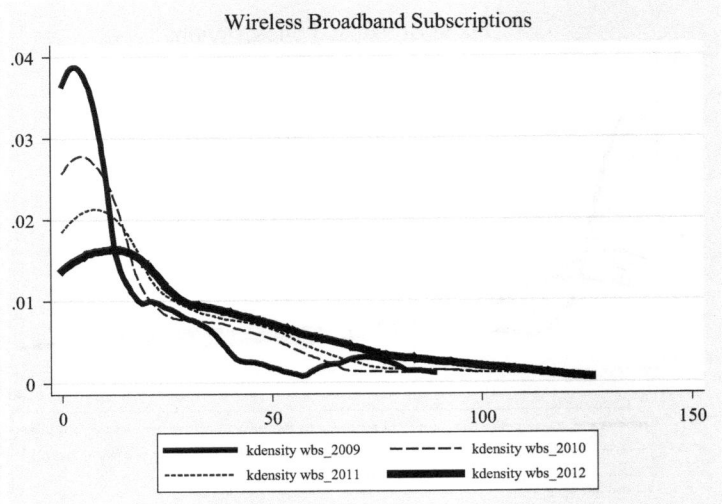

Appendix F. ICT Marginal Growths and Replication Coefficients. 17 Low-Income Economies. Period 1995–2012

© Springer International Publishing Switzerland 2015
E. Lechman, *ICT Diffusion in Developing Countries*,
DOI 10.1007/978-3-319-18254-4

Table F.1 MCS marginal growths and replication coefficients

Year	Bangladesh		Benin		Burkina Faso		Cambodia		Comoros		Eritrea		Ethiopia		Kenya		Madagascar	
	$\Omega_{MCS,i,y}$	$\Phi_{MCS,i,y}$	$\Omega_{MCS,i,y}$	$\Phi_{MCS,i,y}$	$\Omega_{MCS,i,y}$	$\Phi_{MCS,i,y}$	$\Omega_{MCS,i,y}$	$\Phi_{MCS,i,y}$	$\Omega_{MCS,i,y}$	$\Phi_{MCS,i,y}$	$\Omega_{MCS,i,y}$	$\Phi_{MCS,i,y}$	$\Omega_{MCS,i,y}$	$\Phi_{MCS,i,y}$	$\Omega_{MCS,i,y}$	$\Phi_{MCS,i,y}$	$\Omega_{MCS,i,y}$	$\Phi_{MCS,i,y}$
1995	n.a.	n.a.	n.a.	n.a.	n.a.	n.a.	n.a.	n.a.	n.a.	n.a.	n.a.	n.a.	n.a.	n.a.	n.a.	n.a.	n.a.	n.a.
1996	0.00	1.57	0.03	2.50	n.a.	2.78	0.08	1.59	n.a.	n.a.	n.a.	n.a.	n.a.	n.a.	0.00	1.21	0.01	1.71
1997	0.02	6.37	0.02	1.54	0.01	1.77	0.09	1.41	n.a.	n.a.	n.a.	n.a.	n.a.	n.a.	0.01	2.33	0.01	1.73
1998	0.04	2.83	0.03	1.42	0.01	1.79	0.23	1.78	n.a.	n.a.	n.a.	n.a.	n.a.	n.a.	0.01	1.55	0.06	3.02
1999	0.06	1.95	0.01	1.12	0.02	4.87	0.22	1.42	n.a.	n.a.	n.a.	n.a.	n.a.	n.a.	0.04	2.15	0.15	2.71
2000	0.10	1.84	0.69	7.40	0.17	2.93	0.32	1.43	n.a.	n.a.	n.a.	n.a.	0.02	2.56	0.33	5.22	0.17	1.71
2001	0.18	1.83	0.94	2.18	0.42	1.42	0.72	1.68	n.a.	n.a.	n.a.	n.a.	0.01	1.50	1.46	4.59	0.51	2.27
2002	0.40	2.03	1.21	1.69	0.27	2.08	1.20	1.67	n.a.	n.a.	n.a.	n.a.	0.03	1.78	1.73	1.93	0.07	1.07
2003	0.20	1.25	0.13	1.04	0.98	1.62	0.86	1.29	n.a.	n.a.	n.a.	n.a.	0.00	0.99	1.09	1.30	0.67	1.69
2004	0.99	2.01	2.72	1.88	1.16	1.55	2.70	1.70	1.08	4.08	n.a.	n.a.	0.14	2.95	2.62	1.56	0.23	1.14
2005	4.32	3.19	1.49	1.26	1.68	1.56	1.40	1.21	1.15	1.81	0.40	1.94	0.33	2.57	5.58	1.76	0.91	1.48
2006	6.92	2.10	5.22	1.72	2.63	1.77	4.75	1.60	3.40	2.31	0.40	1.48	0.57	2.05	7.08	1.55	2.77	1.99
2007	10.26	1.78	11.06	1.88	5.70	1.58	6.09	1.48	3.83	1.64	0.39	1.31	0.40	1.36	10.09	1.51	5.89	2.06
2008	6.70	1.29	16.84	1.71	7.58	1.23	11.60	1.62	4.31	1.44	0.40	1.25	0.86	1.57	11.99	1.40	12.82	2.12
2009	4.19	1.14	14.07	1.35	4.70	1.45	13.92	1.46	4.28	1.30	0.52	1.26	2.41	2.02	6.58	1.16	6.39	1.26
2010	10.59	1.31	19.93	1.37	11.40	1.31	12.43	1.28	5.79	1.31	0.69	1.27	3.09	1.65	12.41	1.26	5.92	1.19
2011	10.25	1.23	5.00	1.07	11.30	1.26	37.45	1.66	6.71	1.28	0.85	1.26	7.93	2.01	5.78	1.09	3.46	1.09
2012	7.63	1.14	4.25	1.05	12.58	1.26	34.34	1.36	8.60	1.28	0.90	1.22	6.57	1.42	4.36	1.07	-0.67	0.98

	Malawi		Myanmar		Nepal		Niger		Rwanda		Togo		Uganda		Zimbabwe	
	$\Omega_{MCS,i,y}$	$\Phi_{MCS,i,y}$	$\Omega_{MCS,i,y}$	$\Phi_{MCS,i,y}$	$\Omega_{MCS,i,y}$	$\Phi_{MCS,i,y}$	$\Omega_{MCS,i,y}$	$\Phi_{MCS,i,y}$	$\Omega_{MCS,i,y}$	$\Phi_{MCS,i,y}$	$\Omega_{MCS,i,y}$	$\Phi_{MCS,i,y}$	$\Omega_{MCS,i,y}$	$\Phi_{MCS,i,y}$	$\Omega_{MCS,i,y}$	$\Phi_{MCS,i,y}$
1995	n.a.	n.a.	n.a.	n.a.	n.a.	n.a.	n.a.	n.a.	n.a.	n.a.	n.a.	n.a.	n.a.	n.a.	n.a.	n.a.
1996	0.03	9.51	0.01	2.59	n.a.	n.a.	n.a.	n.a.	n.a.	n.a.	n.a.	n.a.	0.01	2.22	n.a.	n.a.
1997	0.03	1.85	0.00	1.15	n.a.	n.a.	n.a.	n.a.	n.a.	n.a.	n.a.	n.a.	0.00	1.21	n.a.	n.a.
1998	0.03	1.46	0.00	0.99	n.a.	n.a.	0.01	13.27	n.a.	n.a.	0.10	2.44	0.11	5.82	0.11	3.26
1999	0.11	2.08	0.01	1.32	n.a.	n.a.	0.01	1.57	0.07	2.01	0.20	2.21	0.11	1.82	1.25	9.04
2000	0.23	2.12	0.00	1.16	0.02	1.82	0.00	0.90	0.32	3.32	0.67	2.87	0.28	2.18	0.73	1.52
2001	0.05	1.11	0.02	1.68	0.03	1.66	0.00	1.00	0.28	1.60	0.88	1.85	0.61	2.16	0.36	1.17
2002	0.24	1.50	0.05	2.10	0.02	1.24	0.47	26.10	0.17	1.24	1.32	1.69	0.39	1.34	0.19	1.07
2003	0.38	1.53	0.04	1.38	0.24	3.68	0.19	1.38	0.52	1.56	1.41	1.44	1.38	1.91	0.19	1.07
2004	0.66	1.60	0.05	1.38	0.13	1.40	0.68	2.02	0.05	1.04	1.53	1.33	1.30	1.45	0.48	1.17
2005	1.49	1.84	0.07	1.38	0.43	1.92	1.10	1.81	0.88	1.59	1.67	1.27	0.38	1.09	1.74	1.52
2006	1.40	1.43	0.17	1.65	3.62	5.02	1.07	1.44	0.89	1.38	4.62	1.59	2.18	1.48	1.58	1.31
2007	3.00	1.64	0.06	1.15	8.08	2.79	2.81	1.80	3.15	1.97	7.95	1.64	6.89	2.02	2.95	1.44
2008	3.00	1.39	0.23	1.47	3.40	1.27	6.54	2.03	6.54	2.02	5.48	1.27	13.27	1.97	3.32	1.35
2009	6.39	1.60	0.26	1.36	5.09	1.32	4.11	1.32	10.13	1.78	9.72	1.38	1.63	1.06	18.02	2.39
2010	3.71	1.22	0.17	1.17	13.16	1.62	6.10	1.36	9.68	1.42	5.67	1.16	9.19	1.32	27.92	1.90
2011	4.80	1.23	1.23	2.08	14.92	1.44	5.64	1.24	7.15	1.22	0.38	1.01	9.76	1.26	9.99	1.17
2012	3.65	1.14	7.93	4.34	10.44	1.21	2.72	1.09	9.77	1.24	8.22	1.20	-2.50	0.95	23.04	1.33

Source: Author's estimations

Table F.2 IU marginal growths and replication coefficients

	Bangladesh		Benin		Burkina Faso		Cambodia		Comoros		Eritrea		Ethiopia		Kenya		Madagascar	
	$\Omega_{IU,i,y}$	$\Phi_{IU,i,y}$	$\Omega_{IU,i,y}$	$\Phi_{IU,i,y}$	$\Omega_{IU,i,y}$	$\Phi_{IU,i,y}$	$\Omega_{IU,i,y}$	$\Phi_{IU,i,y}$	$\Omega_{IU,i,y}$	$\Phi_{IU,i,y}$	$\Omega_{IU,i,y}$	$\Phi_{IU,i,y}$	$\Omega_{IU,i,y}$	$\Phi_{IU,i,y}$	$\Omega_{IU,i,y}$	$\Phi_{IU,i,y}$	$\Omega_{IU,i,y}$	$\Phi_{IU,i,y}$
1995	n.a.	n.a.	n.a.	n.a.	n.a.	n.a.	n.a.	n.a.	n.a.	n.a.	n.a.	n.a.	n.a.	n.a.	n.a.	n.a.	n.a.	n.a.
1996	n.a.	n.a.	n.a.	n.a.	n.a.	n.a.	n.a.	n.a.	n.a.	n.a.	n.a.	n.a.	n.a.	n.a.	n.a.	n.a.	n.a.	n.a.
1997	n.a.	n.a.	0.02	14.56	0.02	19.45	n.a.	n.a.	n.a.	n.a.	n.a.	n.a.	0.00	2.92	0.03	3.89	0.01	3.88
1998	0.00	4.91	0.02	1.94	0.03	2.43	0.01	2.79	n.a.	n.a.	0.00	0.97	0.00	1.95	0.02	1.46	0.05	4.36
1999	0.03	9.82	0.11	3.24	0.02	1.36	0.02	1.96	0.11	3.91	0.02	2.91	0.00	1.30	0.06	2.27	0.11	2.70
2000	0.03	1.96	0.07	1.45	0.02	1.25	0.02	1.47	0.12	1.83	0.11	5.36	0.00	1.22	0.20	2.78	0.03	1.16
2001	0.06	1.83	0.14	1.61	0.08	2.05	0.03	1.64	0.17	1.63	0.02	1.15	0.02	2.43	0.30	1.95	0.03	1.13
2002	0.01	1.08	0.34	1.93	0.04	1.27	0.15	2.95	0.11	1.25	0.07	1.44	0.04	1.95	0.59	1.95	0.12	1.53
2003	0.02	1.17	0.25	1.35	0.17	1.86	0.03	1.15	0.29	1.53	0.02	1.10	0.03	1.46	1.73	2.44	0.08	1.25
2004	0.04	1.21	0.23	1.24	0.03	1.07	0.04	1.15	0.48	1.56	0.04	1.16	0.05	1.47	0.08	1.03	0.10	1.24
2005	0.04	1.21	0.09	1.07	0.07	1.17	0.02	1.06	0.67	1.51	0.04	1.14	0.06	1.41	0.08	1.03	0.04	1.08
2006	0.76	4.14	0.27	1.21	0.16	1.35	0.15	1.48	0.20	1.10	0.05	1.15	0.09	1.41	4.43	2.43	0.04	1.07
2007	0.80	1.80	0.25	1.16	0.12	1.19	0.02	1.05	0.30	1.14	0.03	1.08	0.06	1.19	0.42	1.06	0.04	1.07
2008	0.70	1.39	0.06	1.03	0.17	1.23	0.02	1.04	0.50	1.20	0.06	1.15	0.08	1.22	0.72	1.09	1.00	2.54
2009	0.60	1.24	0.39	1.21	0.21	1.23	0.02	1.04	0.50	1.17	0.07	1.15	0.09	1.20	1.37	1.16	−0.02	0.99
2010	0.60	1.19	0.89	1.40	1.27	2.12	0.73	2.38	1.60	1.46	0.07	1.13	0.21	1.39	3.96	1.39	0.07	1.04
2011	1.30	1.35	0.37	1.12	0.60	1.25	1.84	2.46	0.40	1.08	0.09	1.15	0.35	1.47	14.00	2.00	0.20	1.12
2012	1.30	1.26	0.30	1.09	0.73	1.24	1.84	1.59	0.48	1.09	0.10	1.14	0.38	1.35	4.10	1.15	0.15	1.08

	Malawi		Myanmar		Nepal		Niger		Rwanda		Togo		Uganda		Zimbabwe	
	$\Omega_{IU,i,y}$	$\Phi_{IU,i,y}$	$\Omega_{IU,i,y}$	$\Phi_{IU,i,y}$	$\Omega_{IU,i,y}$	$\Phi_{IU,i,y}$	$\Omega_{IU,i,y}$	$\Phi_{IU,i,y}$	$\Omega_{IU,i,y}$	$\Phi_{IU,i,y}$	$\Omega_{IU,i,y}$	$\Phi_{IU,i,y}$	$\Omega_{IU,i,y}$	$\Phi_{IU,i,y}$	$\Omega_{IU,i,y}$	$\Phi_{IU,i,y}$
1995		n.a.	n.a.	n.a.	0.02	4.88	0.00	1.93	0.00	1.84	0.20	19.32	0.01	2.23	0.02	1.97
1996	0.01	3.87	n.a.	n.a.	0.04	2.93	0.00	1.45	0.01	7.24	0.09	1.45	0.05	6.33	0.05	2.47
1997	0.07	4.83	n.a.	n.a.	0.08	2.28	0.03	9.66	0.06	5.71	0.28	1.93	0.04	1.62	0.08	1.98
1998	0.04	1.45	4E-05	1E+00	0.06	1.40	0.01	1.29	0.00	0.94	0.21	1.36	0.06	1.55	0.24	2.48
1999	0.04	1.29	1E-04	2E+00	0.04	1.17	0.07	2.90	0.18	3.83	0.10	1.13	0.07	1.45	0.40	1.99
2000	0.05	1.31	1E-04	1E+00	0.07	1.30	0.02	1.21	0.05	1.22	0.10	1.11	0.15	1.61	3.19	4.99
2001	0.06	1.30	2E-02	6E+01	0.07	1.22	0.03	1.22	0.06	1.22	0.20	1.20	0.08	1.21	2.40	1.60
2002	0.07	1.25	3E-04	1E+00	0.07	1.18	0.03	1.22	0.07	1.21	0.30	1.25	0.26	1.55	0.17	1.03
2003	0.04	1.11	4E-02	3E+00	0.38	1.84	0.03	1.17	0.13	1.29	0.30	1.20	1.02	2.42	1.45	1.22
2004	0.04	1.11	1E-01	3E+00	0.31	1.38	0.07	1.33	0.77	2.39	0.20	1.11	0.79	1.45	1.78	1.22
2005	0.54	2.27	4E-02	1E+00	0.27	1.24	0.10	1.33	0.79	1.59	0.20	1.10	1.14	1.45	1.06	1.11
2006	-0.27	0.72	3E-03	1E+00	0.32	1.23	0.31	1.79	2.38	2.13	0.20	1.09	4.23	2.15	0.55	1.05
2007	0.37	1.53	0E+00	1E+00	0.24	1.14	0.06	1.09	3.20	1.71	0.20	1.08	1.88	1.24	-0.04	1.00
2008	1.19	2.11	3E-02	1E+00	5.96	4.03	0.07	1.09	0.30	1.04	0.40	1.15	2.72	1.28	0.14	1.01
2009	1.07	1.47	7E-01	4E+00	1.07	1.13	0.47	1.57	-1.00	0.88	0.50	1.17	0.51	1.04	4.20	1.37
2010	1.02	1.31	9E-02	1E+00	2.15	1.24	0.11	1.08	1.02	1.15	0.50	1.14	1.68	1.13	1.39	1.09
2011		n.a.	n.a.	n.a.	0.02	4.88	0.00	1.93	0.00	1.84	0.20	19.32	0.01	2.23	0.02	1.97
2012	0.01	3.87	n.a.	n.a.	0.04	2.93	0.00	1.45	0.01	7.24	0.09	1.45	0.05	6.33	0.05	2.47

Source: Author's estimations

Appendix G. ICT Marginal Growths and Replication Coefficients. 29 Lower-Middle-Income Economies. Period 1995–2012

© Springer International Publishing Switzerland 2015
E. Lechman, *ICT Diffusion in Developing Countries*,
DOI 10.1007/978-3-319-18254-4

Table G.1 MCS marginal growths and replication coefficients

	Armenia		Bolivia		Congo		Egypt		El Salvador		Georgia		Ghana		Guyana		Honduras		India	
	$\Omega_{MCS,Iy}$	$\Phi_{MCS,Iy}$	$\Omega_{MCS,Iy}$	$\Phi_{MCS,Iy}$	$\Omega_{MCS,Iy}$	$\Phi_{MCS,Iy}$	$\Omega_{MCS,Iy}$	$\Phi_{MCS,Iy}$	$\Omega_{MCS,Iy}$	$\Phi_{MCS,Iy}$	$\Omega_{MCS,Iy}$	$\Phi_{MCS,Iy}$	$\Omega_{MCS,Iy}$	$\Phi_{MCS,Iy}$	$\Omega_{MCS,Iy}$	$\Phi_{MCS,Iy}$	$\Omega_{MCS,Iy}$	$\Phi_{MCS,Iy}$	$\Omega_{MCS,Iy}$	$\Phi_{MCS,Iy}$
1992	n.a.	n.a.	0.02	5.13	n.a.	n.a.	0.00	1.07	n.a.	n.a.	n.a.	n.a.	n.a.	n.a.	n.a.	n.a.	n.a.	n.a.	n.a.	n.a.
1993	n.a.	n.a.	0.01	1.67	n.a.	n.a.	0.00	1.38	n.a.	n.a.	n.a.	n.a.	0.01	4.24	0.03	1.22	n.a.	n.a.	n.a.	n.a.
1994	n.a.	n.a.	0.02	1.49	n.a.	n.a.	0.00	1.06	0.06	2.94	n.a.	n.a.	0.01	1.86	0.03	1.21	n.a.	n.a.	n.a.	n.a.
1995	n.a.	n.a.	0.04	1.74	n.a.	n.a.	0.00	0.98	0.15	2.73	n.a.	n.a.	0.02	1.81	0.00	0.99	n.a.	n.a.	n.a.	n.a.
1996	n.a.	n.a.	0.33	4.52	n.a.	n.a.	0.00	0.98	0.17	1.71	0.04	15.57	0.04	2.01	-0.01	0.96	n.a.	n.a.	0.03	4.20
1997	0.15	16.86	1.06	3.47	n.a.	n.a.	0.09	8.74	0.29	1.71	0.56	13.22	0.05	1.67	0.03	1.16	0.21	6.11	0.06	2.64
1998	0.09	1.58	1.45	1.98	n.a.	n.a.	0.04	1.37	1.64	3.39	0.62	2.02	0.11	1.87	0.01	1.03	0.34	2.37	0.03	1.33
1999	0.01	1.05	2.11	1.72	0.05	1.43	0.60	5.22	6.30	3.71	1.54	2.25	0.15	1.64	0.18	1.93	0.70	2.21	0.07	1.55
2000	0.30	2.15	1.81	1.36	2.07	13.63	1.32	2.78	3.85	1.45	1.33	1.48	0.31	1.81	4.97	14.08	1.20	1.94	0.16	1.87
2001	0.27	1.47	2.14	1.31	2.44	2.09	2.10	2.02	1.85	1.15	2.33	1.57	0.57	1.83	4.72	1.88	1.24	1.50	0.27	1.80
2002	1.51	2.81	2.58	1.29	2.08	1.44	2.42	1.58	0.46	1.03	4.46	1.69	0.69	1.55	0.50	1.05	1.29	1.35	0.59	1.96
2003	1.43	1.61	2.61	1.23	3.06	1.45	1.77	1.27	4.28	1.29	4.68	1.43	1.96	2.00	7.73	1.73	0.70	1.14	1.87	2.55
2004	2.95	1.78	5.42	1.38	1.31	1.13	2.48	1.30	11.22	1.59	3.04	1.19	4.22	2.08	4.37	1.24	4.73	1.83	1.62	1.53
2005	3.83	1.57	6.28	1.32	4.63	1.42	8.16	1.75	9.43	1.31	7.62	1.41	5.31	1.65	14.30	1.63	8.12	1.78	3.30	1.70
2006	31.40	3.98	4.34	1.17	9.40	1.60	5.67	1.30	23.46	1.59	12.08	1.46	10.28	1.76	15.28	1.41	13.27	1.71	6.53	1.82
2007	20.81	1.50	3.41	1.11	9.10	1.36	15.88	1.64	37.06	1.59	20.41	1.53	10.03	1.42	17.67	1.34	26.46	1.83	5.63	1.39
2008	-14.33	0.77	17.60	1.52	12.36	1.36	14.15	1.35	12.75	1.13	3.70	1.06	16.31	1.48	-12.21	0.83	26.52	1.45	9.38	1.47
2009	25.40	1.52	13.45	1.26	27.18	1.58	17.41	1.32	9.38	1.08	2.02	1.03	13.71	1.27	4.83	1.08	27.51	1.32	14.59	1.49
2010	56.60	1.77	6.00	1.09	16.65	1.23	18.41	1.26	1.47	1.01	26.19	1.41	8.09	1.13	8.74	1.14	12.39	1.11	18.27	1.41
2011	-22.10	0.83	10.22	1.14	1.50	1.02	14.58	1.16	9.09	1.07	10.63	1.12	13.41	1.19	-4.43	0.94	-21.05	0.83	10.81	1.17
2012	3.58	1.03	9.54	1.12	6.82	1.07	14.84	1.14	4.42	1.03	6.53	1.06	15.72	1.18	1.92	1.03	-10.80	0.90	-3.28	.96

	Indonesia		Lao P.D.R.		Mauritania		Moldova		Mongolia		Morocco		Nicaragua		Nigeria		Pakistan		Paraguay	
	$\Omega_{MCS,i,y}$	$\Phi_{MCS,i,y}$	$\Omega_{MCS,i,y}$	$\Phi_{MCS,i,y}$	$\Omega_{MCS,i,y}$	$\Phi_{MCS,i,y}$	$\Omega_{MCS,i,y}$	$\Phi_{MCS,i,y}$	$\Omega_{MCS,i,y}$	$\Phi_{MCS,i,y}$	$\Omega_{MCS,i,y}$	$\Phi_{MCS,i,y}$	$\Omega_{MCS,i,y}$	$\Phi_{MCS,i,y}$	$\Omega_{MCS,i,y}$	$\Phi_{MCS,i,y}$	$\Omega_{MCS,i,y}$	$\Phi_{MCS,i,y}$	$\Omega_{MCS,i,y}$	$\Phi_{MCS,i,y}$
1992	0.01	1.42	n.a.	n.a.	n.a.	n.a.	n.a.	n.a.	n.a.	n.a.	0.01	2.11	n.a.	n.a.	n.a.	n.a.	0.00	1.55	n.a.	n.a.
1993	0.01	1.48	0.00	1.14	n.a.	n.a.	n.a.	n.a.	n.a.	n.a.	0.01	2.06	n.a.	n.a.	n.a.	n.a.	0.00	1.16	0.04	2.21
1994	0.01	1.44	0.01	1.79	n.a.	n.a.	n.a.	n.a.	n.a.	n.a.	0.03	2.02	0.04	6.58	0.00	1.38	0.01	1.50	0.09	2.21
1995	0.07	2.66	0.02	2.40	n.a.	n.a.	n.a.	n.a.	n.a.	n.a.	0.06	2.11	0.05	1.97	0.00	0.99	0.01	1.62	0.17	2.02
1996	0.18	2.63	0.04	2.41	n.a.	n.a.	0.02	66.19	n.a.	n.a.	0.05	1.43	0.01	1.14	0.00	1.05	0.02	1.62	0.34	2.03
1997	0.17	1.60	0.02	1.27	n.a.	n.a.	0.03	2.41	0.05	2.20	0.11	1.71	0.05	1.45	0.00	1.04	0.05	1.93	1.01	2.51
1998	0.07	1.15	0.03	1.29	n.a.	n.a.	0.11	3.22	0.30	4.48	0.15	1.54	0.21	2.38	0.00	1.30	0.04	1.41	2.83	2.69
1999	0.55	2.05	0.10	1.84	n.a.	n.a.	0.27	2.60	1.07	3.79	0.88	3.12	0.51	2.38	0.00	1.22	0.05	1.32	3.80	1.84
2000	0.68	1.63	0.01	1.03	3.39	7.00	2.95	7.84	4.99	4.43	6.86	6.27	0.89	2.01	0.00	1.17	0.02	1.13	7.03	1.85
2001	1.32	1.75	0.30	2.29	4.64	2.17	2.18	1.65	1.61	1.25	8.28	2.02	1.41	1.80	0.19	8.66	0.29	2.37	5.72	1.37
2002	2.36	1.77	0.45	1.84	3.24	1.38	2.95	1.53	0.78	1.10	4.71	1.29	1.34	1.42	1.00	5.74	0.63	2.24	8.86	1.42
2003	3.04	1.56	1.00	2.01	5.26	1.44	3.69	1.43	4.08	1.46	3.73	1.18	4.26	1.94	1.16	1.96	0.44	1.39	1.23	1.04
2004	5.23	1.62	1.58	1.79	6.60	1.39	8.34	1.68	4.25	1.33	6.40	1.26	4.94	1.56	4.35	2.83	1.66	2.05	−0.96	0.97
2005	7.19	1.52	7.77	3.17	9.04	1.38	8.37	1.41	4.88	1.28	9.86	1.32	6.81	1.50	6.59	1.98	4.85	2.50	1.77	1.06
2006	7.12	1.34	5.77	1.51	9.72	1.30	7.62	1.26	8.22	1.37	11.52	1.28	12.61	1.61	9.24	1.69	13.36	2.65	21.79	1.68
2007	12.41	1.44	7.46	1.44	18.66	1.44	14.71	1.40	15.76	1.52	12.66	1.24	11.59	1.35	4.89	1.22	16.90	1.79	22.89	1.43
2008	19.58	1.48	8.35	1.34	0.95	1.02	15.37	1.30	20.94	1.45	8.39	1.13	10.12	1.23	14.21	1.52	14.36	1.37	16.22	1.21
2009	8.91	1.15	18.67	1.57	14.85	1.24	10.63	1.16	17.19	1.26	7.22	1.10	3.40	1.06	6.30	1.15	2.76	1.05	−4.34	0.95
2010	18.87	1.27	10.99	1.21	12.61	1.16	11.31	1.15	8.38	1.10	20.15	1.25	9.82	1.17	6.70	1.14	1.82	1.03	3.14	1.04
2011	14.67	1.17	21.45	1.34	16.47	1.18	2.23	1.03	14.29	1.15	12.95	1.13	13.63	1.20	3.30	1.06	4.53	1.08	7.67	1.08
2012	11.76	1.11	−19.34	0.77	n.a.	n.a.	11.18	1.12	13.86	1.13	5.95	1.05	4.45	1.05	8.84	1.15	5.25	1.08	2.26	1.02

(continued)

	Philippines		Senegal		Sri Lanka		Swaziland		Syria		Ukraine		Viet Nam		Yemen		Zambia	
	$\Omega_{MCS,i,y}$	$\Phi_{MCS,i,y}$	$\Omega_{MCS,i,y}$	$\Phi_{MCS,i,y}$	$\Omega_{MCS,i,y}$	$\Phi_{MCS,i,y}$	$\Omega_{MCS,i,y}$	$\Phi_{MCS,i,y}$	$\Omega_{MCS,i,y}$	$\Phi_{MCS,i,y}$	$\Omega_{MCS,i,y}$	$\Phi_{MCS,i,y}$	$\Omega_{MCS,i,y}$	$\Phi_{MCS,i,y}$	$\Omega_{MCS,i,y}$	$\Phi_{MCS,i,y}$	$\Omega_{MCS,i,y}$	$\Phi_{MCS,i,y}$
1992	0.03	1.58	n.a.	n.a.	0.00	1.45	n.a.	n.a.	n.a.	n.a.	n.a.	n.a.	n.a.	n.a.	n.a.	n.a.	n.a.	n.a.
1993	0.07	1.79	n.a.	n.a.	0.07	5.50	n.a.	n.a.	n.a.	n.a.	n.a.	n.a.	0.00	4.97	0.03	3.17	n.a.	n.a.
1994	0.10	1.64	n.a.	n.a.	0.08	1.97	n.a.	n.a.	n.a.	n.a.	0.01	77.17	0.01	3.02	0.02	1.51	n.a.	n.a.
1995	0.46	2.81	0.00	1.21	0.12	1.74	n.a.	n.a.	n.a.	n.a.	0.02	2.81	0.01	1.85	0.00	0.97	n.a.	n.a.
1996	0.64	1.90	0.01	11.28	0.11	1.37	n.a.	n.a.	n.a.	n.a.	0.03	2.16	0.06	2.89	0.00	1.03	0.01	1.71
1997	0.50	1.37	0.06	4.80	0.24	1.61	n.a.	n.a.	n.a.	n.a.	0.05	1.92	0.12	2.30	0.02	1.35	0.02	1.63
1998	0.48	1.26	0.22	3.87	0.32	1.51	n.a.	n.a.	n.a.	n.a.	0.12	2.04	0.08	1.37	0.02	1.28	0.04	1.77
1999	1.42	1.61	0.62	3.12	0.44	1.46	0.88	2.92	n.a.	n.a.	0.21	1.89	0.13	1.46	0.06	1.67	0.20	3.32
2000	4.56	2.22	1.62	2.78	0.91	1.66	1.77	2.32	0.16	7.33	1.23	3.81	0.56	2.37	0.02	1.13	0.69	3.42
2001	7.02	1.84	0.44	1.18	1.22	1.54	2.02	1.65	1.01	6.54	2.91	2.74	0.56	1.57	0.64	4.48	0.19	1.20
2002	3.67	1.24	2.34	1.79	1.33	1.38	1.17	1.23	1.16	1.97	3.09	1.67	0.77	1.51	1.80	3.20	0.14	1.12
2003	8.25	1.43	2.00	1.38	2.31	1.48	1.53	1.24	4.50	2.91	5.93	1.77	0.99	1.43	0.91	1.35	0.90	1.69
2004	11.85	1.43	2.89	1.39	4.06	1.57	5.43	1.70	6.42	1.94	15.35	2.13	2.60	1.79	3.99	2.13	1.94	1.88
2005	1.42	1.04	5.13	1.50	5.65	1.50	4.86	1.37	2.97	1.22	34.73	2.20	5.40	1.92	3.78	1.50	4.12	1.99
2006	8.54	1.21	10.40	1.68	10.03	1.60	4.25	1.23	8.62	1.53	41.03	1.64	10.74	1.95	3.10	1.27	5.84	1.71
2007	15.45	1.31	4.75	1.18	12.45	1.46	11.12	1.50	7.01	1.28	13.70	1.13	29.99	2.36	6.12	1.42	7.67	1.54
2008	10.85	1.17	13.54	1.44	14.86	1.38	12.59	1.38	2.81	1.09	1.44	1.01	33.68	1.65	9.16	1.45	6.62	1.30
2009	6.89	1.09	10.80	1.25	24.96	1.46	10.54	1.23	12.97	1.37	-1.09	0.99	25.67	1.30	7.70	1.26	5.95	1.21
2010	6.72	1.08	9.60	1.18	4.48	1.06	4.22	1.07	6.67	1.14	-1.64	0.99	13.93	1.13	11.30	1.30	6.85	1.20
2011	10.11	1.11	5.73	1.09	3.92	1.05	2.41	1.04	4.92	1.09	4.23	1.04	16.31	1.13	1.37	1.03	18.67	1.45
2012	7.42	1.07	13.41	1.19	4.09	1.05	2.16	1.03	0.06	1.00	9.00	1.07	6.06	1.04	8.21	1.16	14.89	1.25

Source: Author's elaboration

Note: Over the period 1984–1991 both $\Omega_{MCS,i,y}$ and $\Phi_{MCS,i,y}$ are negligible, hence the estimates are not reported

Table G.2 IU marginal growths and replication coefficients

	Armenia		Bolivia		Congo		Egypt		El Salvador		Georgia		Ghana		Guyana		Honduras		India	
	$\Omega_{IU,i,y}$	$\Phi_{IU,i,y}$	$\Omega_{IU,i,y}$	$\Phi_{IU,i,y}$	$\Omega_{IU,i,y}$	$\Phi_{IU,i,y}$	$\Omega_{IU,i,y}$	$\Phi_{IU,i,y}$	$\Omega_{IU,i,y}$	$\Phi_{IU,i,y}$	$\Omega_{IU,i,y}$	$\Phi_{IU,i,y}$	$\Omega_{IU,i,y}$	$\Phi_{IU,i,y}$	$\Omega_{IU,i,y}$	$\Phi_{IU,i,y}$	$\Omega_{IU,i,y}$	$\Phi_{IU,i,y}$	$\Omega_{IU,i,y}$	$\Phi_{IU,i,y}$
1996	0.04	1.79	0.13	2.94	n.a.	n.a.	0.03	1.96	n.a.	n.a.	0.03	3.38	0.01	16.24	n.a.	n.a.	0.01	1.19	0.02	1.77
1997	0.02	1.18	0.25	2.28	0.00	0.98	0.03	1.47	0.17	2.97	0.02	1.52	0.02	4.87	0.07	2.00	0.13	3.91	0.02	1.53
1998	0.02	1.15	0.18	1.40	0.00	0.98	0.06	1.64	0.17	1.65	0.04	1.69	0.00	1.17	0.13	2.00	0.13	1.76	0.07	1.96
1999	0.84	7.55	0.36	1.57	0.01	4.92	0.14	1.96	0.42	1.99	0.31	4.05	0.07	3.25	3.70	15.02	0.27	1.90	0.13	1.97
2000	0.33	1.34	0.46	1.47	0.01	1.57	0.35	2.21	0.33	1.39	0.07	1.16	0.05	1.46	2.65	1.67	0.63	2.10	0.25	1.93
2001	0.33	1.25	0.68	1.47	0.01	1.22	0.20	1.31	0.32	1.27	0.51	2.05	0.05	1.30	6.60	2.00	0.21	1.18	0.13	1.25
2002	0.33	1.20	1.00	1.47	0.13	4.88	1.88	3.24	0.40	1.27	0.60	1.60	0.63	4.15	0.09	1.01	1.18	1.84	0.88	2.33
2003	2.61	2.33	0.39	1.13	0.30	2.93	1.32	1.48	0.60	1.32	0.97	1.61	0.36	1.44	0.10	1.01	2.20	1.85	0.15	1.10
2004	0.32	1.07	0.93	1.27	0.62	2.34	7.88	2.95	0.70	1.28	1.33	1.52	0.52	1.44	0.00	1.00	0.80	1.17	0.29	1.17
2005	0.35	1.07	0.79	1.18	0.39	1.36	0.83	1.07	1.00	1.31	2.19	1.56	0.11	1.07	0.10	1.01	0.90	1.16	0.41	1.21
2006	0.38	1.07	0.97	1.19	0.54	1.37	0.91	1.07	1.30	1.31	1.45	1.24	0.89	1.49	0.15	1.01	1.30	1.20	0.42	1.17
2007	0.39	1.07	4.30	1.69	0.75	1.37	2.37	1.17	0.61	1.11	0.73	1.10	1.13	1.41	0.15	1.01	1.60	1.21	1.14	1.41
2008	0.19	1.03	2.00	1.19	1.53	1.55	1.98	1.12	3.97	1.65	1.75	1.21	0.42	1.11	4.40	1.32	0.20	1.02	0.43	1.11
2009	9.09	2.46	4.30	1.34	0.21	1.05	7.68	1.43	2.03	1.20	10.06	2.00	1.17	1.27	5.70	1.31	0.20	1.02	0.74	1.17
2010	9.70	1.63	5.60	1.33	0.50	1.11	5.73	1.22	3.79	1.31	6.83	1.34	2.36	1.43	6.00	1.25	1.29	1.13	2.38	1.46
2011	7.00	1.28	7.60	1.34	0.60	1.12	8.41	1.27	3.00	1.19	9.66	1.36	6.31	1.81	1.10	1.04	4.81	1.43	2.57	1.34
2012	7.16	1.22	4.19	1.14	0.51	1.09	4.24	1.11	6.60	1.35	8.94	1.24	3.00	1.21	2.00	1.06	2.22	1.14	2.51	1.25

(continued)

	Indonesia		Lao P.D.R.		Mauritania		Moldova		Mongolia		Morocco		Nicaragua		Nigeria		Pakistan		Paraguay	
	$\Omega_{IU,i,y}$	$\Phi_{IU,i,y}$	$\Omega_{IU,i,y}$	$\Phi_{IU,i,y}$	$\Omega_{IU,i,y}$	$\Phi_{IU,i,y}$	$\Omega_{IU,i,y}$	$\Phi_{IU,i,y}$	$\Omega_{IU,i,y}$	$\Phi_{IU,i,y}$	$\Omega_{IU,i,y}$	$\Phi_{IU,i,y}$	$\Omega_{IU,i,y}$	$\Phi_{IU,i,y}$	$\Omega_{IU,i,y}$	$\Phi_{IU,i,y}$	$\Omega_{IU,i,y}$	$\Phi_{IU,i,y}$	$\Omega_{IU,i,y}$	$\Phi_{IU,i,y}$
1996	0.03	2.17	n.a.	n.a.	n.a.	n.a.	n.a.	n.a.	0.01	2.06	0.00	1.53	0.05	2.80	n.a.	n.a.	n.a.	n.a.	n.a.	n.a.
1997	0.14	3.44	n.a.	n.a.	n.a.	n.a.	0.02	6.06	0.09	6.21	0.02	3.81	0.12	2.45	0.01	1.95	0.02	9.21	0.08	4.89
1998	0.06	1.31	n.a.	n.a.	0.04	9.73	0.23	9.27	0.03	1.29	0.12	6.58	0.10	1.47	0.01	1.46	0.02	1.60	0.10	1.96
1999	0.19	1.74	0.03	3.91	0.08	2.92	0.34	2.30	0.36	3.49	0.03	1.23	0.19	1.64	0.02	1.63	0.01	1.26	0.19	1.96
2000	0.48	2.08	0.07	2.94	0.07	1.62	0.68	2.14	0.75	2.47	0.52	3.95	0.48	1.97	0.02	1.56	0.61	12.10	0.37	1.96
2001	1.09	2.18	0.07	1.64	0.07	1.36	0.20	1.16	0.40	1.32	0.68	1.98	0.47	1.48	0.03	1.40	0.65	1.97	0.35	1.47
2002	0.12	1.06	0.09	1.47	0.10	1.39	2.30	2.55	0.39	1.23	1.00	1.73	0.27	1.18	0.23	3.56	1.26	1.95	0.70	1.63
2003	0.25	1.12	0.07	1.25	0.06	1.17	3.62	1.96	0.36	1.18	0.98	1.41	0.17	1.10	0.24	1.74	2.46	1.96	0.32	1.18
2004	0.21	1.09	0.03	1.08	0.06	1.14	3.22	1.43	0.50	1.21	8.25	3.46	0.44	1.23	0.73	2.30	1.12	1.22	1.34	1.63
2005	1.00	1.39	0.49	2.35	0.19	1.39	4.00	1.38	0.60	1.21	3.48	1.30	0.25	1.11	2.26	2.76	0.17	1.03	4.45	2.29
2006	1.16	1.32	0.32	1.38	0.31	1.46	4.99	1.34	0.80	1.23	4.69	1.31	0.24	1.09	2.00	1.56	0.17	1.03	0.06	1.01
2007	1.02	1.21	0.47	1.40	0.45	1.46	0.83	1.04	4.70	2.09	1.73	1.09	1.09	1.39	1.22	1.22	0.30	1.05	3.25	1.41
2008	2.13	1.37	1.91	2.16	0.44	1.30	2.94	1.14	0.80	1.09	11.60	1.54	1.40	1.36	9.09	2.34	0.20	1.03	3.06	1.27
2009	−1.00	0.87	2.45	1.69	0.41	1.22	4.11	1.18	0.20	1.02	8.20	1.25	2.00	1.38	4.14	1.26	0.50	1.07	4.63	1.32
2010	4.00	1.58	1.00	1.17	1.72	1.75	4.80	1.17	0.20	1.02	10.70	1.26	2.70	1.37	4.00	1.20	0.50	1.07	0.90	1.05
2011	1.36	1.12	2.00	1.29	0.50	1.13	5.70	1.18	2.30	1.23	1.00	1.02	0.60	1.06	4.43	1.18	1.00	1.13	4.10	1.21
2012	3.08	1.25	1.75	1.19	0.87	1.19	5.37	1.14	3.90	1.31	2.00	1.04	2.90	1.27	4.45	1.16	0.96	1.11	3.18	1.13

	Philippines		Senegal		Sri Lanka		Swaziland		Syria		Ukraine		Viet Nam		Yemen		Zambia	
	$\Omega_{IU,i,y}$	$\Phi_{IU,i,y}$	$\Omega_{IU,i,y}$	$\Phi_{IU,i,y}$	$\Omega_{IU,i,y}$	$\Phi_{IU,i,y}$	$\Omega_{IU,i,y}$	$\Phi_{IU,i,y}$	$\Omega_{IU,i,y}$	$\Phi_{IU,i,y}$	$\Omega_{IU,i,y}$	$\Phi_{IU,i,y}$	$\Omega_{IU,i,y}$	$\Phi_{IU,i,y}$	$\Omega_{IU,i,y}$	$\Phi_{IU,i,y}$	$\Omega_{IU,i,y}$	$\Phi_{IU,i,y}$
1996	0.03	1.96	0.01	16.22	0.05	8.93	n.a.	n.a.	n.a.	n.a.	0.06	2.29	n.a.	n.a.	n.a.	n.a.	n.a.	n.a.
1997	0.08	2.45	0.02	2.43	0.11	1.98	0.04	1.76	n.a.	n.a.	0.10	2.02	n.a.	n.a.	n.a.	n.a.	n.a.	n.a.
1998	0.97	8.06	0.05	2.92	0.13	0.82	0.01	1.09	0.03	1.95	0.10	1.51	0.01	3.28	0.01	1.55	0.02	3.24
1999	0.33	1.30	0.23	3.89	0.05	0.18	0.37	4.90	0.06	1.95	0.10	1.35	0.12	9.86	0.03	2.43	0.12	4.87
2000	0.55	1.38	0.09	1.30	0.30	0.86	0.46	1.97	0.06	1.46	0.31	1.77	0.13	1.97	0.03	1.46	0.04	1.30
2001	0.54	1.27	0.58	2.44	0.15	0.23	0.36	1.38	0.17	1.95	0.52	1.73	1.01	4.98	0.01	1.10	0.04	1.22
2002	1.81	1.72	0.02	1.02	0.26	0.32	0.53	1.42	1.74	5.92	0.64	1.51	0.59	1.47	0.43	5.71	0.24	2.05
2003	0.53	1.12	1.09	2.09	0.41	0.39	0.62	1.34	1.30	1.62	1.27	1.68	1.93	2.04	0.09	1.17	0.50	2.05
2004	0.39	1.08	2.28	2.09	−0.01	−0.01	0.79	1.32	0.92	1.27	0.34	1.11	3.86	2.02	0.28	1.46	1.03	2.05
2005	0.15	1.03	0.40	1.09	0.35	0.24	0.47	1.15	1.33	1.31	0.26	1.07	5.10	1.67	0.17	1.19	0.84	1.42
2006	0.34	1.06	0.83	1.17	0.75	0.42	0.00	1.00	2.18	1.39	0.76	1.20	4.51	1.35	0.20	1.19	1.31	1.46
2007	0.23	1.04	2.09	1.37	1.34	0.53	0.40	1.11	3.67	1.47	2.04	1.45	3.50	1.20	3.76	4.01	0.71	1.17
2008	0.25	1.04	2.90	1.38	1.92	0.49	2.75	1.67	2.50	1.22	4.45	1.68	3.16	1.15	1.88	1.38	0.68	1.14
2009	2.78	1.45	3.90	1.37	2.98	0.51	2.09	1.31	3.30	1.24	6.90	1.63	2.63	1.11	3.07	1.45	0.76	1.14
2010	16.00	2.78	1.50	1.10	3.22	0.37	2.10	1.23	3.40	1.20	5.40	1.30	4.10	1.15	2.39	1.24	3.69	1.58
2011	4.00	1.16	1.50	1.09	3.00	0.25	7.09	1.64	1.80	1.09	5.41	1.23	4.42	1.14	2.56	1.21	1.50	1.15
2012	7.24	1.25	1.70	1.10	3.29	0.22	2.65	1.15	1.80	1.08	4.99	1.17	4.42	1.13	2.54	1.17	1.97	1.17

Appendix H. Mobile Cellular Telephony and Internet Users Penetration Rates: Determinants. Correlation Matrices. Low-Income and Lower-Middle-Income Economies. Period 1997–2012

© Springer International Publishing Switzerland 2015
E. Lechman, *ICT Diffusion in Developing Countries*,
DOI 10.1007/978-3-319-18254-4

Table H.1 Mobile cellular telephony penetration rates—determinants. Correlations and joint sample sizes. 16 low-income Economies (outliers—excluded)

	Time coverage	$MCS_{i,y}$	$GDPPPPpc_{i,y}$	$Call_{i,y}$	$SMS_{i,y}$	$FTL_{i,y}$	$School_{i,y}$	$PopDens_{i,y}$	$Urban_{i,y}$
$MCS_{i,y}$	1997–2012	–							
$GDPPPPpc_{i,y}$	1997–2012	0.46 (213)	–						
$Call_{i,y}$	1997–2012 (with breaks in time series)	−0.23 (186)	−0.27 (181)	–					
$SMS_{i,y}$	1997–2012 (with breaks in time series)	−0.36 (116)	−0.36 (115)	0.61 (114)	–				
$FTL_{i,y}$	1997–2012	0.29 (228)	0.35 (231)	−0.13 (186)	0.21 (116)	–			
$School_{i,y}$	1997–2012 (with breaks in time series)	0.28 (200)	0.27 (202)	−0.10 (164)	−0.10 (105)	0.25 (214)	–		
$PopDens_{i,y}$	1997–2012	0.14 (228)	0.23 (231)	−0.31 (186)	−0.16 (116)	0.04 (246)	0.21 (214)	–	
$Urban_{i,y}$	1997–2012	0.19 (228)	0.37 (231)	0.22 (186)	0.16 (116)	0.23 (246)	0.09 (214)	0.05 (246)	–

Source: Author's calculations
Note: Joint sample sizes reported below coefficients

Table H.2 Mobile cellular telephony penetration rates—affordability. 16 low-income economies (outliers—excluded).

	Time coverage	$MCS_{i,y}$	$MCSIPB_{i,y}$	$CallsMonth_{i,y}$	$SMSMonth_{i,y}$	$MCSChargeMonth_{i,y}$
$MCS_{i,y}$	1997–2012	–				
$MCSIPB_{i,y}$	2008–2012 (with breaks in time series)	−0.36 (61)	–			
$CallsMonth_{i,y}$	1997–2012 (with breaks in time series)	0.52 (181)	−0.61 (60)	–		
$SMSMonth_{i,y}$	1997–2012 (with breaks in time series)	0.59 (115)	−0.60 (59)	0.88 (114)	–	
$MCSChargeMonth_{i,y}$	1997–2012 (with breaks in time series)	0.64 (81)	−0.26 (21)	0.29 (81)	0.46 (61)	–

Source: Author's calculations
Note: Joint sample sizes reported below coefficients

Table H.3 Internet users penetration rates—determinants. Correlations and joint sample sizes. 16 low-income economies (outliers—excluded). 16 low-income countries (outliers—excluded)

	Time coverage	$IU_{i,y}$	$GDPPPPpc_{i,y}$	$FIS_{i,y}$	$FBS_{i,y}$	$WBS_{i,y}$	$FBSCharge\ Month_{i,y}$	$School_{i,y}$	$PopDens_{i,y}$	$Urban_{i,y}$
$IU_{i,y}$	1997–2012	—								
$GDPPPPpc_{i,y}$	1997–2012	0.34 (233)	—							
$FIS_{i,y}$	1997–2012	0.29 (221)	0.02 (211)	—						
$FBS_{i,y}$	2001–2012 (with breaks in time series)	0.45 (130)	0.60 (122)	0.21 (130)	—					
$WBS_{i,y}$	2009–2012 (with breaks in time series)	0.45 (42)	0.23 (39)	0.18 (42)	0.54 (42)	—				
$FBSCharge_{i,y}$	2005–2012 (with breaks in time series)	−0.28 (74)	−0.35 (73)	−0.03 (74)	−0.30 (73)	−0.12 (39)	—			
$School_{i,y}$	1997–2012 (with breaks in time series)	0.31 (214)	0.30 (206)	0.26 (196)	0.25 (115)	0.27 (25)	−0.35 (66)	—		
$PopDens_{i,y}$	1997–2012	0.16 (243)	0.23 (234)	0.10 (221)	0.43 (130)	−0.04 (42)	−0.13 (74)	0.22 (215)	—	
$Urban_{i,y}$	1997–2012	0.09 (243)	0.37 (234)	0.03 (221)	0.03 (130)	−0.27 (42)	−0.17 (74)	0.13 (215)	0.05 (246)	—

Source: Author's calculations

Note: Joint sample sizes reported below coefficients

Table H.4 Internet users penetration rates–affordability. Correlations and joint sample sizes. 16 low-income Economies (outliers—excluded).

	Time coverage	$IU_{i,y}$	$FBSIPB_{i,y}$	$FBSChargeMonth_{i,y}$
$IU_{i,y}$	1997–2012	–		
$FBSIPB_{i,y}$	2008–2012	−0.33 (66)	–	
$FBSChargeMonth_{i,y}$	2005–2012 (with breaks in time series)	0.37 (73)	−0.25 (59)	–

Source: Author's calculations
Note: Joint sample sizes reported below coefficients

Table H.5 Mobile cellular telephony penetration rates—determinants. Correlations and joint sample sizes. 29 lower-middle-income economies (outliers—excluded)

	Time coverage	$MCS_{i,y}$	$GDPPPPpc_{i,y}$	$Call_{i,y}$	$SMS_{i,y}$	$FTL_{i,y}$	$School_{i,y}$	$PopDens_{i,y}$	$Urban_{i,y}$
$MCS_{i,y}$	1997–2012	–							
$GDPPPPpc_{i,y}$	1997–2012	0.44 (434)	–						
$Call_{i,y}$	1997–2012 (with breaks in time series)	−0.20 (434)	−0.18 (347)	–					
$SMS_{i,y}$	1997–2012 (with breaks in time series)	−0.18 (180)	−0.06 (232)	0.45 (228)	–				
$FTL_{i,y}$	1997–2012	0.37 (446)	0.47 (438)	−0.09 (357)	0.11 (239)	–			
$School_{i,y}$	1997–2012 (with breaks in time series)	0.26 (397)	0.40 (388)	−0.10 (320)	0.14 (213)	0.16 (402)	–		
$PopDens_{i,y}$	1997–2012	0.03 (447)	0.13 (439)	−0.31 (358)	−0.19 (240)	0.08 (453)	0.07 (403)	–	
$Urban_{i,y}$	1997–2012	0.26 (447)	0.27 (439)	0.18 (358)	0.11 (240)	0.38 (453)	0.26 (403)	−0.23 (454)	–

Source: Author's calculations
Note: Joint sample sizes reported below coefficients

Table H.6 Mobile cellular telephony penetration rates—affordability. Correlations and joint sample sizes. 29 lower-middle-income economies (outliers—excluded)

	Time coverage	$MCS_{i,y}$	$MCSIPB_{i,y}$	$CallsMonth_{i,y}$	$SMSMonth_{i,y}$	$MCSChargeMonth_{i,y}$
$MCS_{i,y}$	1997–2012	–				
$MCSIPB_{i,y}$	2008–2012	−0.23 (124)	–			
$CallsMonth_{i,y}$	1997–2012 (with breaks in time series)	0.38 (353)	−0.36 (120)	–		
$SMSMonth_{i,y}$	1997–2012 (with breaks in time series)	0.30 (236)	−0.35 (118)	0.41 (224)	–	
$MCSChargeMonth_{i,y}$	1997–2012 (with breaks in time series)	0.29 (125)	No observations to correlate	0.20 (123)	0.15 (23)	–

Source: Author's calculations
Note: Joint sample sizes reported below coefficients

Table H.7 Internet users penetration rates—determinants. Correlations and joint sample sizes. 29 lower-middle-income countries (outliers—excluded)

	Time coverage	$IU_{i,y}$	$GDPPPPpc_{i,y}$	$FIS_{i,y}$	$FBS_{i,y}$	$WBS_{i,y}$	$FBSChargeMonth_{i,y}$	$School_{i,y}$	$PopDens_{i,y}$	$Urban_{i,y}$
$IU_{i,y}$	1997–2012	–								
$GDPPPPpc_{i,y}$	1997–2012	0.43 (438)	–							
$FIS_{i,y}$	1997–2012	0.58 (341)	0.40 (329)	–						
$FBS_{i,y}$	2000–2012 (with breaks in time series)	0.65 (280)	0.31 (271)	0.75 (279)	–					
$WBS_{i,y}$	2007–2012 (with breaks in time series)	0.33 (99)	0.50 (95)	0.09 (99)	0.19 (99)	–				
$FBSCharge_{i,y}$	2005–2012 (with breaks in time series)	–0.18 (142)	–0.04 (138)	–0.19 (142)	–0.18 (142)	–0.12 (92)	–			
$School_{i,y}$	1997–2012	0.13 (402)	0.40 (388)	0.05 (306)	0.00 (255)	0.22 (84)	0.10 (124)	–		
$PopDens_{i,y}$	1997–2012	0.01 (453)	0.13 (439)	0.16 (341)	0.03 (280)	–0.07 (99)	–0.13 (142)	0.07 (403)	–	
$Urban_{i,y}$	1997–2012	0.18 (453)	0.27 (439)	0.17 (341)	0.21 (280)	0.26 (99)	–0.31 (142)	0.26 (403)	–0.23 (454)	–

Source: Author's calculations

Note: Joint sample sizes reported below coefficients

Table H.8 Internet users penetration rates—affordability. Correlations and joint sample sizes. 29 lower-middle-income economies (outliers—excluded)

	Time coverage	$IU_{i,y}$	$FBSIPB_{i,y}$	$FBSChargeMonth_{i,y}$
$IU_{i,y}$	1997–2012	–		
$FBSIPB_{i,y}$	2008–2012	−0.24 (124)	–	
$FBSChargeMonth_{i,y}$	2005–2012 (with breaks in time series)	0.40 (137)	−0.34 (119)	–

Source: Author's calculations
Note: Joint sample sizes reported below coefficients

Appendix I. Mobile Cellular Telephony and Internet Users: Regression Results. Low-Income and Lower-Middle-Income Economies. Period 1997–2012

© Springer International Publishing Switzerland 2015
E. Lechman, *ICT Diffusion in Developing Countries*,
DOI 10.1007/978-3-319-18254-4

Table I.1 Mobile cellular subscriptions. Fixed effects regressions

Explanatory variables	Low-income economies											
	(1)	(2)	(3)	(4)	(5)	(6)	(7)	(8)	(9)	(10)	(11)	(12)
LnGDPPPPpc$_{i,y}$	1.07 (0.96)	**2.79** (**0.86**)	2.07 (2.1)	**2.18** (**0.75**)		**10.2** (**1.60**)						
LnCall$_{i,y}$	0.01 (0.36)		−0.12 (0.57)	−0.08 (0.36)			−1.26 (0.66)					
LnSMS$_{i,y}$		0.27 (0.28)						−**1.60** (**0.49**)				
LnFTL$_{i,y}$	0.37 (0.35)	0.23 (0.12)	**2.17** (**0.65**)		0.58 (0.35)				**3.71** (**0.73**)			
LnSchool$_{i,y}$	1.61 (1.05)	1.07 (1.46)	**6.15** (**1.40**)		**1.72** (**0.74**)					**10.08** (**1.28**)		
LnPopDens$_{i,y}$	**12.09** (**2.87**)	**10.5** (**3.55**)		**13.7** (**2.12**)	**11.7** (**2.46**)						**17.98** (**1.09**)	
LnUrban$_{i,y}$	4.10 (4.68)	1.61 (5.82)		4.44 (4.21)	4.30 (3.50)							**23.05** (**3.58**)
R-sq. (within)	0.89	0.87	0.68	0.88	0.89	0.41	0.10	0.18	0.52	0.40	0.85	0.73
ρ (rho)	0.99	0.99	0.85	0.99	0.99	0.72	0.44	0.46	0.72	0.80	0.99	0.97
mean VIF	1.66	1.50	1.33	1.45	1.40	–	–	–	–	–	–	–
F-test (Prob > F)	190.7 (0.00)	116.5 (0.00)	21.65 (0.00)	118.9 (0.00)	219.5 (0.00)	40.48 (0.00)	3.63 (0.07)	10.41 (0.00)	25.35 (0.00)	61.15 (0.00)	271.5 (0.00)	41.42 (0.00)
# of observations	159	103	159	183	200	215	188	118	230	200	230	230
# of countries	16	16	16	16	16	16	16	16	16	16	16	16

Explanatory variables	Lower-middle-income economies											
	(1)	(2)	(3)	(4)	(5)	(6)	(7)	(8)	(9)	(10)	(11)	(12)
LnGDPPPPpc$_{i,y}$	**4.92** (0.70)	**3.41** (0.72)	**5.62** (0.74)	**5.62** (0.62)		**9.19** (0.81)						
LnCall$_{i,y}$	-0.25 (0.14)		-0.22 (0.14)	-0.31 (0.13)			**-1.24** (0.28)					
LnSMS$_{i,y}$		-0.07 (0.09)						**-1.18** (0.24)				
LnFTL$_{i,y}$	**0.97** (0.32)	0.42 (0.20)	**1.61** (0.35)		**2.25** (0.41)				**3.45** (0.50)			
LnSchool$_{i,y}$	**2.88** (0.91)	2.09 (0.90)	6.01 (1.63)		**5.52** (1.32)					**11.03** (2.26)		
LnPopDens$_{i,y}$	**7.91** (0.14)	**7.16** (1.32)		**10.05** (1.70)	**9.43** (2.43)						**19.17** (1.36)	
LnUrban$_{i,y}$	1.07 (1.86)	17.72 (1.56)		2.92 (1.97)	3.23 (1.96)							**22.67** (.4.41)
R-sq. (within)	0.85	0.87	0.80	0.84	0.73	0.66	0.14	0.22	0.39	0.23	0.59	0.43
ρ (rho)	0.99	0.99	0.96	0.99	0.99	0.93	0.33	0.48	0.83	0.33	0.99	0.96
mean VIF	1.44	1.48	1.48	1.21	1.18	–	–	–	–	–	–	–
F-test (Prob > F)	162.8 (0.00)	56.83 (0.00)	61.48 (0.00)	118.48 (0.00)	132.33 (0.00)	127.99 (0.00)	18.55 (0.00)	23.34 (0.00)	47.61 (0.00)	23.82 (0.00)	197.55 (0.00)	26.33 (0.00)
# of observations	306	204	306	345	396	434	356	240	446	397	447	447
# of countries	29	29	29	29	29	29	29	29	29	29	29	29

Source: Author's estimates

Note: Estimates account for country fixed effects. Panel—unbalanced. Constant included—not reported. Extreme observations—excluded. Robust standard errors—reported below coefficients. In bolds—results statistically significant at 5 % level of significance

Table I.2 Mobile cellular subscriptions. Dynamic panel regressions

Explanatory variables	Low-income economies											
	(1)	(2)	(3)	(4)	(5)	(6)	(7)	(8)	(9)	(10)	(11)	(12)
$LnMCS_{i,y-1}$	**0.70** (**0.15**)	**0.62** (**0.08**)	**0.80** (**0.07**)	**0.78** (**0.07**)	**0.78** (0.11)	**0.87** (**0.01**)	**0.92** (**0.03**)	**0.88** (**0.05**)	**0.91** (**0.02**)	**0.88** (**0.03**)	**0.96** (**0.05**)	**0.90** (**0.01**)
$LnGDPPPPpc_{i,y}$	0.76 (0.85)	**1.79** (**0.56**)		1.29 (19.84)	**1.84** (0.67)	**0.98** (0.38)						
$LnCall_{i,y}$	0.23 (0.16)			0.19 (0.13)	0.19 (0.13)		−0.01 (0.05)					
$LnSMS_{i,y}$		**0.16** (**0.07**)						0.10 (0.07)				
$LnFTL_{i,y}$	0.17 (0.14)	0.10 (0.08)	0.16 (0.10)	0.17 (0.13)					0.11 (0.06)			
$LnSchool_{i,y}$	1.72 (1.4)	**1.13** (**0.52**)	0.80 (0.57)	1.76 (1.6)						0.59 (0.45)		
$LnPopDens_{i,y}$	0.86 (2.5)	0.53 (1.8)			2.06 (2.9)						−0.82 (1.06)	
$LnUrban_{i,y}$	1.7 (1.5)	2.01 (1.94)	1.29 (1.2)		−0.10 (1.2)							0.59 (0.53)
Arellano-Bond test (Prob > z)	−0.87 (0.38)	−0.55 (0.58)	−0.86 (0.38)	−0.85 (0.39)	−1.07 (0.28)	−1.08 (0.27)	−0.90 (0.36)	−0.56 (0.57)	−1.19 (0.23)	−0.91 (0.36)	−1.24 (0.21)	−1.21 (0.22)
# of instruments for differenced equation	81	70	104	79	88	107	85	75	107	101	107	107
# of observations	119	69	166	119	140	180	144	80	194	166	194	194
# of countries	16	16	16	16	16	16	16	16	16	16	16	16

Explanatory variables	Lower-middle-income economies											
	(1)	(2)	(3)	(4)	(5)	(6)	(7)	(8)	(9)	(10)	(11)	(12)
$LnMCS_{i,y-1}$	**0.75** (0.05)	**0.72** (0.13)	**0.78** (0.05)	**0.84** (0.05)	**0.85** (0.03)	**0.89** (0.03)	**0.88** (0.03)	**0.87** (0.04)	**0.87** (0.01)	**0.86** (0.02)	**0.93** (0.03)	**0.90** (0.02)
$LnGDPPPPpc_{i,y}$	0.38 (0.34)	0.87 (0.43)	0.46 (0.33)	0.20 (0.24)		-0.03 (0.29)						
$LnCall_{i,y}$	-0.004 (0.03)		-0.01 (0.03)	0.03 (0.03)			0.03 (0.03)					
$LnSMS_{i,y}$		-0.048 (0.048)						-0.06 (0.04)				
$LnFTL_{i,y}$	**0.22** (0.08)	0.20 (0.12)	**0.24** (0.08)		0.16 (0.09)				**0.14** (0.05)			
$LnSchool_{i,y}$	-0.02 (0.53)	0.84 (0.43)	0.08 (0.58)		-0.27 (0.40)					-0.49 (0.49)		
$LnPopDens_{i,y}$	1.78 (1.1)	0.78 (1.3)		0.66 (0.97)	-1.02 (0.89)						**-1.5** (0.68)	
$LnUrban_{i,y}$	-0.14 (1.4)	-0.80 (1.6)		0.59 (1.3)	0.84 (0.95)							-0.75 (1.04)
Arellano-Bond test (Prob > z)	-0.77 (0.44)	0.24 (0.80)	-0.74 (0.45)	-1.4 (0.14)	-1.4 (0.15)	-2.02 (0.04)	-1.4 (0.16)	-0.37 (0.70)	-2.0 (0.03)	-1.4 (0.16)	-1.9 (0.04)	-1.9 (0.04)
# of instruments for differenced equation	95	87	93	94	110	107	91	83	107	107	107	107
# of observations	220	140	220	263	324	376	271	174	385	326	387	387
# of countries	29	29	29	29	29	29	29	29	29	29	29	29

Source: Author's estimates

Note: GMM estimator applied, one—step results. Panel—unbalanced. Constant included—not reported. Extreme observations—excluded. Robust standard errors—reported below coefficients. In bolds—results statistically significant at 5 % level of significance. Reported Arellano-Bond test is for 2nd order

Table I.3 Mobile cellular subscriptions affordability. Fixed effects regressions

Explanatory variables	Low-income economies				Lower-middle-income economies			
	(1)	(2)	(3)	(4)	(1)	(2)	(3)	(4)
LnMCSIPB$_{i,y}$	−0.34 (0.15)				0.09 (0.11)			
LnCallsMonth$_{i,y}$		2.24 (0.30)				1.54 (0.15)		
LnSMSMonth$_{i,y}$			1.85 (0.28)				10.85 (0.90)	
LnMCSChargeMonth$_{i,y}$				1.04 (0.08)				1.70 (0.45)
R-sq. (within)	0.05	0.45	0.58	0.65	0.01	0.50	0.56	0.39
ρ (rho)	0.83	0.48	0.62	0.55	0.67	0.53	0.58	0.57
mean VIF	–	–	–	–	–	–	–	–
F-test (Prob > F)	4.85 (0.04)	55.8 (0.00)	41.19 (0.00)	139.5 (0.00)	0.61 (0.44)	130.2 (0.00)	143.1 (0.00)	14.02 (0.00)
# of observations	63	183	116	82	124	352	346	125
# of countries	16	16	16	16	29	29	29	29

Source: Author's estimates
Note: Estimates account for country fixed effects. Panel—unbalanced. Constant included—not reported. Extreme observations—excluded. Robust standard errors—reported below coefficients. In bolds—results statistically significant at 5 % level of significance

Table I.4 Mobile cellular subscriptions affordability. Dynamic panel regressions

Explanatory variables	Low-income economies				Lower-middle-income economies			
	(1)	(2)	(3)	(4)	(1)	(2)	(3)	(4)
LnMCS$_{i,y-1}$	0.81 (0.05)	0.91 (0.02)	0.88 (0.05)	0.96 (0.06)	0.73 (0.03)	0.88 (0.04)	0.85 (0.03)	0.80 (0.07)
LnMCSIPB$_{i,y}$	0.05 (0.04)				0.01 (0.03)			
LnCallsMonth$_{i,y}$		0.04 (0.08)				−0.002 (0.03)		
LnSMSMonth$_{i,y}$			−0.03 (0.07)				0.18 (0.23)	
LnMCSCharge Month$_{i,y}$				−0.09 (0.06)				0.24 (0.21)
Arellano-Bond test (Prob > z)	−0.04 (0.96)	−0.90 (0.36)	−0.58 (0.56)	−1.1 (0.26)	−1.2 (0.24)	−1.4 (0.15)	−1.9 (0.05)	−0.75 (0.47)
# of instruments for differenced equation	43	84	74	47	52	91	91	28
# of observations	45	139	82	55	93	268	264	80
# of countries	16	16	16	16	29	29	29	29

Source: Author's estimates
Note: GMM estimator applied, one—step results. Panel—unbalanced. Constant included—not reported. Extreme observations—excluded. Robust standard errors—reported below coefficients. In bolds—results statistically significant at 5 % level of significance. Reported Arellano-Bond test is for 2nd order

Table I.5 Internet users. Fixed effects regressions

Explanatory variables	Low-income economies										
	(1)	(2)	(3)	(4)	(5)	(6)	(7)	(8)	(9)	(10)	(11)
$LnGDPPPPpc_{i,y}$	4.22 (2.74)	5.01 (2.50)		**7.30** **(1.44)**							
$LnFIS_{i,y}$	0.22 (0.13)				**1.14** **(0.10)**						
$LnFBS_{i,y}$			−0.20 (0.18)			**0.38** **(0.03)**					
$LnFBSCharge_{i,y}$		**−0.15** **(0.04)**	**−0.15** **(0.04)**				**−0.48** **(0.07)**				
$LnWBS_{i,y}$	0.02 (0.05)	−0.06 (0.04)	0.00 (0.07)					**0.20** **(0.08)**			
$LnSchool_{i,y}$	−3.26 (2.24)	−3.48 (2.28)	−2.72 (3.16)						**6.74** **(0.83)**		
$LnPopDens_{i,y}$	−0.06 (4.03)	4.31 (2.92)	2.40 (1.81)							**12.19** **(1.06)**	
$LnUrban_{i,y}$	9.01 (6.89)	4.25 (4.87)	**10.98** **(4.1)**								**16.23** **(2.30)**
R-sq. (within)	0.78	0.80	0.71	0.37	0.68	0.53	0.49	0.29	0.37	0.80	0.69
ρ (rho)	0.99	0.99	0.99	0.68	0.51	0.68	0.85	0.83	0.72	0.99	0.96
mean VIF	1.65	2.11	1.84	–	–	–	–	–	–	–	–
F-test (Prob>F)	39.97 (0.00)	202.86 (0.00)	232.75 (0.00)	25.69 (0.00)	110.09 (0.00)	96.61 (0.00)	43.86 (0.00)	5.13 (0.03)	64.61 (0.00)	130.04 (0.00)	49.55 (0.00)
# of observations	34	33	34	233	221	130	74	42	214	243	243
# of countries	16	16	16	16	16	16	16	16	16	16	16

Explanatory variables	Lower-middle-income economies										
	(1)	(2)	(3)	(4)	(5)	(6)	(7)	(8)	(9)	(10)	(11)
$LnGDPPPPpc_{i,y}$	1.23 (0.74)	**1.49** **(0.63)**		**7.89** **(0.90)**							
$LnFIS_{i,y}$	0.04 (0.05)				**1.01** **(0.11)**						
$LnFBS_{i,y}$		−0.06 (0.06)	−0.08 (0.07)			**0.40** **(0.03)**					
$LnFBSCharge_{i,y}$		**−0.33** **(0.07)**	**−0.24** **(0.08)**				**−0.41** **(0.10)**				
$LnWBS_{i,y}$	**0.03** **(0.01)**	**0.04** **(0.01)**	**0.07** **(0.1)**					**0.14** **(0.01)**			
$LnSchool_{i,y}$	−1.41 (0.89)	−0.78 (0.95)	−1.46 (1.02)						**9.38** **(2.19)**		
$LnPopDens_{i,y}$	**4.31** **(1.23)**	**3.15** **(1.14)**	2.85 (1.47)							**17.81** **(1.08)**	
$LnUrban_{i,y}$	2.79 (2.45)	2.91 (2.68)	**6.75** **(2.13)**								**21.71** **(3.87)**
R-sq. (within)	0.82	0.86	0.82	0.60	0.55	0.78	0.26	0.55	0.20	0.65	0.45
ρ (rho)	0.99	0.99	0.99	0.91	0.57	0.65	0.69	0.87	0.31	0.99	0.96
mean VIF	1.51	1.57	1.47	–	–	–	–	–	–	–	–
F-test (Prob>F)	52.76 (0.00)	44.42 (0.00)	40.11 (0.00)	75.76 (0.00)	80.93 (0.00)	153.94 (0.00)	16.51 (0.00)	83.66 (0.00)	18.35 (0.00)	271.63 (0.00)	31.48 (0.00)
# of observations	79	75	83	437	341	280	141	99	401	453	453
# of countries	29	29	29	29	29	29	29	29	29	29	29

Source: Author's estimates

Note: Estimates account for country fixed effects. Panel—unbalanced. Constant included—not reported. Extreme observations—excluded. Robust standard errors—reported below coefficients. In bolds—results statistically significant at 5 % level of significance

Table I.6 Internet Users. Dynamic panel regressions

Explanatory variables	Low-income economies										
	(1)	(2)	(3)	(4)	(5)	(6)	(7)	(8)	(9)	(10)	(11)
LnIU$_{i,y-1}$	–	–	–	**0.80** **(0.03)**	**0.76** **(0.04)**	**0.84** **(0.03)**	**0.64** **(0.08)**		**0.83** **(0.01)**	**0.81** **(0.03)**	**0.76** **(0.04)**
LnGDPPPPpc$_{i,y}$	–	–	–	**0.83** **(0.39)**							
LnFIS$_{i,y}$	–	–	–		**0.02** **(0.09)**						
LnFBS$_{i,y}$	–	–	–			0.02 (0.02)					
LnFBSCharge$_{i,y}$	–	–	–				**-0.12** **(0.04)**				
LnWBS$_{i,y}$	–	–	–								
LnSchool$_{i,y}$	–	–	–						0.05 (0.30)		
LnPopDens$_{i,y}$	–	–	–							0.71 (0.57)	
LnUrban$_{i,y}$	–	–	–								2.33 (1.2)
Arellano-Bond test (Prob > z)	–	–	–	-0.25 (0.80)	0.87 (0.38)	-0.81 (0.41)	-0.45 (0.65)		0.95 (0.33)	0.70 (0.48)	0.71 (0.47)
# of instruments for differenced equation	–	–	–	106	107	94	51		100	107	107
# of observations	–	–	–	199	201	116	53		177	211	211
# of countries	–	–	–	16	16	16	16		16	16	16

(continued)

Explanatory variables	Lower-middle-income economies										
	(1)	(2)	(3)	(4)	(5)	(6)	(7)	(8)	(9)	(10)	(11)
$LnIU_{i,y-1}$	**0.49** (0.08)	–	–	**0.83** (0.02)	**0.83** (0.02)	**0.83** (0.04)	**0.83** (0.05)	**0.67** (0.06)	**0.84** (0.02)	**0.85** (0.03)	**0.86** (0.02)
$LnGDPPPPPpc_{i,y}$	0.41 (0.22)	–	–	**0.61** (0.25)							
$LnFIS_{i,y}$	0.02 (0.02)	–	–		**0.09** (0.03)						
$LnFBS_{i,y}$		–	–			0.03 (0.02)					
$LnFBSCharge_{i,y}$		–	–				0.04 (0.03)				
$LnWBS_{i,y}$	**0.04** (0.01)	–	–					**0.05** (0.01)			
$LnSchool_{i,y}$	**–1.2** (0.59)	–	–						–0.23 (0.41)		
$LnPopDens_{i,y}$	1.2 (0.74)	–	–							0.58 (0.80)	
$LnUrban_{i,y}$	0.94 (1.1)	–	–								0.57 (0.87)
Arellano-Bond test (Prob > z)	1.4 (0.15)	–	–	1.5 (0.12)	0.08 (0.93)	–0.46 (0.64)	0.08 (0.93)	1.4 (0.15)	1.0 (0.31)	1.31 (0.19)	1.31 (0.18)
# of instruments for differenced equation	49	–	–	107	101	101	58	47	107	107	107
# of observations	50	–	–	381	312	251	104	70	331	395	395
# of countries	29	–	–	29	29	29	29	29	29	29	29

Source: Author's estimates

Note: GMM estimator applied, one—step results. Panel—unbalanced. Constant included—not reported. Extreme observations—excluded. Robust standard errors—reported below coefficients. In bolds—results statistically significant at 5 % level of significance. Reported Arellano-Bond test is for 2nd order. In low-income countries—specification (1)–(3)—not applicable—number of instruments exceeds number of observations. In lower-middle-income countries—specifications (2)–(3)—not applicable—number of instruments exceeds number of observations

Table I.7 Internet users affordability. Fixed effects regressions

Explanatory variables	Low-income economies		Lower-middle-income economies	
	(1)	(2)	(1)	(2)
LnFBSIPB$_{i,y}$	**−0.36** **(0.06)**		**−0.41** **(0.06)**	
LnFBSChargeMonth$_{i,y}$		**0.46** **(0.06)**		**0.46** **(0.08)**
R-sq. (within)	0.52	0.59	0.44	0.43
ρ (rho)	0.86	0.77	0.78	0.75
Mean VIF	–	–	–	–
F-test (Prob > F)	26.68 (0.00)	52.48 (0.00)	38.43 (0.00)	33.03 (0.00)
# of observations	61	73	124	136
# of countries	16	16	29	29

Source: Author's estimates

Note: Estimates account for country fixed effects. Panel—unbalanced. Constant included—not reported. Extreme observations—excluded. Robust standard errors—reported below coefficients. In bolds—results statistically significant at 5 % level of significance

Table I.8 Internet users affordability. Dynamic panel regressions

Explanatory variables	Low-income economies		Lower-middle-income economies	
	(1)	(2)	(1)	(2)
LnIU$_{i,y-1}$	**0.64** **(0.10)**	**0.66** **(0.08)**	**0.84** **(0.04)**	**0.82** **(0.07)**
LnFBSIPB$_{i,y}$	**−0.13** **(0.03)**		0.01 (0.04)	
LnFBSChargeMonth$_{i,y}$		**0.12** **(0.03)**		−0.01 (0.04)
Arellano-Bond test (Prob > z)	0.67 (0.49)	−0.79 (0.42)	−0.23 (0.81)	−0.64 (0.52)
# of instruments for differenced equation	42	50	52	58
# of observations	43	51	95	100
# of countries	16	16	29	29

Source: Author's estimates

Note: GMM estimator applied, one—step results. Panel—unbalanced. Constant included—not reported. Robust standard errors—reported below coefficients. In bolds—results statistically significant at 5 % level of significance. Reported Arellano-Bond test is for 2nd order

Appendix J. 'Technological Take-Off' Conditions. Low-Income and Lower-Middle-Income Economies

'MCS-technological take-off ' conditions

Low-income economies (15 countries)				
$Y_{crit,MCS}$—period	2004–2007	'MCS-technological take-off'—period	2005–2011	
$MCS_{i,y}$ in $Y_{crit,MCS}$ (average), per 100 inhab. $MCS_{i,y}$ in $Y_{crit,MCS}$ (median), per 100 inhab.	7.05 6.6	Duration of the diffusion initial phase (average), in years	12 years	
Direct determinants	**Average**	**Median**	**Min. value**	**Max. value**
Mobile-cellular prepaid connection charge, USD	25.00 (16.0—if Burkina Faso is excluded)	5.4	3.03 (Zimbabwe)	114.7 (Burkina Faso)
Mobile-cellular prepaid—price of a 1-min local call, USD	0.19	0.21	0.03 (Bangladesh)	0.37 (Kenya, Niger)
Mobile-cellular prepaid—price of SMS, USD	0.05	0.06	0.01 (Bangladesh)	0.12 (Comoros)
Mobile Cellular Sub-Basket, % of GNI per capita per month	34.3	45.5	3.38 (Bangladesh)	60.0 (Togo)
Indirect determinants	**Average**	**Median**	**Min. value**	**Max. value**
Fixed telephony penetration rate, per 100 inhab.	1.33	1.9	0.16 (Rwanda)	3.09 (Comoros)

(continued)

© Springer International Publishing Switzerland 2015
E. Lechman, *ICT Diffusion in Developing Countries*,
DOI 10.1007/978-3-319-18254-4

Low-income economies (15 countries)

Gross Domestic Product per capita in PPP, US dollars	1,434.6	1,200	794.0 (Malawi)	2,136 (Cambodia)
Economic Freedom index	53.2	52.8	33.5 (Zimbabwe)	63.1 (Uganda)
Investment Freedom index	39.2	35.0	70.0 (Madagascar)	10.0 (Zimbabwe)
Electrification rate, %	18.7	17.0	7.0 (Malawi/ Burkina Faso)	34.0 (Zimbabwe)
Level of competition on telecommunication market	Full competition— 12 Partial competition—1 Monopoly—2			
Country characteristics	**Average**	**Median**	**Min. value**	**Max. value**
Access to improved water, %	67.0	68.5	44.0 (Madagascar)	95.0 (Comoros)
Literacy rate, %	55.9	61.0	23.0 (Burkina Faso)	83.0 (Zimbabwe)
Extreme poverty, %	52.1	53.0	30.0 (Ethiopia)	74.0 (Malawi)
Rural population, %	75.9	82.2	58.3 (Benin)	85.9 (Uganda)
Population density, people per km square	185.3 (119.2—if Bangladesh is excluded)	101.8	32.9 (Zimbabwe)	1,112.9 (Bangladesh)
Country freedom status	Free—1 country Partly free—10 countries Not free—4 countries			
Democracy (Political Freedom)	Score 2—7 countries Score 1—7 countries Score 0—1 country			
Dominant religion	Islam—5 countries Christianity—5 countries Indigenous beliefs— 2 countries Hinduism—1 country Buddhism— 2 countries			

(continued)

Lower-middle-income economies (29 countries)

$Y_{Crit.MCS}$—period	1999–2005	'MCS-technological take-off'—period	2001–2008	
$MCS_{i,y}$ in $Y_{Crit.MCS}$ (average), per 100 inhab. $MCS_{i,y}$ in $Y_{Crit.MCS}$ (median), per 100 inhab.	8.22 6.9	Duration of the diffusion initial phase (average), in years	11.7 years	
Direct determinants	**Average**	**Median**	**Min. value**	**Max. value**
Mobile-cellular prepaid connection charge, USD	9.32	5.4	2.2 (India)	31.1 (Syria)
Mobile-cellular prepaid—price of a one-minute local call, USD	0.17	0.16	0.02 (India)	0.48 (Nicaragua)
Mobile-cellular prepaid—price of SMS, USD	0.05	0.05	0.01 (Indonesia/Pakistan/ Paraguay)	0.27 (Georgia)
Mobile Cellular Sub-Basket, % of GNI per capita per month	7.18	6.15	1.8 (Sri Lanka)	18.5 (Zambia)
Indirect determinants	**Average**	**Median**	**Min. value**	**Max. value**
Fixed telephony penetration rate, per 100 inhab.	7.21	4.5	0.44 (Congo. Rep.)	22.5 (Ukraine)
Gross Domestic Product per capita in PPP, US dollars	4,407.1	3,890	2,049 (Senegal)	8,332 (Egypt)
Electrification rate, %	62.1	36.0	3.0 (Congo Rep.)	98.0 (Egypt)
Economic Freedom index	56.7	53.8	40.6 (Syria)	71.2 (El Salvador)
Investment Freedom index	49.3	50.0	30.0 (Honduras/Lao P.D.R./ Nigeria/Pakistan/ Syria/Viet Nam)	90.0 (Bolivia)
Level of competition on telecommunication market	Full competition— 22 countries Partial competition— 6 countries Monopoly—1 country			

(continued)

Lower-middle-income economies (29 countries)

Country characteristics	Average	Median	Min. value	Max. value
Access to improved water, %	78.6	80.5	44.0 (Mauritania)	97.0 (Ukraine)
Literacy rate, %	76.4	81.0	39.0 (Senegal)	99.0 (Armenia/ Georgia/ Ukraine)
Extreme poverty, %	25.3	19.5	0.50 (Ukraine)	65.0 (Zambia)
Rural population, %	54.9	54.3	32.5 (Ukraine)	84.7 (Sri Lanka)
Population density, people per km square	107.3	76.2	1.5 (Mongolia)	379.1 (India)
Country freedom status	Free— 8 countries Partly free— 13 countries Not free— 8 countries			
Democracy (Political Freedom)	Score 2—18 countries Score 1—5 countries Score 0—6 countries			
Dominant religion	Islam—9 countries Christianity— 15 countries Hinduism— 2 countries Buddhism—3 countries			

Source: Author's elaboration
[a]The evidence is reported exclusively for those countries where the '*MCS-technological take-off*' was identified

'IU-technological take-off' condition. Low-income and lower-middle-income economies

Low-income economies (7 countries)				
$Y_{Crit,IU}$—period	2008–2011	'IU-technological take-off'—period	2008–2012	
$IU_{i.y}$ in $Y_{Crit.\,IU}$ (average), % $IU_{i.y}$ in $Y_{Crit.\,IU}$ (average), %	7.3 8.0	Duration of the diffusion initial phase (average), in years	14.3 years	
Direct determinants	**Average**	**Median**	**Min. value**	**Max. value**
Fixed-narrowband subscriptions, per 100 inhab.	0.45	0.41	0.02 (Kenya)	1.5 (Bangladesh)
Fixed-broadband subscriptions, per 100 inhab.	0.24	0.20	0.10 (Rwanda/ Togo)	0.70 (Zimbabwe)
Wireless-broadband subscriptions, per 100 inhab.	2.39	0.47	0.01 (Kenya)	6.7 (Cambodia)
Fixed (wired)-broadband connection charge, USD	93.6 (64.7—if Zimbabwe id excluded)	50.4	4.2 (Bangladesh)	267.3 (Zimbabwe)
Fixed (wired)-broadband monthly subscription charge, USD	52.07 (if Zimbabwe is excluded)	31.2	4.2 (Bangladesh)	2,672.9 (Zimbabwe)
Fixed-Broadband Sub-Basket, % of GNI per capita per month	166.1 (if Zimbabwe is excluded); 93.8 (if Zimbabwe and Uganda are excluded)	56.3	7.3 (Bangladesh)	1,059 (Zimbabwe); 600 (Uganda)
Indirect determinants	**Average**	**Median**	**Min. value**	**Max. value**
Gross Domestic Product per capita in PPP, US dollars	1,936	1,999	1,189.2 (Rwanda)	2,363.8 (Bangladesh)
Economic Freedom index	53.6	54.6	36.7 (Zimbabwe)	63.8 (Uganda)
Investment Freedom index	39.2	45.0	10.0 (Zimbabwe)	60.0 (Cambodia)
Obstacles to use index	15.6	13.5	10.0 (Kenya)	17.0 (Zimbabwe)
Limits of contents index	15.1	11.0	7.0 (Kenya)	14.0 (Zimbabwe)
Violations of users rights index	21.1	19.5	12.0 (Kenya)	24.0 (Bangladesh)

(continued)

Low-income economies (7 countries)

Electrification rate %	23.6	21.5	9.0 (Uganda)	34.0 (Zimbabwe)
Level of competition in telecommunication market (Fixed Broadband Connections/Internet Services)	**Fixed Broadband Connections** Full competition—7 countries Partial competition—none Monopoly—none **Internet Services** Full competition—7 countries Partial competition—none Monopoly—none			
Country characteristics	**Average**	**Median**	**Min. value**	**Max. value**
Access to improved water, %	74.1	71.3	56.1 (Kenya)	86.3 (Nepal)
Literacy rate, %	69.2	72.0	57.3 (Nepal)	83.6 (Zimbabwe)
Extreme poverty, %	38.3	40.5	18.0 (Cambodia)	63.0 (Rwanda)
Rural population, %	77.5	79.8	65.0 (Zimbabwe)	85.6 (Uganda)
Population density, people per km square	305.4 (158.3 if Bangladesh is excluded)	159.0	32.8 (Zimbabwe)	1,188.4 (Bangladesh)
Country freedom status	Free—none Partly free—4 countries Not free—3 countries			
Democracy (Political Freedom)	Score 2—2 countries Score 1—4 countries Score 0—1 country			
Dominant religion	Islam—1 country Christianity—4 countries Hinduism—1 country Buddhism—1 country			

(continued)

Lower-middle-income economies (26 countries)

$Y_{Crit,IU}$—period	2004–2011	'IU-technological take-off'—period	2004–2012	
$IU_{i.y}$ in $Y_{Crit.IU}$ (average), % $IU_{i.y}$ in $Y_{Crit.IU}$ (median), %	9.52 8.45	Duration of the diffusion initial phase (average), in years	14.4 years	
Direct determinants	**Average**	**Median**	**Min. value**	**Max. value**
Fixed-narrowband subscriptions, per 100 inhab.	1.62	1.3	0.13 (Zambia)	5.6 (Moldova)
Fixed-broadband subscriptions, per 100 inhab.	0.69	0.32	0.01 (Honduras/ Nigeria/Sri Lanka/ Syria)	3.2 (Moldova)
Wireless-broadband subscriptions, per 100 inhab.	5.13	0.26	0.26 (Zambia)	18.6 (Indonesia)
Fixed (wired)-broadband connection charge, USD	131.5 (79.54—if Zambia is excluded)	92.8	3.9 (Sri Lanka)	962.8 (Zambia)
Fixed (wired)-broadband monthly subscription charge, USD	133 (67.0—if Swaziland is excluded)	29.3	3.1 (Viet Nam)	1,781.8 (Swaziland)
Fixed-Broadband Sub-Basket, % of GNI per capita per month	26 (24—if Nigeria and Swaziland are excluded)	22.8	5.1 (India)	890.4 (Nigeria)
Indirect determinants	**Average**	**Median**	**Min. value**	**Max. value**
Gross Domestic Product per capita in PPP, US dollars	5,194	5,206	2,161.7 (Senegal)	10,272.1 (Egypt)
Economic Freedom index	56.4	56.9	48.4 (Guyana)	68.5 (El Salvador)
Investment Freedom index	47.05	50.0	20.0 (Bolivia)	70.0 (El Salvador/ Morocco/ Nicaragua)
Obstacles to use index	13.0	14.0	7.0 (Ukraine)	23.0 (Syria)
Limits of contents index	12.8	25.0	5.0 (Philippines)	26.0 (Viet Nam)
Violations of users rights index	20.1	33.0	8.0 (Philippines)	35.0 (Syria)

(continued)

Lower-middle-income economies (26 countries)

Electrification rate, %	65.1	36.0	33.0 (Senegal)	98.0 (Egypt)
Level of competition on telecommunication market (Fixed Broadband Connections/Internet Services)	**Fixed Broadband Connections**[a] Full competition—16 countries Partial competition—6 countries Monopoly—1 country **Internet Services** Full competition—21 countries Partial competition—5 countries Monopoly—none			

Country characteristics	Average	Median	Min. value	Max. value
Access to improved water, %	83.2	85.6	54.5 (Yemen)	99 (Egypt)
Literacy rate, %	81.2	85.0	51.2 (Nigeria)	99.6 (Armenia)
Extreme poverty, %	21.1	17.0	0.02 (Ukraine)	75.0 (Zambia)
Rural population, %	53.4	50.7	31.7 (Ukraine)	84.6 (Sri Lanka)
Population density, people per km square	115.1	77.2	1.7 (Mongolia)	405.5 (India)
Country freedom status	Free—8 countries Partly free—13 countries Not free—5 countries			
Democracy (Political Freedom)	Score 2—18 countries Score 1—4 countries Score 0—4 countries			
Dominant religion[b]	Islam—6 countries Christianity—14 countries Hinduism—2 countries Buddhism—3 countries Indigenous beliefs—1 country			

Source: Author's elaboration

[a]Not available for all countries

[b]In few countries two or more religions are reported